Primers in Exercise Science Series

Biochemistry Primer for Exercise Science

THIRD EDITION

Michael E. Houston, PhD

Virginia Tech
Blacksburg, Virginia

Human Kinetics

Library of Congress Cataloging-in-Publication Data

Houston, Michael E., 1941-
    Biochemistry primer for exercise science / Michael E. Houston. -- 3rd
ed.
        p. ; cm.
    Includes bibliographical references and index.
    ISBN-13: 978-0-7360-5612-0 (soft cover)
    ISBN-10: 0-7360-5612-2 (soft cover)
    1. Biochemistry. 2. Exercise--Physiological aspects. 3. Metabolism.
    I. Title.
    [DNLM: 1. Exercise--physiology. 2. Metabolism--physiology.
QU 120 H843b 2006]
    QP514.2.H68 2006
    612'.044--dc22

2006001211

ISBN-10: 0-7360-5612-2
ISBN-13: 978-0-7360-5612-0

Acquisitions Editor: Michael S. Bahrke, PhD; **Developmental Editor:** Maggie Schwarzentraub; **Assistant Editors:** Maureen Eckstein and Jillian Evans; **Copyeditor:** Joyce Sexton; **Proofreader:** Sarah Wiseman; **Indexer:** Sharon Duffy; **Permission Manager:** Dalene Reeder; **Graphic Designer:** Fred Starbird; **Graphic Artist:** Denise Lowry; **Cover Designer:** Keith Blomberg; **Art Manager:** Kelly Hendren; **Illustrator:** Jason McAlexander and Terry N. Hayden; **Printer:** Versa Press

Printed in the United States of America   10   9   8   7   6   5   4   3   2

Human Kinetics
Web site: www.HumanKinetics.com

United States: Human Kinetics, P.O. Box 5076, Champaign, IL 61825-5076
800-747-4457
e-mail: humank@hkusa.com

Canada: Human Kinetics, 475 Devonshire Road, Unit 100, Windsor, ON N8Y 2L5
800-465-7301 (in Canada only)
e-mail: orders@hkcanada.com

Europe: Human Kinetics, 107 Bradford Road, Stanningley
Leeds LS28 6AT, United Kingdom
+44 (0) 113 255 5665
e-mail: hk@hkeurope.com

Australia: Human Kinetics, 57A Price Avenue, Lower Mitcham, South Australia 5062
08 8372 0999
e-mail: liaw@hkaustralia.com

New Zealand: Human Kinetics, Division of Sports Distributors NZ Ltd.
P.O. Box 300 226 Albany, North Shore City, Auckland
0064 9 448 1207
e-mail: info@humankinetics.co.nz

To my family, Cheri, Mike Jr., Beth, Pat, Tim, Lee-Ann, and Vanessa,
whose support for me and unwavering dedication
to personal improvement and family
have enriched and inspired me.

Contents

Preface

We are reminded almost constantly that there is an obesity crisis for a growing number of the world's population. We are told that we need to increase our physical activity while at the same time cutting back on our food energy intake. These are simple bioenergetics concepts, yet difficult to accomplish for many. Students trained in traditional exercise physiology can understand the basic energy concepts that underlie our obesity epidemic but may lack the molecular basics on which all human science is founded. We need educated people who not only understand exercise physiology but also are thoroughly grounded in the thermodynamic concepts of human energy metabolism at a more detailed, molecular level. Part I lays a groundwork with chapters on amino acids and proteins and then enzymes, the critical biological catalysts. The major emphasis of this book is part II on metabolism, with all the fundamental biochemical pathways explained and then integrated with information on the exercising human.

The human genome has been mapped in minute detail, and individual genes are being parsed from the enormous information we store in each cell nucleus. We need to interpret the genetic concepts in terms of human metabolism, and we need to understand how what we eat and what we do are governed by our genes but also how they influence the expression of our genes. Traditional chemistry and biochemistry texts provide scientific basics for an understanding of human metabolism, but too often their focus is on principles applicable to a huge range of species. Too often the interpretation of these principles in relation to everyday life is up to the student to acquire. Part III provides the basic concepts of gene expression and protein synthesis, emphasizing how exercise and exercise training modify these processes, allowing us to perform at a higher level.

The rationale for the first edition of *Biochemistry Primer for Exercise Science* was to present the basic biochemistry that a student of exercise science or kinesiology would need in order to understand the molecular aspects of human physical activity. Instructors and reviewers asked for more integration and interpretation of biochemistry and physiology of human physical activity. The second edition was designed to address the integration of biochemistry and exercise physiology, with a more user-friendly approach. That edition received enthusiastic responses, yet readers also requested more.

The third edition builds on the success of the second edition, providing the basic molecular information so essential to understanding the latest in exercise science. New topics have been added, along with more interpretation of biochemistry in the context of the active human. Besides updated information based on new research and ideas from exercise science and molecular biology, more has been added. There is a new section on oxidative stress and its implications for our lifestyle and health. The concept of signaling or signal transduction is integrated into four of the chapters. Greater attention has been given to mechanisms behind the concepts. New review questions have been added, and answers to the questions are provided at the back of the book. Chapters have been reorganized in general around the pattern of the biochemistry basics followed by exercise applications and then more difficult concepts placed at the ends of the chapters. The chapter on bioenergetics has been thoroughly revamped, with a comprehensive overview of energy systems and their role in exercise followed by the more rigorous thermodynamics concepts. Finally, as an instructor in metabolic nutrition, I have tried to bridge the fields of human nutrition, physiology, and biochemistry into a more unifying science.

As before, the third edition builds on the wisdom gained from more than 35 years of teaching exercise science and nutrition to undergraduate students in Canada, Denmark, and the United States. Your comments and suggestions are most eagerly sought.

This book is the first volume in Human Kinetics' Primers in Exercise Science Series, which provides students and professionals alike with a non-intimidating basic understanding of the science behind each topic in the series, and where appropriate, how that science is applied. These books are written by leading researchers and teachers in their respective areas of expertise to present in an easy-to-understand manner essential concepts in dynamic, complex areas of scientific knowledge. The books in the series are ideal for researchers and professionals who need to obtain background in an unfamiliar scientific area or as an accessible basic reference for those who will be returning to the material often.

Acknowledgments

Our success as scientists is based on the quality of the students, teachers, collaborators, and colleagues with whom we interact and from whom we learn. I have been fortunate to work with some outstanding people. Professors Bengt Saltin and Howie Green had enormous influence on me, especially in my early years. Richard Hughson, Jay Thomson, Mike Sharratt, and other fine folks in kinesiology at Waterloo made my 27 years there very fulfilling. Jim Stull of UT Southwestern offered me an opportunity to learn at a truly exciting institution. Tim Inglis, Susan Samuels, Stu Phillips, and Maureen MacDonald were inspiring and fun to be with at UBC. I am grateful for the support of the students and faculty members at Virginia Tech. I owe so much to my graduate students, especially Rob Grange, Peter Tiidus, and Rene Vandenboom. Finally, thanks to the graduate and undergraduate students at Waterloo, UBC, August Krogh, and Virginia Tech, who kept me excited about learning.

PART

I

ESSENTIAL BACKGROUND

Everything we do, everything we are, and everything we will become depend on the action of thousands of different proteins. These large molecules, composed of individual units known as amino acids, regulate everything in our bodies. The arrangement of amino acids in proteins is dictated by the sequence of bases, a part of nucleotides organized into genes within our DNA. Deoxyribonucleic acid base sequence is copied during transcription as working blueprints containing RNA bases. The sequence of bases in RNA, read in groups of three, dictates the sequence of amino acids in each of our thousands of different proteins. This simple story is now understood by high school students. In chapter 1, we review acids, bases, and buffering; we also learn about the nature of amino acids and the ways in which these are linked to make peptides and proteins. Chapter 2 affords a detailed understanding of enzymes and how these key catalytic molecules enable our complex metabolism to function.

CHAPTER

1

AMINO ACIDS, PEPTIDES, AND PROTEINS

Proteins are the molecules responsible for what happens in cells and organisms. We can consider them the body's action molecules: They can be enzymes that catalyze all the chemical reactions in the body; receptor molecules inside cells, in membranes, or on membranes that bind specific substances; contractile proteins involved in the contraction of skeletal, smooth, and heart muscle; or transport molecules that move substances in the blood, within cells, and across membranes. Proteins are also parts of bones, ligaments, and tendons; and in the form of antibodies or receptors on lymphocytes, they help protect us from disease. Peptides are like small proteins that often have specific regulatory roles. Examples of important peptides are hormones, such as insulin, and growth factors such as insulin-like growth factors (IGFs).

Twenty different amino acids are used to make proteins. The genes in our cell nuclei contain the information needed to specify which amino acids are used in making a protein and in what order. The genes in chromosomes are segments of deoxyribonucleic acid (DNA), large molecules containing four different bases. The sequence of bases in a gene spells out the sequence of amino acids for a protein. Because of the huge size of DNA molecules, only small segments of the genes are copied at any time. The copies are in the form of new molecules known as messenger ribonucleic acid (mRNA). The copying process is known as **transcription.** The mRNA information is then used to order the binding of the 100 or more specific amino acids to make a protein. This step, known as **translation,** takes place outside of the nucleus.

Proteins are continually being turned over: Old protein molecules are broken down to their constituent amino acids in a process known as **degradation,** and new protein molecules are synthesized. This continual synthesis and degradation of proteins in cells is known as protein turnover (see figure 1.1). Amino acids released when a protein molecule is broken down can be reutilized. We can express the rate of turnover of an individual protein as the time taken for one-half of the protein molecules to be replaced, that is, the **half-life.** Some proteins have a short half-life, measured in minutes or hours,

Figure 1.1 A simple outline of the lifetime of a protein, begun when its gene is transcribed, making mRNA. The mRNA information is translated into the sequence of amino acids in the protein, followed by the breakdown of the protein to its constituent amino acids.

while others have a half-life that may be expressed in weeks. The content of a particular protein in a cell can reflect the use of that protein. Consider how fast muscles atrophy when a limb is immobilized by a cast. Most of the loss of muscle can be attributed to contractile proteins, due to an increase in contractile protein degradation and a decrease in contractile protein synthesis. Research examining the role of specific exercise training, nutrition, and ergogenic aids in the turnover of skeletal muscle proteins is thriving, both from the perspective of basic science and in order to delineate the practical implications for sport and health.

THE NATURE OF AMINO ACIDS

Amino acids play key roles in the body, especially as building blocks for proteins and peptides. Of the 20 different amino acids needed to synthesize proteins and peptides, at least nine are like vitamins and minerals and must be obtained from our diet. As the name implies, amino acids have characteristics of simple acids, and they also contain at least one group that can act as a base.

Acids, Bases, and Buffers

Our cells operate within a narrow pH range. Our blood pH, although slightly higher than intracellular pH, is carefully regulated by the action of buffers. Deviations above and below normal pH lead to conditions known as **alkalosis** and **acidosis,** respectively. Severe exercise causes a temporary acidosis, first in the active muscle fibers and then in the blood. Acidosis is related to impaired muscle activity or fatigue. Acidosis can also be caused by starvation and diabetes, while alkalosis can occur with uncontrolled vomiting. Amino acids have groups that can behave as both acids and bases. There are other important acids and bases that are generated by our metabolism.

In high school you learned about **acids** and bases, often dealing with strong acids such as hydrochloric or sulfuric acids or strong bases such as sodium and potassium hydroxide. You learned that an acid is a substance that, in water, releases a proton (H^+) and that a base releases hydroxide (OH^-) ions in water. These definitions were originally advanced by Arrhenius, and they serve us well when we consider strong acids and bases. However, the range of acidity and basicity in our bodies is very narrow, hovering primarily around a pH of 7. The *pH scale* is based on the negative logarithm of the hydrogen ion concentration

($-\log [H^+]$) (where square brackets say "concentration of"). A pH of 7 means that the $[H^+]$ is 10^{-7} molar (also written as 1×10^{-7} M). A pH of 7 is also neutral in that there is an equal concentration of protons and hydroxide ions. Acidity (acidosis) denotes a pH less than 7, and basicity or alkalosis denotes a pH larger than 7. Since the pH scale is a logarithmic scale, a difference of one pH unit means there is a 10-fold difference of concentration of protons. In biological terms, extreme acidity is found in the gastric juice of the stomach, with a pH of about 2 (i.e., gastric juice is roughly 250,000 times more acidic than the blood). On the other hand, you have probably used acid solutions stronger than gastric juice or bases much stronger than anything we encounter with living organisms. Therefore, biologists use a different concept for acids and bases.

KEY POINT

A proton is simply a hydrogen atom minus its only electron. As such, it is a very tiny positively charged particle that interacts well with water molecules. You probably learned that a proton is associated with a single water molecule (hydrated) in what is described as a hydronium ion (H_3O^+). It is more likely that a proton in water interacts with four water molecules, existing as $H_9O_4^+$. For our purposes, we will just use H^+, but understand that this naked proton is hydrated with about four water molecules.

The **Brønsted-Lowry** definitions for acids and bases are more effective for understanding acid and base properties of cells and tissues and fluids in our bodies. An acid is a proton donor whereas a base is a proton acceptor. If we have pure water, with a $[H^+]$ of 10^{-7} M (pH = 7), adding an acid to it will result in an increase in the $[H^+]$ because the acid will donate more protons. If we bubble ammonia gas (NH_3) into our pure water, the $[H^+]$ will decrease because the ammonia will act as a base and remove some of the protons from water, generating the ammonium ion (NH_4^+); this will raise the pH. Ammonia is therefore a base by our definition.

We can define the strength of acids and bases in biological systems by using a simple system of measurement, the pK_a. We will illustrate this system by using a general acid that we designate HA. This is an acid because it can donate a proton through a process called **dissociation.** It is not a strong acid, such as HCl, and thus does not dissociate completely. Therefore, in the following equation, incomplete dissociation is shown by a double-headed arrow, indicating that an equilibrium is established in which there exist both undissociated acid (HA) and its conjugate base (A⁻).

$$HA \leftrightarrow H^+ + A^-$$
(acid)   (proton)   (conjugate base)

Because weak acids do not dissociate completely, we can write an equilibrium expression for the reversible dissociation using square brackets to represent concentration. The **acid dissociation constant** K_a reflects the degree to which the acid is dissociated.

$$K_a = [H^+] [A^-] / [HA]$$

The stronger the acid (HA), the more it is dissociated, the lower the pH, and the higher the concentration of the conjugate base (A⁻). Therefore, the larger the numerical value for K_a, the stronger the acid.

Suppose we dissolve HA in a solution containing the sodium salt of A⁻ (i.e., NaA) such that at equilibrium, the concentration of undissociated acid, HA, is equal to the concentration of its conjugate base, A⁻. Now the equilibrium equation is simplified to

$$[H^+] = K_a$$

because the other terms ([HA] and [A⁻]) cancel out since we are creating the circumstances in which they have the same value.

Now we will take the negative logarithm of both sides of this equation; that is,

$$-\log [H^+] = -\log K_a,$$

or

$$pH = pK_a$$

By definition, pH is the negative logarithm of [H⁺]. Thus pK_a is the negative logarithm of K_a, the acid dissociation constant.

The pK_a for an acid is the pH of a solution in which the acid is one-half dissociated; that is, 50% of the molecules are dissociated and 50% are not dissociated. Therefore, the smaller the value of pK_a for an acid, the stronger the acid. Let's illustrate this concept by using two substances, acetic acid (CH_3COOH—a carboxylic acid) and ammonia, each dissolved in a separate beaker. We know that acetic acid will dissociate as shown in the following.

$$CH_3COOH \leftrightarrow CH_3COO^- + H^+$$
(acetic acid)   (acetate)

Acetic acid is a weak acid, and we can write an equilibrium expression for it. Note that the K_a for acetic acid is 1.8×10^{-5}.

$$1.8 \times 10^{-5} = [CH_3COO^-] [H^+] / [CH_3COOH]$$

Using the definition for pK_a as just shown, the pK_a for acetic acid would be $-\log 1.8 \times 10^{-5}$ or $-(\log 1.8 + \log 10^{-5})$, which turns out to be 4.74. Assume we start out with a solution of sodium acetate and acetic acid so that at equilibrium the concentration of undissociated acetic acid equals the concentration of acetate ions. The pH where this would occur would be 4.74—that is, the pH of a solution in which the acetic acid is 50% dissociated.

When the gas ammonia is bubbled into water, some of the ammonia removes a proton from water, forming the ammonium ion (NH_4^+). There would also be free ammonia (NH_3) in the solution. Using our description for acids as proton donors, we could say that the ammonium ion is an acid and the ammonia is its conjugate base. Therefore, we could write the dissociation of the acid NH_4^+ as follows:

$$NH_4^+ \leftrightarrow NH_3 + H^+$$

and from this write an equilibrium expression:

$$K_a = [NH_3] [H^+] / [NH_4^+]$$

Looking up the value for the acid dissociation constant for the ammonium ion, we get the value 5.6×10^{-10}. Recall that the larger the K_a, the stronger the acid; we can see that ammonium ion is a much weaker acid than acetic acid. Of course, this means that ammonia is a much stronger base than acetate. During very severe exercise, ammonia is produced by the deamination (removal of an amino group) of adenosine monophosphate or AMP. The ammonia can remove a proton from the cytoplasm of the exercising muscle cell to help maintain muscle cell pH. Can you determine the pK_a for ammonium ion?

Buffers are chemicals that resist changes in the pH of a solution. They play enormously important roles in our bodies given the very narrow pH range at which our metabolism operates. They do this by maintaining the [H⁺] within a very narrow range through absorbing added protons or reducing the effects of added OH⁻ ions. A buffer usually consists of a mixture of a weak acid and its salt such as acetic acid and sodium acetate. Buffers also work best in the pH range just above and just below the pK_a for the acid. Consider the preceding equation for the dissociation of the weak acid acetic acid. With a pK_a of 4.74, acetic acid would function best as a buffer just above and below pH 4.74, where there would be a roughly equal concentration of CH_3COOH (undissociated acetic acid) and its conjugate base acetate (CH_3COO^-). Addition of H^+ ions would result in their being attached to the acetate, forming undissociated acetic acid. Addition of hydroxide ions would result in their being combined with H^+ ions to form water, and more acetic acid would dissociate to maintain the H^+ balance. Similarly, a solution with an equal concentration of NH_3 (ammonia) and NH_4^+ (ammonium ion) would buffer best at the pK_a for ammonium ions (9.26).

Physiological buffers need to operate at roughly pH 7, since this is the internal pH of most cells. **Inorganic phosphate,** often denoted by the abbreviation Pi, is a mixture of dihydrogen ($H_2PO_4^-$) and monohydrogen (HPO_4^{2-}) ions. The $H_2PO_4^-$ is a weak acid whose dissociation is shown in the following:

$$H_2PO_4^- \leftrightarrow H^+ + HPO_4^{2-}$$

The pK_a for dissociation of the dihydrogen phosphate ion is 7.2, so a mixture of the dihydrogen and monohydrogen phosphates is a good buffer in the physiological pH range. Inside cells, the free concentration of Pi (the mixture of the two phosphates) is at least 1 millimolar (1 mM or 1 mmol/L). In addition, there are other phosphate-containing compounds, such as ATP (adenosine triphosphate), whose phosphate groups can also participate in pH control of the cell. Can you predict which of the two species of Pi would dominate in a skeletal muscle cell when the pH is reduced by severe exercise? Or if the pH were increased above 7 by alkalosis?

Before leaving this important topic, let's revisit the dissociation of the general weak acid, HA.

$$HA \leftrightarrow H^+ + A^-$$

From this, write the equilibrium expression:

$$K_a = [H^+] [A^-] / [HA]$$

Rearrange this to produce:

$$[H^+] = K_a [HA] / [A^-]$$

Take negative logs of each side:

$$-\log [H^+] = -\log K_a - \log [HA] / [A^-]$$

or

$$pH = pK_a + \log [A^-] / [HA]$$

This last equation is known as the **Henderson-Hasselbalch equation** and is a very useful way of expressing the relationship between the pH of a solution, the pK_a of a weak acid, and the relative concentrations of the weak acid and its conjugate base. This equation will prove to be very useful when we talk about the acid–base characteristics of the amino acids.

Structure of Amino Acids

The genetic code provides information for 20 different amino acids that make up proteins. Figure 1.2 illustrates what is typically shown for the general structure of an amino acid. In this figure, each amino acid has a carboxyl group—a weak acid, an amino group—a weak base, a carbon atom identified as the alpha (α)-carbon (also known as carbon 2), and a variable group known as the side chain and indicated by the letter R. In fact, 19 of the 20 common amino acids have this general structure. What makes each amino acid unique is its side chain or R group. Each amino acid has (a) a specific side chain (designated as an R group), (b) a name, (c) a three-letter designation, and (d) a single capital letter to represent it. Differences in the properties of the 20 common amino acids are based on their side chains.

The general amino acid structure depicted in figure 1.2 never occurs exactly as shown because of

Figure 1.2   The general structure of an amino acid, showing its weak acid (carboxyl) group and weak base (amino) group. These are attached to a central or alpha (α)-carbon along with a hydrogen atom and its variable part, given by R.

the acid–base properties of the amino and carboxyl groups. Figure 1.3 shows the general structure of an amino acid as it would exist at pH 7. This is often described as the **zwitterion** form, in which there is both a positively and a negatively charged group. Figure 1.4 shows each of the 20 amino acids, including the name, three-letter short form, single-letter identification, and the overall structural formula with the specific side chain. Amino acids are typically organized on the basis of the side chain structure and how this will interact with water. Charged polar and uncharged polar side chains will interact with polar water molecules.

Figure 1.3 The general structure of an amino acid as it would exist at cellular pH of 7.

> **KEY POINT**
>
> The 20 amino acids used to make proteins are the same ones used by all animal and plant life. Diversity in the structure of amino acids helps produce the diversity in the nature and function of proteins.

Acid–Base Properties of Amino Acids

All amino acids have at least one acid group (proton donor) and one basic group (proton acceptor). Therefore, amino acids may have both acid and base properties; that is, they are **amphoteric.** In amino acids, the major acid groups are the carboxyl (–COOH) and the ammonium (–NH_3^+) groups. These protonated forms can each give up a proton. The major base groups are the carboxylate (–COO^-) and amino (–NH_2) groups, which are unprotonated and can accept a proton.

Now let us look at equations describing the ionization of the protonated (acid) forms of the groups:

$$-COOH \leftrightarrow -COO^- + H^+$$

will have a K_a value and hence a pK_a.

$$-NH_3^+ \leftrightarrow -NH_2 + H^+$$

will also have a K_a and thus a pK_a.

The carboxyl group (–COOH) is a stronger acid than ammonium (–NH_3^+) and will have a lower pK_a value (or its K_a value will be larger). Conversely, the amino group will be a better proton acceptor (base) than the –COO^- and will have a larger pK_a.

Most of the amino acids have two groups that can act as acids, each with a pK_a, usually identified as pK_1 (the lower value) and pK_2 (the higher value). Figure 1.5 shows the structure of the amino acid alanine when it is a zwitterion. In this form, alanine has no net charge, but it has one positive and one negative charge. The pH at which a molecule has an equal number of positive and negative charges (and therefore no net charge) is its **isoelectric point,** designated pI. For amino acids like alanine, the pI is one-half the sum of pK_1 and pK_2.

Monoamino-dicarboxylic amino acids, such as aspartic acid and glutamic acid (see figure 1.4), have two groups that can be carboxyl or carboxylate and only one group that can be amino or ammonium. We call such amino acids acidic amino acids; and there will be three pK_a values for these, identified from lowest to highest as pK_1, pK_2, and pK_3. As shown in figure 1.6, the isoelectric point (pI) of aspartic acid will be the pH where each molecule has one positive and one negative charge, yet no net charge. The actual value for the pI will be one-half the value of pK_1 and pK_2, that is, 2.8. Because aspartic and glutamic acids always have a negative charge at cellular pH, they are almost always described as aspartate and glutamate, respectively.

Amino acids such as lysine (see figure 1.4) have two amino/ammonium groups and only one carboxyl/carboxylate group and are often called basic amino acids. Figure 1.7 shows the structure of lysine at its pI. Notice that lysine has no net charge, due to an equal number of positive and negative groups. The pI value for lysine will be one-half the sum of pK_2 and pK_3, that is, 9.8.

In addition to the major ionizable groups discussed so far (i.e., the amino and carboxyl groups

Amino acids with uncharged polar side chains

Serine
Ser
S

$$H - \underset{\underset{NH_3^+}{|}}{\overset{\overset{COO^-}{|}}{C}} - CH_2 - OH$$

Threonine
Thr
T

$$H - \underset{\underset{NH_3^+}{|}}{\overset{\overset{COO^-}{|}}{C}} - \underset{\underset{OH}{|}}{\overset{\overset{H}{|}}{C}} - CH_3$$

Asparagine
Asn
N

$$H - \underset{\underset{NH_3^+}{|}}{\overset{\overset{COO^-}{|}}{C}} - CH_2 - C \overset{O}{\underset{NH_2}{}}$$

Glutamine
Gln
Q

$$H - \underset{\underset{NH_3^+}{|}}{\overset{\overset{COO^-}{|}}{C}} - CH_2 - CH_2 - C \overset{O}{\underset{NH_2}{}}$$

Tyrosine
Tyr
Y

$$H - \underset{\underset{NH_3^+}{|}}{\overset{\overset{COO^-}{|}}{C}} - CH_2 - \bigcirc - OH$$

Cysteine
Cys
C

$$H - \underset{\underset{NH_3^+}{|}}{\overset{\overset{COO^-}{|}}{C}} - CH_2 - SH$$

Amino acids with charged polar side chains

Lysine
Lys
K

$$H - \underset{\underset{NH_3^+}{|}}{\overset{\overset{COO^-}{|}}{C}} - CH_2 - CH_2 - CH_2 - CH_2 - NH_3^+$$

Arginine
Arg
R

$$H - \underset{\underset{NH_3^+}{|}}{\overset{\overset{COO^-}{|}}{C}} - CH_2 - CH_2 - CH_2 - NH - C \overset{NH_2}{\underset{NH_2^+}{}}$$

Histidine
His
H

$$H - \underset{\underset{NH_3^+}{|}}{\overset{\overset{COO^-}{|}}{C}} - CH_2 - [imidazole ring]$$

Aspartic acid
Asp
D

$$H - \underset{\underset{NH_3^+}{|}}{\overset{\overset{COO^-}{|}}{C}} - CH_2 - C \overset{O}{\underset{O^-}{}}$$

Glutamic acid
Glu
E

$$H - \underset{\underset{NH_3^+}{|}}{\overset{\overset{COO^-}{|}}{C}} - CH_2 - CH_2 - C \overset{O}{\underset{O^-}{}}$$

Amino acids with nonpolar side chains

Glycine
Gly
G

$$H - \underset{\underset{NH_3^+}{|}}{\overset{\overset{COO^-}{|}}{C}} - H$$

Alanine
Ala
A

$$H - \underset{\underset{NH_3^+}{|}}{\overset{\overset{COO^-}{|}}{C}} - CH_3$$

Valine
Val
V

$$H - \underset{\underset{NH_3^+}{|}}{\overset{\overset{COO^-}{|}}{C}} - CH \overset{CH_3}{\underset{CH_3}{}}$$

Leucine
Leu
L

$$H - \underset{\underset{NH_3^+}{|}}{\overset{\overset{COO^-}{|}}{C}} - CH_2 - CH \overset{CH_3}{\underset{CH_3}{}}$$

Isoleucine
Ile
I

$$H - \underset{\underset{NH_3^+}{|}}{\overset{\overset{COO^-}{|}}{C}} - \underset{\underset{H}{|}}{\overset{\overset{CH_3}{|}}{C}} - CH_2 - CH_3$$

Methionine
Met
M

$$H - \underset{\underset{NH_3^+}{|}}{\overset{\overset{COO^-}{|}}{C}} - CH_2 - CH_2 - S - CH_3$$

Proline
Pro
P

$$COO^- \quad [pyrrolidine ring with H_2, CH_2, CH_2, N, H]$$

Phenylalanine
Phe
F

$$H - \underset{\underset{NH_3^+}{|}}{\overset{\overset{COO^-}{|}}{C}} - CH_2 - \bigcirc$$

Tryptophan
Trp
W

$$H - \underset{\underset{NH_3^+}{|}}{\overset{\overset{COO^-}{|}}{C}} - CH_2 - [indole ring]$$

Figure 1.4 Structures of the 20 amino acids used to make proteins, with their names, three-letter abbreviations, and single-letter identifiers.

$$COO^-$$
$$^+H_3N - C - H$$
$$CH_3$$

Figure 1.5 Structure of alanine at its isoelectric point.

$$COO^- \leftarrow pK_1 = 1.9$$
$$^+H_3N - C - H$$
$$pK_3 = 9.6 \quad CH_2$$
$$COOH \leftarrow pK_2 = 3.7$$

Figure 1.6 Structure of aspartate (aspartic acid) at its isoelectric point. The pK_a values for the three dissociable groups are shown in order of increasing magnitude.

$$COO^- \leftarrow pK_1 = 2.2$$
$$H_2N - C - H$$
$$pK_2 = 9.0 \quad (CH_2)_4$$
$$NH_3^+ \leftarrow pK_3 = 10.6$$

Figure 1.7 Structure of lysine at its isoelectric point. The pK_a values for the dissociable groups are shown.

attached to the α-carbon and side chain), others can be quite important (see figure 1.8). Histidine is often found at the active site of many enzymes, for it influences their catalytic ability.

KEY POINT

The pH of intracellular and extracellular fluids must be carefully regulated since even a small change in pH can add or remove a proton from an amino acid. Such a change may alter the structure and thus the function of proteins, altering our metabolism.

Stereoisomerism of Amino Acids

As shown in figure 1.4, all amino acids except glycine have four different groups attached to the α-carbon. Because of this, the α-carbon is a chiral center, and the amino acid is chiral or asymmetric, with two different ways of arranging these groups; that is, the amino acid has two different configurations. Figure 1.9 shows the groups around the α-carbon to be three-dimensional; the dashed lines mean that the bonds are going into the paper, and the wedges mean that the bonds are coming out of the plane of the paper toward you. When the carboxylate group is at the top and going into the paper, the L and D refer to the position of the ammonium group, that is, on the left side (L or *levo*) or right side (D or *dextro*). These **stereoisomers** (or space isomers) are also enantiomers—pairs of molecules that are nonsuperimposable mirror images of each other. They may be compared to a right and left hand. When you hold your left hand to a mirror, the image is that of your right hand. Similarly, holding a D-amino acid to a mirror gives an L-amino acid as the image. All naturally occurring amino acids are in the L-configuration.

$-CH_2SH \longleftrightarrow -CH_2S^- + H^+ \qquad pK_a = 8.2$

$-CH_2-\text{(phenyl)}-OH \longleftrightarrow -CH_2-\text{(phenyl)}-O^- + H^+ \quad pK_a = 10.1$

$-CH_2-\text{(imidazole, }^+NH) \longleftrightarrow -CH_2-\text{(imidazole, N)} + H^+ \quad pK_a = 6.0$

Figure 1.8 Equations to show the acid dissociation characteristics for the side chains of the amino acids cysteine, (top), tyrosine (middle), and histidine (bottom). The numerical value for the pK_a for each dissociation is shown.

Figure 1.9 The two stereoisomeric forms for the general structure for an amino acid. The amino acids in our bodies are in the L-configuration.

You may have learned another way to describe the arrangement of four different groups about a chiral center. When the groups are assigned a priority number, based on atomic number and arranged in descending priority, the lowest-priority group (hydrogen) points into the page, away from the viewer. If the group arrangement, from highest to lowest priority, runs counterclockwise, the configuration is said to be S (for *sinister* or left). If the configuration instead runs clockwise, it is said to be R (for *rectus* or right). Generally, the L-amino acids have an S absolute configuration.

CHARACTERISTICS OF PEPTIDES

Peptides are formed when amino acids join together via their ammonium and carboxylate groups in a specialized form of the amide bond known as a **peptide bond.** Because the body's amino acids are in an environment with a pH around 7, the amino group is protonated and is an ammonium group, whereas the carboxyl group is unprotonated and is a carboxylate group. Figure 1.10 shows how a peptide is formed from two amino acids. Combination of the ammonium and carboxylate groups yields the peptide bond, and a water molecule is eliminated. The peptide bond is rigid and planar. Peptides are

drawn by convention starting with the free ammonium group and ending with the free carboxylate group. We describe these as the N-terminus and the C-terminus, respectively.

The amino acids in a peptide are known as *amino acid residues*. The prefixes used to characterize the number of amino acid residues in a peptide are *di*—two, *tri*—three, *tetra*—four, *penta*—five, *hexa*—six, *hepta*—seven, *octa*—eight, *nona*—nine, and *deca*—ten. Thus a nonapeptide consists of nine amino acid residues. An *oligopeptide* is one that contains roughly 10 to 20 amino acid residues. The term **polypeptide** refers to a large peptide that contains more than 20 amino acids; indeed, some polypeptides have more than 4,000 amino acid residues. A protein and a polypeptide can be the same thing if there is only one polypeptide in the molecule. However, many proteins contain more than one polypeptide chain. In this case the protein is not a polypeptide, but a molecule containing two or more specific polypeptides. Since a peptide has only one N-terminus and one C-terminus, its properties are defined by the properties of the individual side chains of its constituent amino acids.

The peptide shown in figure 1.10 is a dipeptide. We would name it glycylalanine, starting from the amino acid at the N-terminus. We could also refer to this as gly-ala, using the three-letter short form, or GA, using the single-letter amino acid designations. There is a very important tripeptide in all cells that plays a key role in maintaining our cellular oxidation–reduction (redox) status. Known as glutathione, it is made up of the amino acids glutamate, cysteine, and glycine. The first peptide bond between glutamate and cysteine is via the side chain carboxyl of glutamate—not the usual α-carboxyl group. This is the gamma (γ)-carbon atom of glutamate. We thus designate glutathione as γ-glutamylcysteinylglycine (γglu-cys-gly or γECG).

Many important hormones are peptides, such as insulin, glucagon, and growth hormone. Growth factors have specific, potent effects on tissues, and these

Figure 1.10 The combination of two amino acids, glycine and alanine, to form the dipeptide glycylalanine.

are peptides. The hypothalamus, a key regulator of the overall function of our body, releases a number of peptide factors that control the secretions from other glands, particularly the pituitary gland. As we will see, even fat cells produce and secrete important peptide signaling molecules.

> ## ▶ K E Y P O I N T ◀
>
> Peptides are linear (as opposed to branched) molecules. This is a natural consequence of the linear sequence of bases within a gene in DNA, which is responsible for specifying the linear sequence of amino acids in the polypeptide chain. The **primary structure** of a peptide is the sequence of amino acids starting from the N-terminus.

STRUCTURE OF PROTEINS

Proteins are composed of one or more polypeptide chains. Many also contain other substances such as metal ions (e.g., hemoglobin contains iron), carbohydrates (e.g., glycoproteins contain sugars attached to amino acids; these can be found on cell membranes, pointing into the fluid surrounding cells), and fat or lipid (e.g., blood lipoproteins transport fat such as cholesterol and triacylglycerol). The precise biological structure of a protein, and hence its function, is determined by the amino acids it contains.

Primary Structure of Proteins

As mentioned earlier, the primary structure of a protein (or peptide) is the sequence of amino acids starting from the N-(amino) terminus. With 20 different amino acids, the number of different possible sequences for even a small polypeptide is enormous. However, a specific protein in your body must have its proper amino acid sequence. Even replacing one amino acid with another could make a protein totally useless and possibly dangerous. For example, sickle cell anemia, characterized by short-lived erythrocytes of unusual shape, results from the replacement of just one amino acid (valine) with another (glutamic acid) in one chain of the hemoglobin molecule. As shown in figure 1.4, valine has an uncharged hydrophobic side chain, whereas the side chain of glutamic acid carries a negative charge. Even this modest change can have a profound influence on the properties of the hemoglobin molecule. The

cystic fibrosis transmembrane conductance regulator protein (CFTR) allows chloride ions to move across membranes. This protein contains 1,480 amino acids. The disease associated with this protein, cystic fibrosis, is due to the absence of a single phenylalanine residue (phe-508).

The amino acid sequences of many thousands of proteins in humans and other species are now known. The sequence can be learned through actual sequencing of the amino acids in each polypeptide or learned from the base sequence of the gene for that polypeptide. The sequence always begins at the N-terminus of each polypeptide chain if the protein has more than one. The N-terminus is designated number 1, with succeeding amino acids given higher numbers. Since we know the sequence of so many polypeptides, the specific amino acid is described along with its position in the polypeptide. For example, serine-283 (ser-283) is the 283rd amino acid residue in the polypeptide, and it is serine. Amino acid sequence determines the three-dimensional structure for the protein. Moreover, with computers, amino acid sequences, and known three-dimensional structures, one can make computer-guided predictions about the three-dimensional structure of a new protein given its sequence. Scientists in biotech companies throughout the world are designing peptides to improve human function. The study of large compilations of proteins, as those expressed in a single cell, is known as **proteomics.** Researchers in this area are proceeding at a rapid pace, in many cases looking for specific amino acid sequence differences that underlie particular diseases.

Bonds and Interactions Responsible for Protein Structure

Proteins have a very specific structure in the cell that is essential for a protein to perform its unique function. As we have seen, the primary structure, or the sequence of amino acids in the polypeptide chain, is maintained by peptide bonds. These are strong, rigid bonds joining the individual amino acids together in the protein. However, the three-dimensional structure of proteins is very important to their function. The three-dimensional structure is maintained by a variety of strong and weak bonds. Figure 1.11 illustrates these bonds and how they are responsible for the overall three-dimensional shape of a protein.

Disulfide bonds are covalent bonds that join together the side chains of two cysteine residues ($-CH_2SH$) in the same polypeptide or different

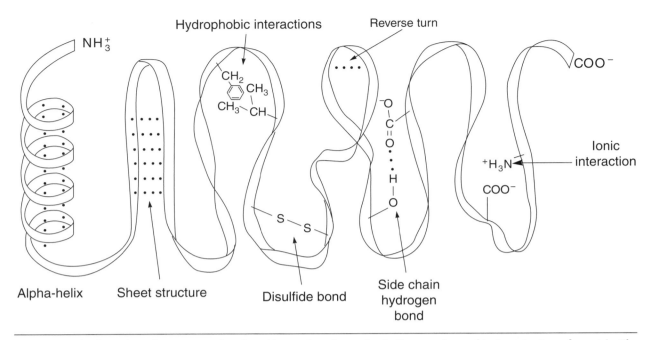

Figure 1.11   A schematic to show common bonds and interactions that maintain the secondary and tertiary structure of a protein. The ribbon shows the peptide backbone. The dotted lines represent hydrogen bonding.

polypeptides, resulting in disulfide (S–S) bonds and the loss of two hydrogen atoms. Disulfide bonds in the same polypeptide chain are intramolecular bonds, whereas those joining different polypeptide chains are intermolecular bonds. For example, the polypeptide hormone insulin consists of two peptide chains held together by two intermolecular disulfide bonds.

Hydrogen bonds are weak electrostatic attractions between an electronegative element, such as oxygen or nitrogen, and a hydrogen atom that is covalently bonded to another electronegative element. Hydrogen bonds between individual water molecules, for example, are responsible for some of the unique properties of water. Figure 1.11 shows the hydrogen bond between the components of a peptide bond; such bonds are important in maintaining overall protein structure. Hydrogen bonds also form between polar parts of some amino acid side chains and those in the protein molecule and water (see polar side chains of amino acids in figure 1.4). Although hydrogen bonds are weak, they are important because so many of them are involved in the structure of a protein. Figure 1.11 shows *ionic interactions* between oppositely charged groups, as found in proteins, such as the N- and C-termini and the charged carboxylate or ammonium groups in the side chains.

Amino acids are primarily found in a polar environment because they are in an aqueous medium with charged ions and polar molecules. Proteins spanning membranes are an exception because the interior of a membrane is hydrophobic. Water is a polar molecule and interacts well with polar and charged groups on amino acids. This means that inside cells and in the blood, polar and charged amino acids are most likely to be exposed to their aqueous environment. However, 9 of the 20 common amino acids are **hydrophobic** or nonpolar because their side chains are composed of carbon and hydrogen atoms and thus have no affinity for water (see figure 1.4). In fact, they can disrupt the relatively organized structure of liquid water. These side chains tend to cluster in the interior of the protein molecule, not because they have an affinity for each other but because they cannot interact with water. Fat globules also form tiny spheres in water to keep their exposed surface as small as possible. We call the clustering of hydrophobic groups on amino acids hydrophobic interactions (see figure 1.11).

Secondary Structure of Proteins

So far, a one-dimensional protein structure has been described, that is, the sequence of amino acids. As already mentioned, proteins actually have a three-dimensional structure. Imagine a peptide backbone with the side chains of each amino acid radiating out. The backbone consists of three elements—the α-carbon, the nitrogen group involved in the peptide bond, and the carbonyl group. The **secondary structure** is the spatial path taken by these three elements of the peptide backbone. Side chains (or R groups) of individual amino acids are not considered in secondary structure. Two common, recognizable structural features of the polypeptide backbone, the **alpha-helix** and the **beta-sheet,** are found in many proteins. These result from the way the polypeptide backbone organizes itself. They are stabilized by hydrogen bonding between neighboring elements of the peptide bond, as shown in figure 1.11 with dotted lines. The reverse turn, also shown in figure 1.11, involves a sharp turn in the polypeptide backbone and is also maintained by hydrogen bonding.

Tertiary Structure of Proteins

The tertiary structure is the overall three-dimensional arrangement of a polypeptide chain, including the secondary structure and any nonordered interactions involving amino acid residues that are far apart. Figure 1.12 shows the three-dimensional structure of the small oxygen-binding protein, myoglobin. This contains a single polypeptide chain and a single heme group. Most proteins are found in aqueous

mediums where they assume a compact globular structure maintained by the forces already described. Examples include proteins in the blood and enzymes within cells. The hydrophobic side chains are buried in the interior of the protein, away from the aqueous medium, with **hydrophilic** side chains and the N- and C-termini exposed on the surface where they can interact with water or other polar molecules and ions. The tertiary structure also involves the spatial position of ions or organic groups that are part of the makeup of many proteins. The term **amphipathic** is often used to describe a part of a protein that must interact with both a hydrophobic and a hydrophilic region. *Amphipathic* means that the section of the protein will have a polar region able to interact, for example, with the cytoplasm of the cell, and a nonpolar region that may interact with the hydrophobic part of the cell membrane.

KEY POINT

The sequence of bases in a gene specifies the amino acid sequence of a polypeptide, which determines secondary and tertiary structure. The biological function of the polypeptide is therefore based on its overall structure.

The secondary and tertiary structure of a polypeptide chain depends on the kind and sequence of its amino acids. In the cell, or wherever a protein is found, the overall structure or conformation of a protein must be maintained, because proteins must recognize and interact with other molecules. Biochemists describe a protein's three-dimensional form as its fold. However, the conformation of a protein in vivo is not fixed but changes in subtle ways as it carries out its particular function. For example, the cross bridges in the contractile protein myosin alter their conformation to generate force during muscle contraction. Membrane transport proteins alter their conformations as they move substances from one side of a membrane to the other.

Quaternary Structure of Proteins

Many proteins consist of more than one polypeptide chain, each containing its own unique primary, secondary, and tertiary structures. We call these chains subunits. **Quaternary structure** refers to

Figure 1.12 Three-dimensional view of the small oxygen-binding protein myoglobin. In the upper middle portion, the heme group can be seen with its central iron ion (sphere). Much of the secondary structure of myoglobin is alpha-helices.

Reprinted, by permission, from J.M. Berg, J.L. Tymoczko, and J.L. Stryer, 2002, *Biochemistry*, 5th ed. (New York: W.H. Freeman), 31.

the arrangement of the individual subunits with respect to each other. The subunits in an oligomeric protein (i.e., one containing more than one subunit) are held together with noncovalent bonds, that is, hydrogen bonds, electrostatic interactions, and hydrophobic interactions. Oligomeric proteins are common because several subunits allow subtle ways of altering the protein's function. For example, hemoglobin A, the adult form of hemoglobin, is a tetramer consisting of four subunits, two α (with 141 amino acids per subunit) and two β (with 146 amino acids per subunit). Each subunit contains one heme group that binds one oxygen molecule. The quaternary structure of hemoglobin refers to the way the two α and two β subunits interact with each other. The oligomeric nature of hemoglobin enhances its role in loading oxygen in the lung and releasing it at the cell level. In the lungs, where the oxygen concentration is high, binding of one oxygen molecule to one subunit facilitates the binding of oxygen molecules to the other subunits. This helps make hemoglobin saturated with oxygen in the lungs, maximizing its ability to transport oxygen. At the tissue level, where the oxygen concentration is much lower, the subunit interactions facilitate the unbinding of oxygen so that it is available for diffusion into adjacent cells.

Myosin, a contractile protein, is a hexamer (i.e., consists of six subunits). There are two heavy chains, each with a molecular weight of about 220,000 (often expressed as 220,000 daltons or 220 kilodaltons, kD). The two chains wind around each other, form-

ing a single long chain for most of their length. At one end, they each form separate globular regions, known as heads. In each globular head there are two light chain polypeptides, each with a molecular weight of about 18 to 25 kD. One is described as a regulatory light chain and the other is known as an essential light chain (see figure 1.13). The globular heads act as the cross bridges that attach to the protein actin to generate force when a muscle contracts. We will discuss the role of different heavy and light chains in terms of myosin function and fiber typing in chapter 3.

The term **domain** refers to part of a protein molecule, usually globular in shape, that has a specific function, such as binding or catalysis. A single protein may contain a number of domains. For example, each heavy chain in the contractile protein myosin has an actin-binding domain and an ATP-binding domain. These domains act in concert to utilize the energy of ATP to generate a force and produce movement.

The noncovalent forces that maintain the secondary, tertiary, and quaternary structures of proteins are generally weak. **Denaturation** refers to a disruption in the overall conformation of a protein with loss of biological activity. When denatured, proteins usually become less soluble. Denaturation can be caused by heat, acids, bases, organic solvents, detergents, agitation (e.g., beating egg whites to produce a meringue), or specific denaturing agents such as urea. We seldom encounter such harsh conditions that our body proteins are denatured. However, burning the skin or the reflux of acid from the

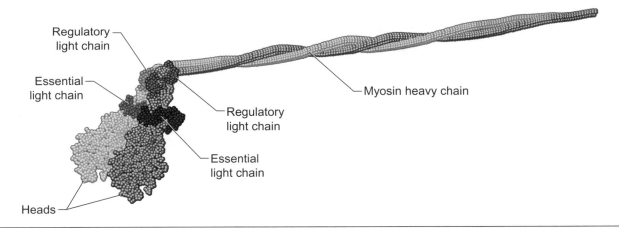

Figure 1.13 The contractile protein myosin. Myosin is a hexamer, composed of six subunits. Two heavy chains run the length of the molecule, terminating in two globular heads that act as cross bridges. Four myosin light chains are also part of the myosin molecule, two per cross bridge head; one is found in the neck region where each globular head begins and another is located further into each globular head.

Reprinted, by permission, from R. Vandenboom, 2004, "The myofibrillar complex and fatigue: A review," *Canadian Journal of Applied Physiology* 29(3): 330-356.

stomach into the esophagus may cause some protein denaturation. Before proteins are broken down in their normal turnover process, they are denatured by enzyme-catalyzed chemical modification, then broken down to their constituent amino acids. In high concentrations, urea, a normal product of amino acid breakdown in our cells, can be used as a laboratory denaturant.

SUMMARY

Maintenance of protein acid–base balance is critical to life. Acids are substances that can donate protons, and bases are proton acceptors. Buffers are chemicals that can help maintain pH in the face of added acid or base. Proteins are the action molecules of life, made from 20 different amino acids. Differences among amino acids are based on their characteristic side chains, known as R groups. Amino acids exist in all cells of the body and in the fluids outside the cells. Natural amino acids are in the L-configuration. Because each has at least one group that acts as an acid and one that acts as a base, they are said to be amphoteric. All amino acids have at least one charged group, and at the neutral pH associated with life they have at least two charged groups. When an amino acid has no net charge, it is said to be a zwitterion. The pH where this occurs is known as the isoelectric point.

Peptides are formed when amino acids are joined together by peptide bonds. In general, proteins are large peptides (polypeptides), usually containing 100 or more amino acids. The primary structure of a peptide is the sequence of amino acid residues starting from the end containing the free amino group, known as the N-terminus. Because of their large size, proteins have higher levels of three-dimensional structure, known as secondary and tertiary structures. If a protein consists of more than one polypeptide, the structural relationship of the polypeptide subunits is described as a quaternary structure. The forces such as hydrogen bonds and hydrophobic interactions maintaining the secondary, tertiary, and quaternary structures are generally weak.

REVIEW QUESTIONS

1. Using the information found in figure 1.4, answer the following questions about the peptide shown in shorthand: Gly-Arg-Cys-Glu-Asp-Lys-Phe-Val-Trp-Cys-Leu.

   a. How many amino acid residues are there?

   b. Identify the N- and C-terminal amino acids.

   c. What would be the net charge on this peptide at pH 7.0?

   d. Identify the amino acids that have a polar, but uncharged, side chain.

   e. Identify the nonpolar amino acids.

   f. What are the hydrophobic amino acids?

   g. Is it possible for this amino acid to have an intramolecular disulfide bond? Explain.

   h. Write the primary sequence using single-letter amino acid abbreviations.

2. Using the Henderson-Hasselbalch equation, predict the pH of a solution in which cysteine carries a net negative charge.

3. Write the dominant molecular formula for the constituent of Pi that exists at pH of 6.0 and then 8.0.

CHAPTER
2

ENZYMES

Enzymes are proteins that catalyze the thousands of different chemical reactions that constitute our metabolism. As proteins, they are large molecules, made from the 20 different amino acids. As catalysts, enzymes speed up chemical reactions without being destroyed in the process. As proteins, their lifetime is relatively short, so the concentration of most enzymes at any time reflects their relative use. This chapter begins with a review of catalysts, the role enzymes play as catalysts, and key factors that are critical to enzyme action in our cells. Many enzymes function with the aid of helpers or cofactors, especially in reactions where oxidation and reduction take place. We conclude with an overview of how enzymes are regulated and how we can measure enzyme action.

ENZYMES AS CATALYSTS

Enzymes speed up chemical reactions by lowering the energy barrier to a reaction—called the energy of activation—so that the reaction takes place at the low temperature of an organism (37 °C for a human). In the laboratory, chemical reactions may take place because we heat the reacting substances, overcoming the energy barrier that prevents the reaction from occurring at a lower temperature. Because our bodies function at a relatively constant, but low, temperature (37 °C), enzymes are essential to reduce the energy barrier or energy of activation. In this way, the thousands of chemical reactions that reflect our metabolism can take place speedily. In fact, enzymes are so effective that they can increase the speed of reactions by millions of times the rate of an uncatalyzed reaction.

The molecule (or molecules) that an enzyme acts on is known as its **substrate,** which is then converted into a product or products. For example, when we digest dietary protein (the substrate) with protein-digesting enzymes in our small intestine, we produce amino acids and small peptides as products. Enzymes are highly specific, catalyzing a single reaction or type of reaction. Left to themselves, many chemicals can interact over long periods of time to reach an equilibrium in which there is no net change in substrate or product concentrations. The enzyme allows the reaction to reach equilibrium much faster than it would if the enzyme were absent, but the position of equilibrium remains the same.

A part of the large enzyme molecule will reversibly bind to the substrate (or substrates), and then a specific part or parts of the enzyme will catalyze the detailed change necessary to convert the substrate into a product. The enzyme has amino acid residues that bind the substrate, as well as those that carry out the actual catalysis. There is thus a **binding site** and a **catalytic site,** although the term **active site** often represents both the binding and catalytic domains of the enzyme protein. Active sites are typically clefts in the three-dimensional enzyme structure, where amino acid residues from close or distant parts of the molecule can act on the substrate. The bonds and interactions we discussed in the previous chapter maintain the conformation of the active site and interact with the specific substrate molecule(s).

We can write a general equation to describe a simple enzyme-catalyzed reaction in which a single substrate S is converted into a single product P. The reaction could be irreversible, with all substrate molecules converted into product molecules and indicated by an arrow with only one head pointing toward the product. Irreversible reactions are typically described as nonequilibrium reactions. Alternatively, the reaction could be reversible, such that, given time, it establishes an equilibrium, with the ratio of product concentration to substrate concentration a constant known as the **equilibrium constant** (K_{eq}). Reversible reactions are commonly called equilibrium reactions, and the two terms mean the same thing. An equilibrium reaction will be shown with a double-headed arrow, meaning that the product is also a substrate for the reverse reaction. Figure 2.1 illustrates these two types of reactions and outlines their properties.

- Irreversible reaction
- Large energy change
- Nonequilibrium reaction
- Less common

- Reversible reaction
- Small energy change
- Equilibrium reaction
- Product also a substrate for the reverse reaction
- More common

Figure 2.1 The properties of irreversible (left) and reversible (right) reactions, catalyzed by enzyme E, in which a substrate (S) is converted to product (P).

Most enzyme-catalyzed reactions are considered to take place in discrete steps (see figure 2.2). Enzyme E first combines reversibly with substrate S to form an initial enzyme-substrate complex ES. This complex is then converted to an enzyme-intermediate complex EI, which then changes to an enzyme-product complex EP. Enzyme-product complex then dissociates into free product, and the enzyme is released unchanged.

Figure 2.2 Enzyme-catalyzed reactions. These reactions are considered to take place in steps. Substrate S first combines with enzyme (E) to form an ES complex. This subsequently becomes an enzyme-intermediate complex (EI), which proceeds on to the enzyme-product complex (EP) until product is released.

RATES OF ENZYMATIC REACTIONS

Consider the case of an enzyme-catalyzed reaction in which substrate S is converted to product P, catalyzed by enzyme E. This reaction is reversible, so that P is a substrate for the reverse reaction. Our goal is to measure the rate of this reaction in the direction S↔P. If we begin with only S plus the enzyme, P will be formed. As the concentration of P increases and that of S decreases, the reverse reaction will take place, becoming more important as the concentration of P increases. Therefore the reaction rate in the direction toward P decreases over time. Accordingly, the rate of the forward reaction must be measured quickly before any appreciable amount of P is formed. We call this quickly measured forward reaction rate the initial velocity. Imagine a barrel filled with water. Now punch a hole in it. The water will come shooting out, forcefully at first, but with the force gradually decreasing as the level of water in the barrel falls. This is what happens to the rate of conversion of S to P as the concentration of S decreases.

Effect of Substrate Concentration

Let us carry out an experiment in which we set up 10 test tubes, each containing a solution at 25 °C, pH 7.0, and a fixed concentration of enzyme E. We add a specific amount of substrate S to each test tube, mix it, and quickly measure the rate of the reaction, either by measuring the rate of disappearance of S or the rate of appearance of P. Let us assume that the concentration of S in test tube 1 is 2 micromolar (2 μM), with higher and higher concentrations in the other tubes such that the concentration of S in tube 10 is 500 μM. After getting our initial velocities, expressed in units of micromoles of S disappearing per minute (or micromoles of P appearing per minute), we plot the initial velocity as a function of substrate concentration.

Figure 2.3 shows that initial velocity is higher as substrate concentration is increased. Note that the relationship is not linear but hyperbolic. That is, at low substrate concentration the velocity increases linearly with increasing substrate concentration, but at a higher substrate concentration the velocity flattens out, approaching a **maximum velocity** called **Vmax.** When this is reached, increasing substrate concentration will not produce any increase in the rate of the reaction. At this point, each enzyme molecule is working as fast as it can, converting S to P. If we determine the value of the maximum velocity, divide this in half, then determine what substrate concentration will produce one-half of Vmax, we get a concentration known as the **Michaelis constant** and represented by the abbreviation K_m. The K_m is defined as the substrate concentration needed to

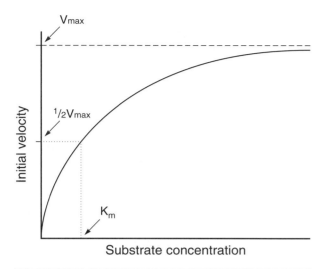

Figure 2.3 Relationship between the initial velocity of an enzyme-catalyzed reaction and the substrate concentration. Vmax is the maximum velocity that can be achieved using a fixed amount of enzyme. K_m is the Michaelis constant, the substrate concentration needed to generate one-half maximum velocity.

produce one-half the maximal velocity of an enzyme-catalyzed reaction. The K_m has units of concentration because it represents an intercept on the X-axis. Vmax is the limiting rate for the velocity of an enzyme-catalyzed reaction at fixed enzyme concentration. It occurs when enzyme active sites are so totally saturated with substrate that as a substrate molecule is converted to product and leaves the enzyme, another substrate molecule immediately binds.

Vmax and K_m are known as **kinetic parameters** for an enzyme-catalyzed reaction. K_m is said to reflect the affinity of an enzyme for its substrate; the smaller the value of K_m, the greater the affinity an enzyme has for its substrate. In other words, if the enzyme has a high affinity for its substrate, it can convert the substrate to product even when the [S] is low. The reverse reaction will have its own kinetic parameters, because

the product becomes the substrate for the backward reaction. The kinetic parameters for the reverse reaction are unlikely to be identical to those for the forward reaction. Enzymes that exhibit hyperbolic kinetics, as shown in figure 2.3, are said to exhibit Michaelis-Menten kinetics. Not all enzymes behave this way, as we shall discover later in this chapter.

In the cell, the substrate concentration is generally equal to or less than the value for its K_m. This offers two advantages: (a) A substantial fraction of enzyme catalytic ability is being used, and (b) the substrate concentration is low enough that the enzyme can still respond to changes in substrate concentration because it is on the steep part of the curve (see figure 2.3). If the substrate concentration is much greater than K_m we will get efficient use of the enzyme, but it will respond less effectively to changes in substrate levels because it is on the flat part of the curve (see figure 2.3). K_m values for substrates thus reflect in general the concentrations of these substrates in the cell.

KEY POINT

Tissues can adjust the amount of enzyme molecules by altering the rate of expression of the gene for specific enzymes, leading to higher or lower cellular content of the enzymes. The liver can vary the amount of enzymes involved in metabolizing carbohydrates and fats by adjusting the rate of expression of genes for key enzymes in response to changes in our dietary intake of carbohydrates and fats. In this way, the amount of enzyme reflects in general the amount of its substrate.

KEY POINT

Which direction a reaction actually takes is based on the relative concentrations of substrates and products compared to the individual K_m values. This direction is driven by the release of energy. A reaction taking place is analogous to a ball rolling down a hill. Balls roll down hills, not up. Reactions take place because energy is released. At equilibrium, nothing happens and no energy is released.

There are enzymes that catalyze a given reaction in different tissues, but the enzymes have different kinetic parameters for the substrate(s). Often products of different genes, they are known as **isozymes** or isoenzymes. When glucose enters a cell, the first thing that happens is that a phosphate group is attached to it. This reaction is catalyzed by four isozymes known as hexokinase I, II, III, and IV. Three of these isozymes have low K_m values for glucose (20-120 μM); the fourth (hexokinase IV, also known as glucokinase and found in the liver) has a high K_m for glucose (5-10 mM). The low-K_m isozymes can phosphorylate glucose even when the blood concentration is low. This is especially important to the brain, which depends solely on glucose as its fuel under normal circumstances. The high-K_m isozyme is found in the liver, where glucose is stored when blood concentration is elevated, for example following a meal. The liver isozyme thus readily responds to glucose as a substrate only when blood glucose is elevated. Figure 2.4 shows the Michaelis-Menten kinetics for a low-K_m hexokinase and for glucokinase.

Figure 2.4   Relative velocities of a low-K_m and a high-K_m isozyme of hexokinase. The high-K_m hexokinase is known as glucokinase.

Determining reliable values for the kinetic parameters K_m and Vmax from the velocity–substrate concentration curve shown in figure 2.3 is difficult. A wide range of substrate concentrations must be tested to ensure a reasonable value of Vmax because we also need this value to determine K_m. An easier way, shown in figure 2.5, is a **Lineweaver-Burk plot.** Here, the reciprocals of velocity and substrate concentration are plotted, which allows one to determine accurately the kinetic parameters for an enzyme-catalyzed

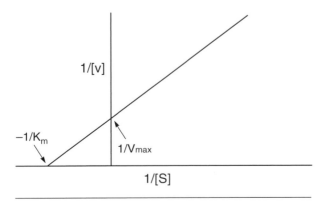

Figure 2.5   Lineweaver-Burk plot showing the determination of reliable values for K_m and Vmax.

reaction because these values are determined from intercepts on the horizontal and vertical axes. One can thus obtain more accurate values for the kinetic parameters from fewer data points.

Protein Transporters

The cell membrane and internal membranes (such as the mitochondrial membrane) act as barriers to the movement of molecules and ions. Since cells can function only if there is a two-way traffic of ions and molecules across their membranes, specific transport proteins, embedded in or traversing the membrane, are needed to make this happen. The membrane transport protein may go by such names as translocase, porter, carrier, or transporter. The substances carried may be small ions (Na^+, Mg^{2+}, Ca^{2+}, etc.) or nutrient molecules such as glucose, amino acids, or fats.

The membrane protein may function as a simple conduit that allows an ion or molecule to flow down its concentration gradient, either into or out of the cell or organelle (mitochondrion). We call this **facilitated diffusion.** It is diffusion because the diffusing substance is moving from a higher to a lower concentration. The membrane transporter facilitates the process. Alternatively, the transport process requires the diffusing substance to move against its concentration gradient. Moving a molecule or ion from a lower to a higher concentration is similar to a ball rolling up a hill. This can take place only if driven by a release of energy. The energy release may come from the breakdown of ATP (adenosine triphosphate) or by the simultaneous movement of another molecule down a steeper concentration gradient. We call this energy-requiring membrane transport process **active transport.** For example, creatine is synthesized in the liver or taken in our diet from the meat we eat. Creatine circulates in the blood and enters cells that have a specific creatine transporter. Creatine is especially concentrated in muscle cells (fibers), at levels far above that in the blood. The reason muscle cells can take up creatine is that sodium ions (Na^+) are simultaneously transported into the muscle fiber down a steep gradient. The energy released in the flow of sodium allows the creatine to move up its gradient. In essence, the sodium drags the creatine into the fiber (figure 2.6).

Although membrane transporters are not technically enzymes, catalyzing reactions in which a substrate is converted into a product, they function in a way that is consistent with the kinetics of enzymes. They are similar to enzymes because the transport-

Blood   Sarcolemma   Cytosol

[Na$^+$] 142 mM   [Na$^+$] 10 mM
[Cr]   0.05 mM   [Cr]   8 mM

Na$^+$

Cr

Creatine transporter

Figure 2.6 An example of active transport. Creatine (Cr) transport, across the sarcolemma from the blood and into skeletal muscle cells, takes place against a concentration gradient because of the simultaneous transport of sodium ions down their gradient.

ers recognize only specific transport substances and move them in a particular direction across the membrane. Their kinetics are similar to those of many enzymes in that we can describe the transport kinetics with Vmax and K_m. The transporters may also be subject to inhibition by other substances just as enzymes can be inhibited.

▶ K E Y   P O I N T ◀

Transport of substances across cell membranes and intracellular membranes is critical to our metabolism. Transport proteins are selective in terms of the ions or molecules they recognize and the direction of transport of these. In all cases, transport across a membrane by either diffusion or active transport occurs with the release of energy.

Effect of Enzyme Concentration

If an enzyme is saturated with substrate and thus working as fast as possible, adding more enzyme increases the reaction velocity. Vmax is thus proportional to enzyme concentration. However, changing enzyme concentration increases only Vmax and has no influence on K_m. This is illustrated in figure 2.7, where a 60% increase in enzyme content produces a 60% increase in Vmax but no change in K_m. We can use the relationship between enzyme concentration and maxi-

mum velocity to determine how much of a particular enzyme is present in a tissue or fluid (e.g., blood). When we undertake an exercise training program, the amount of some enzymes in the trained muscles may increase by a factor of 2 or more. This adaptation is in response to the increased demand placed on certain enzymes during training. The muscle cell responds by stimulating the synthesis of more enzyme proteins to reduce the stress of the training.

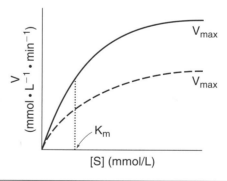

Figure 2.7 Enzyme velocity as a function of substrate concentration for two different concentrations of enzyme molecules. The solid curve is based on a 67% increase in enzyme concentration. Note that while Vmax is 67% higher when 67% more enzyme is present, the K_m is not changed.

Effect of pH

Most enzymes have a *pH optimum,* that is, a particular pH or narrow pH range where enzyme activity is maximal. As shown in figure 2.8, values of pH on either side of optimum produce lower reaction rates. The reason is that some of the forces holding an enzyme in its native conformation depend on charged groups. A change in pH can alter these charges, thus reducing enzyme function. The change may directly influence the active site or some other part of the enzyme that indirectly affects the active site. A change in pH may also alter the substrate for an enzyme, which could also influence rate. For most enzymes, their pH optimum reflects where they are active in the body. As an example, the stomach enzyme pepsin has a pH optimum around 2 because the stomach is acidic. Severe muscle fatigue is due in part to the acidification of the muscle. This can diminish the activity of certain enzymes, thus compromising muscle function.

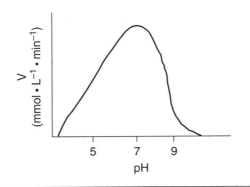

Figure 2.8   The effect of pH on enzyme-catalyzed reaction velocity.

Effect of Temperature on Enzyme Reactions

Like all chemical reactions, enzyme-catalyzed reactions increase in rate if the temperature is increased. However, since the forces holding the three-dimensional conformation of an enzyme are generally weak, heating too much disrupts the conformation and decreases enzyme activity. Thus, if we plot the velocity of an enzyme-catalyzed reaction as a function of temperature, the curve rises with increasing temperature until about 50 °C, at which point most enzymes start to denature and the velocity drops quickly (figure 2.9). Biochemists describe the relationship between temperature and reaction rate by the quotient Q_{10}, which describes the fold increase in reaction rate for a 10 °C rise in temperature. For many biological processes, the rate of a reaction approximately doubles for each 10 °C increase in temperature. This means the Q_{10} is about 2. Can you

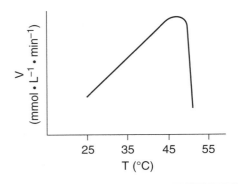

Figure 2.9   The effect of temperature on enzyme-catalyzed reaction velocity.

see how warming up muscles prior to competition can increase the rates at which the energy-yielding reactions take place by increasing the activities of the enzymes?

Turnover Number (k_{cat})

Turnover number, also known as the *catalytic constant* (k_{cat}), is defined as the maximum number of molecules of substrate converted to product per enzyme active site per unit of time (usually 1 s). It is a measure of how fast an enzyme can convert substrate to product. Since turnover number is considered a measure of the maximum catalytic activity for an enzyme, the enzyme must obviously be saturated with substrate for this to occur. Some enzymes are extremely fast as catalysts. For example, catalase, an antioxidant enzyme that breaks down hydrogen peroxide, has a k_{cat} of approximately 10 million. Carbonic anhydrase, which combines carbon dioxide with water to make carbonic acid, has a k_{cat} of approximately 1 million. However, most enzymes do not operate under conditions in which they are constantly saturated with substrate. A better way to express their catalytic efficiency in vivo is to use the expression k_{cat}/K_m, which gives a truer picture of enzyme function under physiological conditions. The k_{cat} directly reflects the speed with which an enzyme turns substrate into product, whereas the K_m is inversely related to the affinity of the enzyme for its substrate. Thus the ratio k_{cat}/K_m provides a measure for the catalytic efficiency of the enzyme, that is, how fast the enzyme can work given its physiological substrate concentration. While there may be large differences in values for k_{cat} among enzymes, the ratio k_{cat}/K_m provides far less diversity since most enzymes with large k_{cat} values generally have large K_m values.

Enzyme Inhibition

Enzymes can be inhibited by a variety of substances. We describe these inhibitors according to how they influence the enzyme. **Competitive inhibitors** resemble the normal substrate for an enzyme, for they bind to the active site, but they cannot be changed by the enzyme into product. In this sense they act as mimics, but because of subtle differences between them and the normal substrate they are not chemically altered by the enzyme. The inhibitor simply occupies the active site, binding, leaving, binding, and leaving, in a reversible fashion. Competitive inhibitors thus compete with the normal substrate for a place on the active site of the enzyme. This inhibition can be overcome by the addition of excess substrate. Accordingly, a competitive inhibitor will not affect the Vmax of the enzyme. It will, however, increase the K_m because more substrate will be needed to overcome the effects of the competitive inhibitor. Figure 2.10 show the effects of a competitive inhibitor, using both a normal velocity–substrate curve and a Lineweaver-Burk plot.

A **noncompetitive inhibitor** does not resemble the normal enzyme substrate and does not bind to the active site. However, when bound to the enzyme, it interferes with its function, taking that enzyme molecule out of commission. Hence, noncompetitive inhibitors lower Vmax but do not alter K_m. Examples of noncompetitive inhibitors are heavy metal ions (e.g., Hg^{2+}, Pb^{2+}), known to cause many health problems because of their tight binding to reactive side chains of amino acids in proteins. Figure 2.11 shows the effects of a noncompetitive inhibitor using both a standard velocity–substrate relationship and a Lineweaver-Burk plot.

Many drugs are developed based on the principles of enzyme inhibition, with specific reactions in bacteria, viruses, or tumors targeted by antibiotics or antiviral drugs. With the sequences and three-dimensional structure known for so many enzymes, researchers can design drugs to target specific active sites of enzymes by mimicking their substrates. Reduction of the effects of inflammation and arthritis by targeting key enzymes such as cyclooxygenase-1 (COX-1) and cyclooxygenase-2 (COX-2) represents a key example of using competitive inhibition of enzymes to improve health.

Figure 2.10 *(a)* The effects of a fixed concentration of a competitive inhibitor (dashed line) on the relationship between reaction velocity and substrate concentration for an enzyme-catalyzed reaction. In *(b)*, the same data are shown using a Lineweaver-Burk plot.

Figure 2.11 *(a)* The effects of a fixed concentration of a noncompetitive inhibitor (dashed line) on the relationship between reaction velocity and substrate concentration for an enzyme-catalyzed reaction. In *(b)*, the same data are shown using a Lineweaver-Burk plot.

PROVISION OF REACTIVE GROUPS BY COFACTORS

Enzymes have reactive groups in the form of side chains of amino acids as well as the N- and C-termini. However, they may need other reactive groups not available on amino acids, called *cofactors,* in order to carry out their functions as catalysts. Cofactors may be metal ions, such as Zn^{2+}, Mg^{2+}, or Mn^{2+}, or they may be organic molecules called **coenzymes.** The polypeptide part of the enzyme is called the *apoenzyme;* when it is combined with the cofactor, we have the **holoenzyme,** illustrated as follows:

apoenzyme *(inactive)* + cofactor = holoenzyme *(active)*

A cofactor tightly bound to the enzyme at all times is called a **prosthetic group.** An example of a prosthetic group is heme, bound tightly to hemoglobin and myoglobin (see figure 1.12 on p. 13). If the cofactor is not tightly bound but combines with the enzyme and the other substrate during the reaction, we can consider it a second substrate.

There are eight B vitamins that we need because they form the basic components for coenzymes. Humans and many animals eating a plant-and-animal diet have lost the ability to synthesize the B vitamins, so these must come from the diet. Table 2.1 lists the eight B vitamins, the coenzymes they form, and their short-form names. The deficiency diseases associated with inadequate intake of specific B vitamins are due to insufficient catalytic power of enzymes because they lack their coenzymes.

Table 2.1　The B Vitamins, the Coenzymes They Form, and Their Common Abbreviations

B vitamin	Coenzyme	Abbreviation
Thiamine (B1)	Thiamine pyrophosphate	TPP
Riboflavin (B2)	Flavin adenine dinucleotide	FAD
Niacin (B3)	Nicotinamide adenine dinucleotide	NAD^+
Vitamin B6	Pyridoxal phosphate	PLP
Pantothenic acid	Coenzyme A	CoA
Folate (folacin)	Tetrahydrofolic acid	THFA
Biotin	Biotin	–
Vitamin B12	Methyl cobalamin	–

> **KEY POINT**
>
> The family of B vitamins performs various functions as coenzymes or critical components of coenzymes. Because they are so important to our overall metabolism and health, and because they are water soluble, we need B vitamins on a daily basis.

People often stress the importance of vitamins but neglect the minerals. We need many more minerals in our diet than vitamins. Many minerals affect the function of enzymes, either as tightly bound components of prosthetic groups (copper, iron, manganese, selenium, and zinc) or because they are needed at the time of a reaction (calcium and magnesium). Zinc is an especially useful mineral in that it is an essential component of more than 300 different proteins, including enzymes for synthesizing RNA (ribonucleic acid) and DNA, pancreatic digestive enzymes, and enzymes involved in carbohydrate, fat, protein, and alcohol metabolism. Magnesium (Mg^{2+}) is associated with the energy-rich ATP molecule, so it is involved in virtually all aspects of our metabolism. Minerals not only regulate the activity of many enzymes in a cofactor role; some minerals, such as calcium, act in signal transduction, and some, such as sodium and potassium, are critical to the chemical composition of intracellular and extracellular fluids.

CLASSIFICATION OF ENZYMES

Thousands of different enzymes are involved in our metabolism. The overwhelming preponderance of these enzymes are present in other organisms as well. Creating some form of organization of enzyme names is a function of the International Union of Biochemistry and Molecular Biology (IUBMB). This classification system is shown in table 2.2. Six major classes of enzymes have been identified, and within each of these there are subclasses and then sub-subclasses. Each enzyme is assigned two names: The recommended name is suitable for everyday use, and its systematic name is formal and used to remove any ambiguity. Each enzyme has a four-number designation as well. For example, lactate is reduced through loss of hydrogens and associated electrons (dehydrogenation) facilitated by a specific enzyme. The recommended name for this enzyme is *lactate dehydrogenase;* this is the name used by almost all

Table 2.2 Enzyme Classes, Including Subclasses We Will Encounter

Enzyme class: number and name	Some common subclasses
1. Oxidoreductases	Dehydrogenases
	Reductases
	Oxidases
	Peroxidases
	Catalase
2. Transferases	Kinases
	Phosphomutases
3. Hydrolases	Phosphatases
	Thiolases
	Phospholipases
	Deaminases
	Ribonucleases
	Esterases
4. Lyases	Decarboxylases
	Aldolase
	Hydratases
	Dehyratases
	Synthases
	Lyases
5. Isomerases	Isomerases
	Mutases (but not all)
6. Ligases	Synthetases
	Carboxylases

biochemists. The systematic name is *lactate:NAD+ oxidoreductase.* The classification is EC 1.1.1.27, where EC represents Enzyme Commission and the last number (27) refers to the specific enzyme.

In general, recommended names identify first the substrate, then the reaction type with the ending *ase* added. These are the enzymes we will encounter, for the most part. There are enzymes, however, with names that give no clue to either their substrate or the type of reaction they catalyze. For example, catalase, an enzyme we will encounter in chapter 4, is also a member of the oxidoreductase class, just like lactate dehydrogenase (LDH). One would not know this from the name.

OXIDATIONS AND REDUCTIONS

Oxidation–reduction reactions, or **redox reactions,** are extremely important for all organisms. In these reactions, something gets oxidized and something gets reduced. You may have encountered in earlier chemistry courses the term **oxidation,** which means that something loses electrons, and **reduction,** which means that something gains electrons. The familiar memorization tools *LEO* (loses electrons oxidation) *says GER* (gains electrons reduction) and *OIL* (oxidation is losing) *RIG* (reduction is gaining) have been used by students for many years. Redox reactions are absolutely critical to life and underlie all aspects of metabolism. In all cases the electron gain (reduction) is directly connected with electron loss (oxidation), as seen in the reaction

$$Fe^{3+} + Cu^+ \rightarrow Fe^{2+} + Cu^{2+}$$

In this example, the ferric iron (Fe^{3+}) is reduced because it gains an electron from the copper to form ferrous iron (Fe^{2+}), whereas the copper (Cu^+) is oxidized by losing an electron to the iron, thereby becoming Cu^{2+} or cupric copper ion.

In the cell, there are redox reactions in which ions are oxidized and reduced as shown in the equation. However, in many important redox reactions it is not easy to see that electrons are lost from one molecule and gained by another. For example, figure 2.12 shows two dehydrogenation reactions in which two hydrogens are lost from each of the two partial structures. The important thing is that when the hydrogens leave, they exit with electrons. In the first example, two hydrogen atoms are lost, and a carbon-to-carbon double bond is generated. In the second reaction, two hydrogens are lost from a secondary alcohol, generating a ketone group. In the former case, we consider that the hydrogen atoms each contain an electron. In the latter case, one hydrogen comes off as a hydride

Figure 2.12 Dehydrogenation reactions. These reactions involve electron loss as hydrogen atoms (top reaction) or a hydride ion (bottom reaction).

> ▶ **K E Y   P O I N T** ◀
>
> Most of the energy needed by our bodies to grow and survive arises when electrons on fuel molecules are transferred to the oxygen we breathe. Energy released during this process is captured through the formation of ATP, which is subsequently used to carry out the energy-requiring reactions and processes needed for us to live.

ion and the other as a proton. The **hydride ion** is simply a hydrogen atom (with one proton and one electron) plus an additional electron so that it has a negative charge. In summary, dehydrogenation reactions are oxidation reactions in which two electrons are lost as part of hydrogen atoms or a hydride ion.

Figure 2.13 shows dehydrogenation, but these reactions are also accompanied by hydrogenation reactions, which means that a molecule is reduced because it accepts two hydrogen atoms (each with an electron) or a hydride ion (with two electrons). Two major coenzymes are involved in most cell redox reactions. The oxidized form of the coenzyme **FAD (flavin adenine dinucleotide;** see table 2.1) can accept two hydrogen atoms to become FADH$_2$ (the reduced form of FAD). The FAD coenzyme is involved in redox reactions in which two hydrogen atoms are removed from chemical structures, such as the top reaction in figure 2.12. Those shown in the lower part of figure 2.12 involve the **NAD$^+$ (nicotinamide adenine dinucleotide)**

coenzyme, which accepts a hydride ion, becoming NADH. NAD$^+$ is the oxidized form, whereas **NADH** is the reduced form of this coenzyme.

In the top reaction of figure 2.13, part of the citric acid (tricarboxylic acid or Krebs) cycle, succinate, is oxidized to fumarate through the loss of two hydrogen atoms, forming the carbon-to-carbon double bond. The two hydrogen atoms, each carrying an electron, are picked up by the coenzyme FAD, forming FADH$_2$. Thus, succinate is oxidized to fumarate, and FAD is simultaneously reduced to FADH$_2$ in this redox reaction. The coenzyme (FAD or FADH$_2$) is a tightly bound coenzyme for the enzyme succinate dehydrogenase. In the lower example, lactate loses a hydride ion and a proton when oxidized to pyruvate. At the same time, NAD$^+$ accepts the hydride ion, becoming the reduced form of the coenzyme NADH. In this latter reaction, catalyzed by the enzyme lactate dehydrogenase, the coenzyme (NAD$^+$ or NADH) binds to the enzyme only at the time of the reaction. Therefore, as the reaction proceeds from left to right as shown in figure 2.13, NAD$^+$ is a substrate and NADH is a product. As shown in the two examples in figure 2.13, most redox reactions are reversible, so pyruvate can be reduced to lactate or fumarate reduced to succinate.

The net direction of redox reactions depends on the relative concentrations of the oxidized and reduced forms of the substrates and coenzymes. During exercise, when the rate of pyruvate formation increases in muscle, the enzyme lactate dehydrogenase attempts to maintain equilibrium, producing lactate. The lactate can travel from the muscle cell to the blood,

Figure 2.13 Two biologically important redox reactions, showing substrates, products, and coenzymes.

where its concentration is often used as an indicator of the exercise intensity. Notice in figure 2.13 that we write L-lactate because the middle carbon is chiral, and the L refers to the absolute configuration of the lactate. Lactate comes from lactic acid, a modestly strong acid (pK$_a$ about 3.1). Thus, at the neutral pH where this reaction occurs, lactic acid exists as the ion lactate because it has lost its proton. The same can be said for pyruvate, from pyruvic acid.

Redox reactions represent the molecular basis for energy generation in the cell. Electrons from substrates, generated from the foods we eat, are transferred to coenzymes NAD$^+$ and FAD, generating NADH and FADH$_2$. Subsequently, these electrons are transferred from the reduced coenzymes through a series of carriers embedded in the inner membrane of mitochondria until they are passed to oxygen (O$_2$). Reduction of oxygen through accepting these electrons and H$^+$ to form water molecules is the final step (see the following equation). During the passage of electrons from food-derived substrates to oxygen, a tremendous amount of energy is released and captured through formation of ATP. The ATP is then used to drive energy-requiring processes in the body. This is the basis of the process called **oxidative phosphorylation,** which we will cover in detail in chapter 4.

$$O_2 + 4e^- + 4\,H^+ \rightarrow 2\,H_2O$$

▶ K E Y P O I N T ◀

The harder we exercise, the more ATP we need. This means we must accelerate the rate of transfer of electrons from fuel molecules (dehydrogenation reactions) to oxygen. If we continue to increase the intensity of our exercise, a weak link in the electron transport process is reached. This limitation may be availability of oxygen to accept electrons, or the rate at which we can remove electrons from the fuels, or the actual kind and amount of fuel available.

REGULATION OF ENZYME ACTIVITY

Each chemical reaction in a cell is catalyzed by a specific enzyme. As discussed previously, the rate of an enzyme-catalyzed reaction depends on the concentration of substrate. Reaction rates can also be controlled by the amount of enzyme protein, as well as by the enzyme's location within the cell. In subsequent chapters, we will encounter a variety of examples in which the activity of existing enzymes in the cell must be modified so that cellular metabolism is appropriate. Simple enzymes, obeying the Michaelis-Menten kinetics shown in figure 2.3, are very common but are rarely involved in controlling metabolism. For this, enzymes with more complex kinetics or properties are needed. There are two major ways in which the activity of existing enzymes can be controlled, modification by effector molecules and covalent modification.

Allosteric Enzymes

In one group of enzymes, their activity is dependent not only on substrate and product concentrations but also on the presence of positive or negative effectors. Such enzymes are usually composed of multiple subunits with multiple active sites and are typically placed in metabolic pathways where they can control the overall pathway rate. We call these enzymes **allosteric enzymes,** and the positive and negative effectors positive and negative *allosteric effectors*. The term *allosteric*, derived from the Greek word *allo* (or other), is used because these enzymes have sites, other than the active site, to which effectors can bind. Such binding affects the overall rate of the enzyme-catalyzed reaction by facilitating or diminishing velocity for a given substrate concentration. Allosteric effectors (often just called effectors) can be substrates or products of the allosteric enzyme or other molecules whose concentration provides a message about how active the allosteric enzyme should be.

Molecules that bind to large molecules are known as **ligands.** The ligands may be substrates or products that bind to the active site, as well as allosteric effectors that bind to **allosteric sites.** Therefore, we could classify enzymes as (a) those whose ligands are only substrates and products that bind to the active site and (b) those that bind substrates and products as well as one or more ligands that bind at allosteric sites. Ligands are also molecules, such as hormones, that bind to other proteins such as membrane receptors.

In general, allosteric enzymes are composed of subunits; that is, they have quaternary structure. The kinetics of allosteric enzymes are more complicated than the simple Michaelis-Menten kinetics shown in figure 2.3. Allosteric enzymes typically demonstrate sigmoidal (S-shaped) kinetics, as shown in figure 2.14. Here, the presence of positive or negative

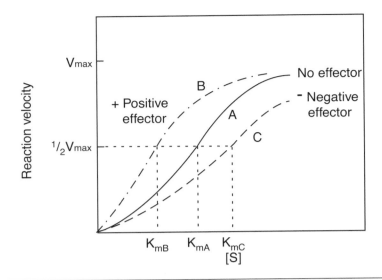

Figure 2.14   The effect of substrate concentration on reaction velocity for an allosteric enzyme *(A)* when neither positive nor negative allosteric effector is present, *(B)* when a positive allosteric effector binds at its allosteric site, and *(C)* when a negative allosteric effector binds to its allosteric site. In the example shown, Vmax is the same for the three cases, but the positive allosteric effector reduces K_m (i.e., K_{mB}) while the negative allosteric effector increases the value of K_m (i.e., K_{mC}).

allosteric effectors greatly modifies the response of the enzyme to its typical substrate. For example, a much higher concentration of substrate would be needed to achieve one-half of Vmax in the presence of a negative effector, but a much smaller amount of substrate is needed to achieve one-half of Vmax with a positive effector bound to its allosteric site. Thus allosteric effectors markedly influence K_m values. Enzymes displaying sigmoid kinetics do so because **cooperativity** in substrate binding exists at the active site in each subunit. This means that the binding of a substrate molecule to the active site of one of the subunits influences subsequent binding of the substrate molecule at the active sites of the other subunits.

Allosteric enzymes play an important role in the regulation of cell metabolism. For example, in a muscle fiber at rest, the breakdown of carbohydrate is very low, primarily due to negative effectors binding to the allosteric enzyme phosphofructokinase (PFK). Phosphofructokinase is located near the beginning of the pathway of glycolysis and acts to regulate the breakdown of carbohydrate. When the fiber becomes active in an exercise situation, the rise in the concentration of positive effectors and the decline in the concentration of negative effectors enormously increase the activity of PFK. On the other hand, under rest conditions, PFK activity is reduced by low concentrations of positive allosteric effectors and by higher levels of two negative allosteric effectors. This forces rested muscle to consume more fat as its fuel. We will discuss the role of PFK in

carbohydrate metabolism in chapter 5. Hexokinase, the enzyme that first interacts with glucose upon entering a cell, is a dimer—that is, composed of two polypeptide subunits. Its substrates are glucose and ATP. Its product, glucose 6-phosphate, can inhibit the activity of the enzyme if its concentration increases. This is an example of **feedback inhibition,** which helps to regulate metabolic pathways by sensing an oversupply of product.

Covalent Modification of Enzymes

Allosteric regulation is more a "fine-tuning" type of enzyme activity modulation. On the other hand, enzyme activity can be rapidly turned on or off by the covalent modification of specific amino acid residues in the enzyme protein. One example of this is acetylation (addition of an acetyl group) to specific residues in a class of DNA-binding proteins known as histones. Acetylation of histones is associated with the enhancement of gene transcription, whereas removal of the acetyl groups (deacetylation) decreases transcription of closely associated genes.

The most common way to modify the activity of enzymes or receptors is to attach a phosphate group to the hydroxyl part of the side chains of the amino acids serine, threonine, and tyrosine (see figures 1.4 and 2.15). This process is known as **phosphorylation,** specifically **protein phosphorylation** since the phosphate is added to part of a protein. Removal of the phosphate group is **dephosphorylation.** The addition of a phosphate group in place of a single hydrogen atom drastically alters a protein. Not only is the phosphate group far larger, but it contains two negative charges, thus making a major change in protein conformation. Of course, the drastic change in protein conformation following removal of the phosphate group reverses the effect. Protein phosphorylation-dephosphorylation is a common and effective mechanism to alter rapidly and reversibly the activity of key enzymes. The source of the phosphate group is ATP (adenosine triphosphate), a molecule we will encounter throughout all subsequent chapters. Protein phosphorylation is catalyzed by a class of enzymes known as **protein kinases,** which are specific for serine, threonine, or tyrosine side chains. For example, there are protein serine kinases, protein tyrosine kinases, and protein ser/thr kinases; the latter can phosphorylate both serine and threonine side chains. A class of enzymes known as **phosphoprotein phosphatases** catalyzes removal of the phosphate groups by hydrolysis. These phosphatases are specific to the amino acid side chain as in the case of protein tyrosine phosphatases. A summary of this process is shown in figure 2.15.

It is not unusual for multiple phosphorylation sites to exist on the same enzyme, allowing for graded modulations in the activity of the enzyme. The enzyme catalyzing the storage of glycogen in liver and muscle, glycogen synthase, has 10 sites where it can be phosphorylated. The more numerous the attached phosphates, the less active the glycogen synthase.

KEY POINT

Protein phosphorylation-dephosphorylation works like a light switch in that it has an all-or-none effect. The enzyme molecule is either active or inactive. Allosteric modification of enzyme activity acts like a dimmer switch, grading the activity of the enzyme.

Control of enzyme activity by phosphorylation plays a critical role in controlling and integrating our metabolism. For example, when we begin to exercise, the hormone epinephrine (adrenaline) is released from the adrenal medulla and binds to a membrane receptor on muscle and fat cells. In active muscle cells, epinephrine binding leads to the phosphorylation and activation of glycogen phosphorylase, the enzyme that breaks down glycogen in muscle. In fat cells, adrenaline binding to its receptor leads to the phosphorylation and thus activation of hormone-sensitive lipase, the enzyme catalyzing the breakdown of stored fat. In many other cases, the protein kinases are activated when external molecules (e.g., a hormone) bind to specific sites on the membrane of the cell. The hormone insulin, on binding to the insulin receptor on the cell membrane, results in a cascade of phosphorylation reactions involving multiple downstream proteins. In skeletal muscle, this cascade leads to glucose uptake. For every phosphorylation reaction, there must be a dephosphorylation—just

Figure 2.15 Modification of enzyme activity. Enzyme activity can be turned on or off by the covalent attachment of a phosphate group from ATP to the OH part of the side chain of the amino acids serine, threonine, and tyrosine. A class of enzymes known as protein kinases attach the phosphate group, transferring the terminal phosphate from ATP. Hydrolysis of the phosphate group by a class of enzymes known as phosphoprotein phosphatases reverses the effect of phosphate group addition.

as for every switching on of a light, the light must eventually be turned off.

We shall see in subsequent chapters that protein phosphorylation is a powerful strategy to control enzymes, other proteins, gene transcription, and protein synthesis. Virtually all reactions and processes that take place in our bodies are regulated by protein phosphorylation-dephosphorylation. Indeed, problems associated with these processes underlie many diseases, especially cancers. The integration of cells in complex organisms such as humans requires carefully regulated signaling throughout the body. You are familiar with this because you have learned about particular hormones in earlier studies. Besides hormones, there is a growing list of special signaling proteins known as **cytokines**—many of these are described as **growth factors.** These are secreted from a wide variety of cells and may interact in different parts of the body by traveling in the blood in an endocrine manner. Alternatively, they may act locally, on cells in their immediate environment (**paracrine** effect) or even on themselves (**autocrine** effect). This sequence in which an external molecule such as a hormone or growth factor binds to its cell surface receptor and thereby influences events within the cell is known as **signal transduction** or signaling. In chapters 8 and 9 we will examine in more detail how signal transduction influences gene expression, protein synthesis, and protein degradation.

> ### ▶ K E Y   P O I N T ◀
>
> Signal transduction or signaling is the process that provides the total network of communication among the various cells, tissues, and organs of the body. Cells cannot survive without communication, but with it, the normal pattern of cell growth, differentiation, and metabolism can be regulated.

MEASUREMENT OF ENZYME ACTIVITY

We sometimes need to know how many functional enzyme molecules exist in a fluid (e.g., blood) or tissue. Because the number of molecules of functional enzyme is proportional to the Vmax, measurement results are in units of reaction velocity (i.e., change in substrate or product concentration per minute)

per unit weight of tissue, per unit amount of protein, or per volume of fluid. Examples are micromoles of product formed per minute per gram of tissue, millimoles of substrate disappearing per minute per milligram of protein, or micromoles of product formed per minute per milliliter of blood, respectively. The expression of the Vmax or, as it is commonly called, *enzyme activity*, is important in a variety of physiological and clinical conditions. For example, the activity of mitochondrial enzymes can almost double given the appropriate exercise training stimulus. This means that the maximal flux or traffic of substrate through the enzyme is twice what it was before training. We may want to determine if a particular exercise training program alters the metabolism of a muscle by measuring the activity of selected enzymes. We can also tell if a particular tissue is damaged by measuring the activity of isozymes specific for that tissue that are released to the blood due to cell membrane damage.

> ### ▶ K E Y   P O I N T ◀
>
> If we double the activity of enzymes involved in dehydrogenating fuel molecules, we can potentially double the ability to transfer electrons to oxygen and form the energy molecule ATP.

When measuring the activity of an enzyme, it is necessary to establish and rigidly follow certain principles. First, we need to make the measurements at a concentration of substrate (or substrates) high enough to generate a true Vmax. The pH of the reaction and the temperature at which it is measured should also be standardized so that meaningful comparisons can be made. Finally, we need to have a simple method for measuring the disappearance of substrate or appearance of product. This determination becomes possible if the substrate or product is colored or can be made to generate a colored complex. For example, phosphate appearance can be readily measured because it forms a colored complex with a number of agents. One useful technique takes advantage of two properties of the coenzyme NADH. First, NADH absorbs light at a wavelength of 340 nanometers (nm), so we can measure its rate of formation or disappearance with a spectrophotometer. The relationship is as follows: A 0.1 mmol solution of NADH has an absorbance of 0.627. Second, NADH fluoresces when bombarded with light of a specific wavelength; thus we can measure its appearance or disappearance with a fluorometer. If the specific reaction does not actually involve NADH, it can be connected to a reaction that does. The rate of the reaction in question then dictates the rate of a connection reaction in which NADH is formed or lost.

Biochemists use the term **international unit** (IU) to express the activity of an enzyme. One inter-national unit of enzyme activity is the amount of enzyme that converts one micromole of substrate to product in one minute. Thus if an enzyme has an activity of 15 international units per gram, 15 μmol of product forms per minute per gram of tissue. Since enzyme activity is so sensitive to temperature, the composition of the medium, and pH, one must specify these conditions when describing enzyme activity in international units.

SUMMARY

Enzymes are biological catalysts—specialized proteins that enormously speed up reactions in cells. Highly specific, they catalyze reactions involving single substrates or a closely related group of substrates. Enzymes have a Michaelis constant, K_m, which is the substrate concentration needed to produce one-half the maximal velocity (Vmax) of the enzyme reaction. The K_m, a characteristic constant for an enzyme–substrate pair, reflects inversely the affinity of the enzyme for its substrate. The maximal velocity of an enzyme-catalyzed reaction, or Vmax, is proportional to the amount of enzyme present and can be determined only when the enzyme is saturated with its substrate. Measurement of the Vmax thus determines the amount of enzyme present. An international unit is defined as the amount of enzyme needed to convert one micromole of substrate to product in one minute. The action of enzymes can be hindered by the presence of inhibitors—specific substances that resemble the normal substrate and compete with it (competitive inhibitors) or that irreversibly alter the structure of the enzyme (noncompetitive inhibitors). In the cell, regulation and integration of the thousands of chemical reactions can be affected through modulation of the activity of a subset of key enzymes. The activity of allosteric enzymes can be changed by the binding of ligands, including substrates or products, to

specific sites known as allosteric sites. Binding of these effector molecules may increase or decrease enzyme activity by changing the enzyme's response to a particular substrate concentration. A more profound change in enzyme activity accompanies the covalent attachment of a phosphate group to an enzyme, catalyzed by a class of enzymes known as protein kinases. Dephosphorylation by phosphoprotein phosphatases reverses the change in activity accompanying phosphorylation.

Membrane transport is carried out by a class of protein molecules with properties similar to those of enzymes. Membrane transport can be described by kinetic constants, Vmax and K_m, and can be subject to competitive and noncompetitive inhibition. Many enzymes require the presence of nonprotein substances to function. These cofactors can be organic molecules, that is, coenzymes, or they can be metal ions. Most coenzymes are derived from the B vitamins in our diet, while our need for many specific mineral nutrients relates to their role as enzyme cofactors. Isoenzymes (isozymes) are closely related, but different, enzyme molecules that catalyze the same reaction but differ in certain properties, such as K_m. Some of the most important enzymes are the dehydrogenases, which add or remove electrons from their substrates. They play a major role in producing energy in the process of oxidative phosphorylation.

REVIEW QUESTIONS

1. Using the following data generated by measuring the rates of an enzyme-catalyzed reaction at constant enzyme concentration, determine the K_m and Vmax.

Substrate concentration	Reaction rate
1 mM	2 mmol · ml^{-1} · min
2 mM	4 mmol · ml^{-1} · min
4 mM	7 mmol · ml^{-1} · min
10 mM	15 mmol · ml^{-1} · min
50 mM	25 mmol · ml^{-1} · min

2. If the data in question 1 were obtained at 25 °C, approximately what would be the Vmax at 35 °C?

3. One lab reports that the Vmax for the mitochondrial enzyme malate dehydrogenase in the quadriceps of a group of elite cyclists averages 40 μmol of substrate per gram wet weight of tissue per minute. Your lab, on the other hand, measures the same enzyme in a similar group of trained subjects, yet your average enzyme activity is double that of the other lab. What are some possibilities that could contribute to this discrepancy between two apparently similar groups of subjects?

4. Change the units for the activity of citrate synthase from 20 μmol per gram per 120 s to (a) millimoles per kilogram per minute, (b) international units per gram.

5. Some labs determine the activity of muscle enzymes on freeze-dried samples of muscle. If the water content of a muscle sample is 75%, what would be the activity of the citrate synthase in micromoles per gram dry weight per minute if the activity is 20 international units per gram wet weight?

6. The experiment described in question 1 was repeated with a known inhibitor of the enzyme and using the same substrate concentrations. The data are paired as follows using the same substrate and reaction velocity units: S/V—1/1.7; 2/3.2; 4/5.8; 10/10.5; 40/21.1 What number did you get that was different? What kind of inhibitor was it?

7. The Vmax you measure in the lab is 20 μmol per minute per 100 mL of solution. You are asked to report this Vmax as IU/L per active site. You have some reservations about this, but go ahead anyway. What value do you report to the lab instructor? Why do you have reservations about this exercise? You will need to do a little research on LDH first.

II

METABOLISM

Money is the central commodity that allows modern societies to function. Individuals work for their money and use this money to buy things. If you operate like most folks, the amount of money that comes in as salary balances the money that goes out for housing, food, and other expenses.

Living organisms operate in the same way, except that the currency is ATP, not money. We create ATP by breaking down fuels such as carbohydrates and fats obtained from our diet. We use this ATP to build proteins, use our muscles, transport things into cells and cell organelles, and generally grow and maintain our bodies. All the reactions that create and use ATP are collectively called metabolism. That branch of metabolism in which ATP is generated is known as catabolism, and the branch in which ATP is used to make things is known as anabolism. The section on metabolism focuses on the detailed processes for the breakdown of the major fuel molecules (fats, carbohydrates, and amino acids from proteins) to make ATP. We will also look at how ATP is used, especially in the formation and storage of fuels.

Throughout these chapters, we emphasize not only the details of the processes, but also the mechanisms that regulate them, including the roles played by diet and various forms of exercise. We begin (in chapter 3) with an overview of metabolism, outlining first the three energy systems that provide ATP to contracting skeletal muscles. This is followed by a section that illustrates how we can quantify energy. Oxidative phosphorylation is the dominant process generating ATP in the body, and this is the theme of chapter 4. For this process to function, carbohydrates (chapter 5), lipids or fats (chapter 6), and amino acids (chapter 7) are degraded, and their chemical energy is converted into ATP.

CHAPTER

3

ENERGY SYSTEMS AND BIOENERGETICS

The ability to use energy is the hallmark of all living organisms. This energy, in the form of **ATP (adenosine triphosphate)**, is found in all cells, but in small amounts, ranging from about 3 mmol per kilogram in most tissues to about 6 mmol per kilogram of skeletal muscle. The content of ATP remains the same even though it is used constantly, because it is being regenerated at a similar rate. Adenosine triphosphate is used to drive a number of processes in cells, such as making cell proteins, storing fuels, synthesizing RNA (ribonucleic acid) molecules, and transporting substances into cells and the organelles within cells, as well as in signaling to regulate cellular processes.

Skeletal muscles, especially in elite athletes, are incredible systems. Consider that an elite marathon runner will cover 42 km (26.2 miles) in less than 130 min but will

- expend approximately 3,000 kilocalories (12,600 kilojoules),
- oxidize almost 700 g of carbohydrates,
- oxidize more than 30 g of fat,
- utilize about 600 L of oxygen, and
- break down and reform more than 150 mol of ATP, weighing about 63 kg (139 lb).

In this chapter, we will look at the energy requirements of skeletal muscle and the role of ATP, an energy-rich phosphate compound that directly drives muscle contraction. The ATP concentration of muscle is very small; we will discuss three energy systems responsible for maintaining the ATP concentration in cells. The last part of the chapter deals with quantitative aspects of cellular energy metabolism.

ENERGY REQUIREMENTS OF SKELETAL MUSCLE

Skeletal muscle is a special tissue that can suddenly increase its rate of ATP use to more than 100 times that of rest, as in sprinting. Skeletal muscle can maintain isometric contractions, as in holding a baby, or can be used at a high level for long periods of time in the case of a marathon. We will start with the structure of skeletal muscle and learn how skeletal muscle contracts.

Structure of Skeletal Muscle

Skeletal muscle, the largest tissue in the body, consists of a number of elongated cells known as fibers, which contain a number of parallel *myofibrils* (see figure 3.1). Muscle fibers have a cell membrane, known as the **sarcolemma,** and mitochondria arranged under the sarcolemma and between myofibrils (figure 3.2). Muscle fibers differ from other cells because they are so large, running from one end of a muscle to the other. This large size explains the need for many nuclei in each skeletal muscle cell. Neurons, emerging from the spinal cord, innervate each fiber at the *neuromuscular junction*, and nerve impulses crossing

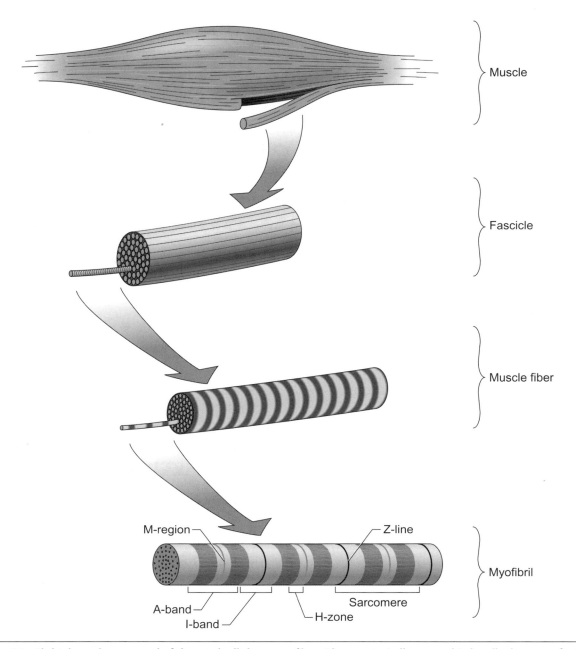

Figure 3.1 Skeletal muscle, composed of elongated cells known as fibers. These are typically arranged in bundles known as fascicles. Within a muscle fiber are myofibrils, aligned in parallel with the fiber long axis. Striations in the fiber and myofibril are due to the arrangement of contractile proteins, creating bands in the sarcomere.

Adapted, by permission, from B. MacIntosh, P. Gardiner, and A.J. McComas, 2004, *Skeletal muscle: Form and function*, 2nd ed. (Champaign, IL: Human Kinetics), 3.

from the neuron to the sarcolemma activate the fiber to make it contract. Contraction is initiated when calcium ions are released from intracellular storage vesicles known as **sarcoplasmic reticulum.**

The striations seen in muscle fibers are due to the arrangement of proteins aligned in parallel in two kinds of filaments in each myofibril. Figure 3.3 shows parts of three thin filaments and a section of

a thick filament. The **thick filament** is composed of individual myosin molecules arranged in overlapping parallel arrays. Three myosin molecules, each with two protruding heads known as cross bridges, can be seen in the thick filament section in figure 3.3. Calcium ions released when the fiber is activated bind to the thin filament protein troponin, allowing strong, force-generating interactions between myosin

Figure 3.2 Schematic of a muscle fiber showing longitudinal and cross-sectional perspective. A myelinated motor neuron, originating from the spinal cord, attaches on the cell membrane (sarcolemma) at the neuromuscular junction. Muscle fibers have mitochondria located under the sarcolemma and between individual myofibrils. Sarcoplasmic reticulum is a network of filaments and sacs that store and release calcium ions. Because of their large size, muscle fibers have numerous nuclei.

Reprinted, by permission, from J.H. Wilmore and D.L. Costill, 2004, *Physiology of sport and exercise*, 3rd ed. (Champaign, IL: Human Kinetics), 36.

and actin, driven by ATP hydrolysis, to take place. The sum of the forces from millions of myosin cross bridges, each attached to an actin molecule, leads to shortening of the overall muscle.

A **thin filament** contains three kinds of proteins. Individual molecules of **actin** (actin monomers),

> **K E Y P O I N T**
>
> Troponin is the name given to a protein containing three subunits. Troponin C can bind reversibly with calcium when the calcium concentration increases, as it does when the fiber is activated by its nerve. Calcium binding by troponin C results in alterations in the two other subunits, troponin I and troponin T, leading to a change in the position of tropomyosin on the thin filament and better access by myosin cross bridges to actin-binding sites.

shown as spheres in figure 3.3, align themselves into two strands that form a helix. Arrayed along the actin fibers are the proteins *tropomyosin* and troponin. When **troponin** binds calcium, it can alter the position of the tropomyosin filaments, exposing sites on each actin monomer that can bind a myosin cross bridge.

Myosin and Muscle Contraction

A myosin molecule consists of six polypeptide chains. Two of these polypeptide chains are very large and are twisted together from one end (the tail) to the other end where they form the heads or **cross bridges.** These are the **myosin heavy chains** (MHC). There are four other polypeptides, called *myosin light chains* (MLC). Two of these are associated with each head or cross bridge. The light chains in the neck region are known as the *regulatory light chains* because each of these can be phosphorylated by accepting a phosphate group from ATP. Phosphorylation of

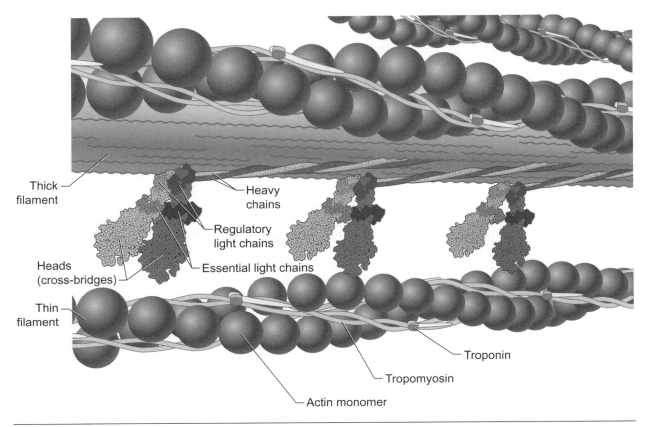

Figure 3.3 Close-up view showing parts of three thin filaments and part of a thick filament in a myofibril. The two cross bridges on three individual myosin molecules in a thick filament can each interact with individual actin monomers (shown as spheres) in the thin filament when the muscle fiber is activated.

Adapted, by permission, from R. Vandenboom, 2004, "The myofibrillar complex and fatigue: A review," *Canadian Journal of Applied Physiology* 29(3): 330-356.

myosin regulatory light chains causes subtle changes in the performance of the myosin crosses bridge. The other two light chains, located farther into each cross bridge, are known as the *essential light chains.*

Myosin has two important properties. Each cross bridge can bind to a single actin molecule in the thin filament. The second property is that each head can act as an enzyme to hydrolyze ATP as is shown in the following equation. The word "hydrolysis" is used because it refers to breaking the covalent bonds of a molecule by the addition of water.

$$ATP + H_2O \rightarrow ADP + Pi$$

This ability to break down ATP into the products **ADP (adenosine diphosphate)** and inorganic phosphate (Pi) is called *myosin ATPase* activity, because it is myosin acting as the **ATPase** enzyme.

The characteristics of a myosin molecule are dominated by the heavy chain properties. Each cross bridge or head contains an actin (monomer)-binding site and an ATP-binding site. The rate at which a myosin

cross bridge can hydrolyze ATP is determined by the heavy chain. By itself, myosin ATPase activity is low, but it can be increased by about 100-fold when actin binds to a myosin cross bridge. Actin serves as a powerful activator for myosin ATPase activity, so we describe this as *actin-activated myosin ATPase* activity or *actomyosin ATPase* activity. Actomyosin ATPase activity is strongly correlated with the maximum speed at which a muscle or muscle fiber can contract (shorten).

There are four different kinds of skeletal muscle myosin heavy chains, identified as *MHC I, MHC IIA, MHC IIX,* and *MHC IIB.* The Roman numeral I is used to designate myosin that has a lower actin-activated myosin ATPase activity; sometimes the word "slow" is used instead of the Roman I. The Roman II means fast. The addition of A, X, or B is intended to subtype the particular fast MHC. Human skeletal muscles have three main myosin heavy chains, MHC I, MHC IIA, and MHC IIX. Smaller animals have MHC IIB, which has the fastest actomyosin ATPase activity.

Two additional MHCs are found in embryonic and neonatal muscles and in adult muscles undergoing regeneration.

We have focused on the MHCs because these dominate the contraction characteristics of the myosin molecule. Nonetheless, myosin is still a molecule with six subunits (two MHCs and four MLCs). Accordingly, exercise scientists talk about *myosin isozymes* (also known as myosin isoforms or isomyosins), different molecular forms of the same enzyme (myosin) that catalyze the same reaction but with different speeds of contraction. The properties of the muscle fibers that contain different myosin isozymes can be illustrated by how fast these fibers shorten or how fast they can undergo a twitch in response to a single electrical stimulus. Thus there are slow-twitch or type I fibers, dominated by the MHC I, and fast-twitch fibers, dominated by more of the fast MHCs. Table 3.1 summarizes the major MHCs and the corresponding fiber types for mammalian skeletal muscles. As an added note, skeletal muscle fibers have many nuclei, each with the ability to express MHC genes. Therefore, a single muscle fiber may have two or more different MHCs, although one tends to dominate, thus leading to a single fiber type characterization.

As already mentioned, it is the action of myosin cross bridges exerting force on actin monomers in the thin filament that leads to muscle shortening and therefore movement. Each cross bridge undergoes a sequence of binding to an actin monomer, ATP hydrolysis, force generation, and then detachment from the actin. This cross-bridge cycle is shown in figure 3.4. In figure 3.4*a*, one of the two cross bridges of a myosin molecule attaches loosely (weakly) to an actin monomer. An ATP molecule is split into the products ADP and Pi, but these remain together so that the actual energy of this splitting is not released. This step can be pictured as a spring compressed between two fingers. Next, some of the energy of the ADP-Pi couple is released, and a strong binding conformation is attained. In figure 3.4*b*, the Pi is released as a consequence of the strong binding, and the power stroke occurs. This is in the nature of a change in the head-neck region of the myosin molecule that strains elastic elements in this region. The strain on the elastic elements exerts a force on the actin monomer. The actual attachment of the myosin head to actin does not change during the power stroke. In figure 3.4*c*, the myosin head, acting through the strain on the elastic elements in the head-neck region, exerts a force on the actin monomer. At the end of this force-generating stroke, ADP is released; a new ATP binds; and as shown in figure 3.4*d*, the previously attached, force-generating cross bridge detaches from actin. In the detached state, ATP is hydrolyzed to ADP and Pi and the cross bridge is available to bind weakly to actin. The sum of millions of cross bridges, each acting on its actin monomer, results in actual movement of the thick filament with respect to the thin filament, muscle fiber, and hence overall muscle shortening.

Figure 3.5 summarizes the overall process, which includes the following steps:

1. Acetylcholine released by the neuron at the neuromuscular junction diffuses across the gap between neuron and sarcolemma, binds

KEY POINT

Many people describe the fastest-contracting human skeletal muscle fiber type as type IIB when in fact it is really IIX. One of the reasons is that the IIX myosin heavy chain was identified much later as distinct from the IIB myosin heavy chain. Accordingly, much of the early literature mistakenly identified IIX human fibers as IIB fibers. You can still read of IIB human muscle fibers.

Table 3.1 Fiber Types in Mammalian Skeletal Muscle, Their Dominant Myosin Heavy Chain Type, and Relative Speed of Contraction

Fiber type*	Myosin heavy chain	Relative speed of contraction
Type I (ST)	MHC I	Slow
Type IIA (FTA)	MHC IIA	Fast
Type IIX (FTX)	MHC IIX	Faster
Type IIB (FTB)**	MHC IIB	Fastest

* Type I and ST (slow-twitch) and type II and FT (fast-twitch) are used synonymously.

** Type IIB (FTB) fibers with IB MHC are found in skeletal muscles of small mammals.

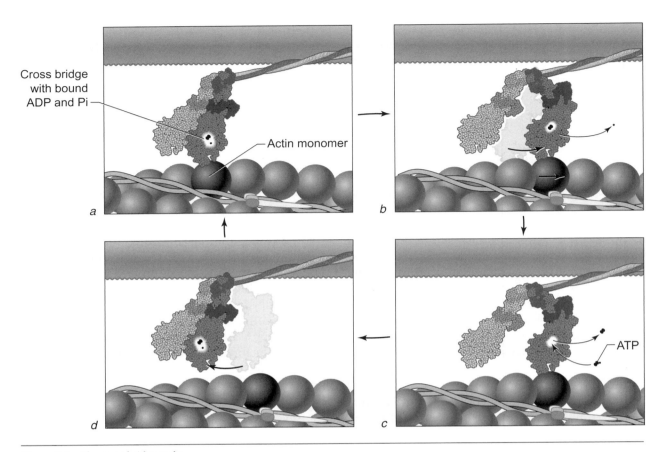

Cross bridge with bound ADP and Pi

Actin monomer

a   *b*   *c*   *d*

Figure 3.4   The cross-bridge cycle.

Adapted, by permission, from R. Vandenboom, 2004, "The myofibrillar complex and fatigue: A review," *Canadian Journal of Applied Physiology* 29(3): 330-356.

to its receptor, and causes depolarization of the sarcolemma.

2. A wave of depolarization passes over the sarcolemma and down into the interior of the fiber via surface invaginations known as **T-tubules.** The depolarization is associated with sodium ions (Na⁺) moving into the fiber and some potassium ions (K⁺) moving out of the fiber.

3. T-tubule depolarization is linked to calcium ion (Ca²⁺) release from a specialized form of endoplasmic reticulum known as sarcoplasmic reticulum or SR.

4. The calcium ion concentration in the fiber interior increases from about 10^{-7} M to about 10^{-5} M (a 100-fold increase).

5. The calcium ions bind to the protein troponin in the thin filament in the sarcomere.

6. This binding allows cross bridges to bind to individual actin monomers in the thin filament.

7. Using the energy released from ATP, the bound myosin cross bridges exert force on the actin in the thin filament.

8. The force exerted by millions of cross bridges in many fibers causes the overall muscle to shorten and do work.

9. The Ca²⁺ is pumped back into the SR in a process requiring energy from ATP. A special membrane ATPase known as SERCA (sarcoplasmic endoplasmic reticulum calcium ATPase) uses the energy from ATP hydrolysis to pump two calcium ions inside the reticulum of the SR or endoplasmic reticulum in other cells.

10. At the same time, a sarcolemma sodium-potassium ATPase (Na⁺-K⁺ ATPase) pumps three sodium ions from the inside of the muscle fiber to the outside while simultaneously moving two potassium ions back inside. This restores the normal polarized sarcolemma.

Figure 3.5 An overview of the steps in muscle contraction. A nerve impulse causes depolarization of the sarcolemma when acetylcholine (Ach) binds to its receptor. A wave of depolarization spreads over the surface of the muscle fiber and into the interior through the T-tubules, resulting in the release of Ca^{2+} ions from the sarcoplasmic reticulum (SR). Ca^{2+} binding to troponin on the thin filament allows myosin cross bridges to bind to actin, creating force when ATP is broken down. In relaxation, Ca^{2+} is transported into the SR by the SR-ATPase using energy from ATP. The Na^+-K^+ ATPase restores the balance of Na^+ and K^+ across the sarcolemma.

ENERGY-RICH PHOSPHATES

We have already discussed ATP as the energy currency in the cell and illustrated how it is used to drive the contraction of muscle. As we will learn in the next section, most of the ATP in the cell is produced by the breakdown of fuels such as fat, carbohydrate, and amino acids (from protein). These fuel molecules are oxidized to simple products such as CO_2 and H_2O, and the energy released drives the phosphorylation of ADP to make ATP as shown in the following equation.

$$ADP + Pi \rightarrow ATP + H_2O$$

Note that the formation of ATP in this reaction is opposite to that of ATP hydrolysis shown earlier in the chapter. Adenosine triphosphate synthesis in this reaction is accomplished when energy released from the **catabolism** (breakdown) of fuels is harnessed through ADP phosphorylation. The energy released by the hydrolysis of ATP drives a process that would not ordinarily take place without an input of energy, such as the cross-bridge cycle in muscle.

The Pool of Phosphates in the Cell

Adenosine triphosphate and ADP are examples of a class of molecules known as **nucleotides.** As we will discuss in chapter 8, nucleotides are molecules that consist of three kinds of components: either a pyrimidine or purine base (adenine in the case of ATP), the sugar ribose, and one or more phosphate groups. When the base adenine is combined with the sugar ribose, we get a new molecule known as a **nucleoside,** specifically adenosine. If we add one phosphate group to the 5' position of the sugar ribose in **adenosine,** we get a nucleoside monophosphate (NMP), specifically adenosine 5'-monophosphate (abbreviated 5'-AMP). If we add another phosphate group to the one existing in 5'-AMP, we get adenosine

5'-diphosphate or 5'-ADP—a nucleoside diphosphate (NDP). Another phosphate added to 5'-ADP gives us the **nucleoside triphosphate** adenosine 5'-triphosphate or 5'-ATP (an NTP). Normally we drop the 5' from these molecule names and simply call them **AMP (adenosine monophosphate),** ADP, and ATP, respectively. The molecule ATP is shown in figure 3.6 as it would exist in the cell, associated with a magnesium ion (Mg^{2+}). Each of the three phosphate groups is identified by a Greek letter. The bonds between the α and β phosphates and the β and γ phosphates are **anhydride bonds**. Note that there are negative charges on the phosphate groups. The combination of the anhydride bonds plus the negative charges in close proximity means that hydrolysis of ATP according to the two reactions shown in figure 3.6 yields a lot of energy. In reaction 2 in figure 3.6, one of the products is PPi or **inorganic pyrophosphate.** The bond between the two phosphate groups in PPi is also an anhydride; and its splitting also yields additional energy, accomplished in the cell by a ubiquitous enzyme, *inorganic pyrophosphatase.*

(1) ATP + H_2O ⟶ ADP + Pi

(2) ATP + H_2O ⟶ AMP + PPi

PPi + H_2O ⟶ 2 Pi

Figure 3.6 Adenosine triphosphate found in the cell in association with magnesium ions. The three phosphate groups in ATP are identified by Greek letters. Hydrolysis of ATP can take place between the β and γ phosphate groups (reaction 1) or between the α and β phosphate groups (reaction 2).

> **KEY POINT**
>
> To understand the fact that ATP hydrolysis releases a lot of energy, consider the analogy of a spring, compressed between your thumb and second finger. Keeping the coils of the spring together requires force. If you quickly release the pressure on the spring, it rapidly pops free, reaching a longer length and releasing energy.

In addition to the nucleoside mono-, di-, and triphosphates represented by AMP, ADP, and ATP, respectively, there are three other important kinds of nucleotides shown as nucleoside triphosphates: *GTP* or guanosine triphosphate contains the purine base guanine; *UTP* or uridine triphosphate contains the pyrimidine base uracil; and *CTP* or cytidine triphosphate includes the pyrimidine base cytosine. Their corresponding nucleoside di- and monophosphates are GDP and GMP, UDP and UMP, and CDP and CMP, respectively. Hydrolysis of the anhydride bonds in GTP, UTP, and CTP generates large energy releases, just the same as with ATP.

The hydrolysis of ATP drives most of the reactions or processes that cannot take place by themselves and need a source of energy. Guanosine triphosphate is specifically involved in the synthesis of proteins and signaling at the plasma membrane of cells; and UTP is used to make glycogen, the storage form for carbohydrates in the cell. As well, ATP, GTP, UTP, and CTP are used to make RNA molecules as we will discuss in chapter 8. All four of these nucleoside triphosphates yield large quantities of energy when they are hydrolyzed. As a result, we say that they are energy-rich molecules.

As we will discuss in more detail later in this chapter, the hydrolysis of ATP or any other nucleoside triphosphate results in the release of a lot of energy. Reactions in which a great deal of energy is released tend to go to completion, so that the initial reactant (ATP) virtually disappears. In fact, if we were to put ATP in a test tube solution containing a buffer to keep the pH near 7, plus Mg^{2+} in quantity sufficient to form an ionic interaction with the phosphates on ATP as shown in figure 3.6 and the ATP-hydrolyzing enzyme myosin, within a few minutes virtually no ATP would remain. We would have a solution with concentrations of ADP and Pi equivalent to that of ATP at the start. There would be no equilibrium established such that we would have a mixture with roughly equal portions of ATP, ADP, and Pi. Reactions such as ATP hydrolysis release so much energy that they go almost completely to the right. In fact, if one performs the calculations to see what the concentration of ATP would be in our test tube, assuming that the end point could be called an equilibrium point, we would have a very difficult time finding any ATP molecules not hydrolyzed—because for every ATP, there would be more than 10 million ADP molecules.

Adenosine triphosphate would be useless as an energy currency in the cell if the ATP were allowed to be converted completely into its products, ADP and Pi. At equilibrium, there would be no net ATP hydrolysis and no energy release because chemical reactions at equilibrium release no energy. For ATP to be the important energy currency it is, the concentration in the cell is kept very far from equilibrium so that the concentration ratio of ATP divided by free ADP is very high—almost 500 in a skeletal muscle cell. Moreover, under most conditions, when ATP is used as an energy source it is replenished at the same rate, so that its concentration does not decrease. As we will see in the last part of this chapter, maintaining a high ATP to ADP ratio ensures that the energy released when ATP is hydrolyzed is very high. During very severe exercise, the concentration of ATP does decline, but by no more than 50% of what it was at rest.

> ## ▶ K E Y   P O I N T ◀
>
> Because more than 90% of all of the ADP in muscle is bound with proteins, especially actin, we use the term *free ADP* to describe that pool of ADP that is not in a protein complex but is free to be involved in chemical reactions.

The adenine nucleotides (i.e., ATP, ADP, AMP) are primarily involved in coupling anabolic and catabolic reactions in the cell. Adenosine triphosphate is formed from ADP when fuel molecules are broken down, and ATP hydrolysis drives most energy-requiring processes. As mentioned earlier, GTP and UTP have their own special roles in metabolism, and all nucleoside triphosphates are necessary to make the various molecules of RNA. When the four nucleoside triphosphates are used as energy sources and to make RNA, the products are NDPs (ADP, GDP, UDP, and CDP) or NMP molecules (AMP, GMP, UMP, and CMP) or both. Since there is a continual synthesis and breakdown of RNA molecules in the cell, the

need to maintain a balanced level of NTP molecules is evident. This means that there must be specific reactions to interconvert the various components of the nucleoside phosphate pool. For example, nucleoside diphosphates must convert to nucleoside triphosphates, and nucleoside monophosphates to nucleoside diphosphates. For the first interconversion, a nonspecific enzyme known as nucleoside diphosphate kinase transfers phosphate groups to and from ATP and ADP, and NTP and NDP, as follows:

$$NDP + ATP \leftrightarrow NTP + ADP$$

In this freely reversible reaction, an NDP (such as UDP, GDP, or CDP) is phosphorylated to NTP by accepting a phosphate group from ATP. Because the reaction is freely reversible, its net direction depends on the relative concentrations of ATP, ADP, and the other NTPs and NDPs. The overall effect of the nucleoside diphosphate kinase reaction is to maintain a balance in the ratio of NTP to NDP in a cell.

To convert NMP to NDP, nucleotide-specific enzymes transfer a phosphate group from ATP to the nucleoside monophosphate, making ADP and a nucleoside diphosphate. They are said to be specific because one exists for each nucleoside monophosphate. For example, *adenylate kinase* catalyzes the following reaction:

$$AMP + ATP \leftrightarrow ADP + ADP$$

Uridine monophosphate kinase catalyzes a similar reaction:

$$UMP + ATP \leftrightarrow UDP + ADP$$

These freely reversible reactions also need magnesium ions because these ions are bound to the ATP molecules. The net effect of the nonspecific nucleoside diphosphate kinase and the specific nucleoside monophosphate kinase (e.g., AMP kinase and UMP kinase) is to maintain a balance between the NTP, NDP, and NMP in the cell.

Phosphagens

The ATP concentration in most tissues is fairly low, about 3 to 8 mmol per liter of cell water or about 2 to 6 mmol of ATP per kilogram of tissue. Since ATP represents the immediate energy source to drive energy-requiring processes, problems could arise if ATP is needed at a rapid rate and therefore is quickly used up. In cells with a slow acceleration of ATP-consuming reactions, ATP concentration can be easily maintained by a gradual acceleration of ATP-producing reactions, such as fuel oxidation. However, in muscle, this could be a big problem because during the transition from rest to maximal exercise, the rate of energy expenditure in a human muscle can increase by more than 100 times. The rate of energy turnover in a rested muscle is about 1 mmol of ATP per kilogram of muscle per minute. During sprinting, an elite athlete is able to turn over ATP at a rate of about 4 mmol of ATP per kilogram of muscle per second (240 mmol · kg^{-1} · min). For normally active individuals, achieving 75% of the maximum rate of ATP turnover of an elite athlete is certainly feasible. During peak activity, all of the muscle cell's ATP could be consumed in about 2 s if it were not regenerated.

> ### ▶ **K E Y   P O I N T** ◀
>
> Although a muscle's ATP can be turned over at an astonishingly high rate, during most kinds of physical activity, except for extreme activities such as sprinting, its content seldom decreases much below the resting level.

As we will see in the next section, ATP in cells is regenerated from ADP through the breakdown of fuel molecules using aerobic or anaerobic catabolic processes. However, ATP-regenerating processes cannot produce ATP at the same rate at which it is hydrolyzed to drive muscle contraction during sprinting. Moreover, these processes take time to gear up to maximum speed, whereas at the start of a sprint, the rate at which ATP is hydrolyzed is about maximal. To prevent muscle cells from using up their ATP at the start of maximal or near-maximal contractions, an alternate energy-rich molecule, known as a **phosphagen,** is capable of regenerating ATP at a very high rate. In vertebrate muscle the phosphagen is **phosphocreatine** (abbreviated PCr), also called creatine phosphate (abbreviated CP). In some invertebrate muscles, the phosphagen is arginine phosphate. In humans, 92% to 96% of the total body's PCr is found in skeletal muscle; the remainder is in cardiac muscle, brain, and testes.

Phosphocreatine (or creatine phosphate) has its phosphate group transferred to ADP to yield ATP and creatine, in a reaction catalyzed by an enzyme known as creatine kinase:

$$ADP + PCr + H^+ \leftrightarrow ATP + Cr$$

This reaction is freely reversible. During muscle contraction, the forward direction is favored in order to regenerate ATP. During recovery, when PCr is much reduced and free Cr is elevated, the backward reaction is favored to regenerate PCr. Because a little more energy is released when PCr is hydrolyzed, compared to when ATP is hydrolyzed, the reaction just presented is tilted a little more to ATP formation (and thus PCr disappearance) during rapid rates of ATP hydrolysis. Note that a proton (H$^+$) is consumed as a substrate when the creatine kinase reaction proceeds toward ATP formation. This is due to the acid–base characteristics of the phosphate groups. Consumption of a proton is advantageous since it can partially reduce the acidification of muscle during very vigorous exercise. We note this later in this chapter.

The actual concentration of PCr in muscle is about three to four times that of ATP (about 18-20 mmol per kilogram of muscle)—not that much considering how fast ATP can be used in very intense muscle activity. However, the extremely high activity of creatine kinase ensures that ATP is regenerated almost as fast as it is broken down near the beginning of sprint-type activities. Although limited in quantity, there is enough PCr to act as a temporary ATP buffer until other ATP-regenerating processes reach maximal rates. In general, both chemical analysis and the use of phosphorus nuclear magnetic resonance spectroscopy reveal that the content of PCr and ATP is higher in muscle or muscle fibers with the highest rates of ATP hydrolysis. The ATP level in muscle must not be allowed to drop very much, for if it does, a condition known as rigor occurs, which can be damaging to muscle cells. In fact, rigor mortis is caused by the loss of muscle ATP some time after death, as a result of the inability of the muscle cells to regenerate ATP.

In skeletal muscle, the activity of creatine kinase exceeds that of all other enzymes. This means that the creatine kinase reaction is very important in muscle for sprinting or burst-type activities such as bounding. Animals in which the gene for creatine kinase is knocked out demonstrate very poor performance during high-intensity exercise. Because skeletal muscle contains more creatine kinase than any other enzyme, damage to a muscle cell membrane, typically through unaccustomed exercise or hard eccentric exercise, results in creatine kinase in the blood. Damage to the heart muscle can also result in the

---

appearance of the cardiac isozyme of creatine kinase in blood. Determination of the activity and isozyme type can reveal the nature and extent of damage to cardiac and skeletal muscle.

ENERGY SYSTEMS

Cells use ATP or related nucleoside triphosphates such as GTP and UTP to do everything that requires an input of energy. This can range from transporting substances into the cell or across internal membranes to signaling information to the interior of the cell, storing fuels, and synthesizing protein and RNA. Muscle cells use ATP for these roles, but during contractile activity, ATPase enzymes become dominant in the breakdown of ATP. The ATPases represent the actomyosin ATPase, the ATPase associated with the sarcolemma to pump sodium and potassium ions, and the ATPase that pumps calcium ions back into the SR. The next equation shows the hydrolysis of ATP in complete detail.

$$MgATP^{2-} + H_2O \rightarrow MgADP^- + n\ H_2PO_4^-$$
$$+ 1\text{-}n\ HPO_4^{2-} + 1\text{-}n\ H^+$$

In the cell, all of the ATP is associated with magnesium ions. It is believed that much of the ADP is also in a complex with magnesium ions. The other product of ATP hydrolysis, which we have simply described as Pi, is actually a mixture of the dihydrogen ($H_2PO_4^-$) and monohydrogen phosphate (HPO_4^{2-}) ions. This is the case because the dihydrogen phosphate ion is a weak acid with an acid dissociation constant (pK_a) just below 7. This means that at the pH of a typical cell (approximately 7), Pi is a mixture of the acid ($H_2PO_4^-$) and its conjugate base (HPO_4^{2-}) in almost equal proportions. It also means that for every monohydrogen phosphate produced by ATP hydrolysis, a proton (H^+) is generated

so that ATP hydrolysis has an acidifying effect in a cell at neutral pH.

We have emphasized that there is not a lot of ATP in any cell and that therefore when ATP is used, it must be immediately regenerated. In skeletal and cardiac muscle, three energy systems are responsible for maintaining ATP concentrations. These are illustrated in figure 3.7 and can be identified as the phosphocreatine system, oxidative phosphorylation, and anaerobic glycolysis. In most cell types, oxidative phosphorylation is dominant in ATP regeneration. In erythrocytes or red blood cells, the only source for ATP regeneration is glycolysis. Skeletal and cardiac muscles use all three systems. In the next sections, we will briefly consider these energy systems. Detailed discussions of oxidative phosphorylation and anaerobic glycolysis are presented in chapters 4 and 5, respectively.

Figure 3.7 Three ATP-hydrolyzing enzymes (ATPases) in muscle, able to use ATP at a rapid rate. The ATP concentration is maintained, or diminishes only slightly, due to the rapid response of the creatine kinase reaction and the aerobic (oxidative phosphorylation) and anaerobic (glycolysis) breakdown of fuel molecules. Although the creatine kinase reaction is freely reversible, during muscle activity it is driven in the direction to regenerate ATP; and during recovery, the reverse reaction is favored to make PCr.

Oxidative Phosphorylation

Oxidative phosphorylation is the fancy name for a process that many people know as the aerobic system or the fuel oxidation system. Biologists often call it cellular respiration or simply respiration. The term *oxidative phosphorylation* is becoming more dominant because it describes the overall process better. We can define this process as the formation of ATP from ADP and Pi in association with the transfer of electrons from fuel molecules to coenzymes to oxygen. Products of oxidative phosphorylation are H_2O and CO_2, as well as ATP.

Figure 3.8 outlines the overall scheme of oxidative phosphorylation. In this scheme, SH₂ represents fuel molecules such as carbohydrates, fats, amino acids, or alcohol. Electrons associated with the hydrogen atoms are transferred from SH₂ to coenzymes (represented by nicotinamide adenine dinucleotide, NAD⁺), and then the electrons on the now-reduced coenzyme (NADH) are transferred on to oxygen, forming H₂O. During this process enough energy is generated and captured to phosphorylate ADP with Pi to make ATP. In chapter 2, electron transfer using dehydrogenase enzymes was illustrated. We devote the next chapter to the steps involved in oxidative phosphorylation. For the present, we can summarize it using figure 3.8:

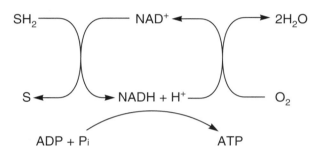

Figure 3.8 A brief summary of oxidative phosphorylation. Electrons in the form of hydrogen atoms are removed from reduced substrates (SH₂) and transferred to the oxidized coenzyme NAD⁺ to make it NADH. Through a series of carriers, electrons on NADH are transferred to oxygen, creating water. Energy released during the transfer of electrons from SH₂ is harnessed to phosphorylate ADP with Pi to make ATP.

The following equation is based on the figure:

$$2 \; SH_2 + O_2 + 5 \; ADP + 5 \; Pi \rightarrow 2 \; S + 2 \; H_2O + 5 \; ATP$$

This very simple equation illustrates that the transfer of electrons to oxygen in the form of H atoms on fuels reduces the oxygen to water, and the energy is used to make ATP. As a summary equation, it does not show the role of the coenzymes NAD⁺ or flavin adenine dinucleotide (FAD) during oxidation and reduction, which we discussed in chapter 2. Although it does reflect the true stoichiometry (the ratio in which molecules react with each other) of ATP production, it does not reveal that during oxidative phosphorylation CO_2 is produced from the carbon atoms of fuel molecules, represented by SH₂.

Oxidative phosphorylation requires fuel molecules that we obtain from our diets, principally in the form of carbohydrates and fats. The oxygen comes from the air we breathe. Oxygen enters the lungs, diffuses to hemoglobin molecules in the blood, and

is transported throughout the body using large and small arteries and arterioles, finally reaching every cell by way of capillaries. Oxygen diffuses from capillaries to cells to the mitochondria of cells, where it accepts electrons to form water and ATP.

It is fairly easy to measure the rate of oxidative phosphorylation by measuring the disappearance of the substrate oxygen. Indeed, this is the method employed by exercise physiologists when they talk about **oxygen consumption,** measured at the mouth with special breath-sampling techniques. The resulting V̇O₂ is used as an index of whole-body metabolism. During exercise, V̇O₂ increases in proportion to the rate of whole-body energy expenditure. With gradually increasing loads of exercise, for example treadmill running or pedaling on a cycle ergometer, V̇O₂ reaches a maximum or peak that is considered to be the highest rate of oxidative phosphorylation for the individual during that activity. If a large enough muscle mass is activated, this peak rate of oxidative phosphorylation is the maximal rate the individual can achieve or V̇O₂max, which can be described in absolute units, liters of oxygen consumed per minute, or relative to body weight in milliliters of oxygen per kilogram body weight per minute.

For many activities, oxidative phosphorylation can provide all of the ATP needed by active skeletal muscles. When exercise level and V̇O₂ are constant over time, oxidative phosphorylation is providing all of the ATP needed by the activity. Such a condition is known as a metabolic **steady state.** From a simple bioenergetics perspective, we can use the approximate relationship in which a V̇O₂ of 1.0 L per minute is equivalent to an energy expenditure of about 5 kcal per minute (21 kJ/min). The precise relationship between V̇O₂ in liters per minute and energy expenditure in kilocalories per minute depends primarily on the relative proportions of carbohydrate and fat that

 K E Y   P O I N T

Oxidation of carbohydrate produces more ATP per unit of oxygen consumed than does the oxidation of fat. In addition, the maximum rate of electron transport from fat to oxygen is approximately one-half the rate reached when carbohydrate is the sole fuel. These facts explain why carbohydrate is the preferred fuel for exercise at an intensity beyond 50% of maximal oxygen consumption.

are being catabolized. The value is slightly higher for pure carbohydrate oxidation compared to pure fat.

Oxidative phosphorylation is considered a low-power process because it relies on oxygen from the atmosphere as the final acceptor of electrons in fuels. With the steps in oxygen transport from the air, to the lungs, to the blood, to the individual cells, and finally to the mitochondria within cells, oxidative phosphorylation cannot quickly reach a maximal rate. Indeed, the time needed to double the rate of oxidative phosphorylation is approximately 15 to 20 s. Although its power is limited in most cases by the oxygen transport system, oxidative phosphorylation is considered to have a high capacity (i.e., large fuel tank), because a major fuel for oxidative phosphorylation is fat. Even a very lean person has enough stored fat to provide fuel for days of oxidative phosphorylation.

Glycolysis

Glycolysis is a metabolic pathway, a sequence of enzyme-catalyzed reactions in which a starting substrate is converted into a product after undergoing a number of steps. From the perspective of muscle, there are two starting substances for the glycolytic pathway, glucose and glycogen. **Glycogen** is simply a polymer consisting entirely of glucose units joined together. It is readily available in muscle and can quickly be used to fuel glycolysis. The products of glycolysis are the same whether the starting substrate is glucose or glycogen. The only difference is that a glucose unit from glycogen produces one more ATP than a free glucose unit. The following are summary reactions for glycolysis beginning with glucose and then with glycogen.

$$glucose + 2\ ADP + 2\ Pi + 2\ NAD^+ \rightarrow 2\ pyruvate + 2\ ATP + 2\ NADH + 2\ H^+$$

$$glycogen_n + 3\ ADP + 3\ Pi + 2\ NAD^+ \rightarrow glycogen_{n-1} + 2\ pyruvate + 3\ ATP + 2\ NADH + H^+$$

For the second reaction, we consider glycogen to consist of n glucose units. When one of these is removed from glycogen and converted to pyruvate, what is left is shown as $glycogen_{n-1}$. The ATP produced can be used to drive any energy-requiring process. The NADH is a source of electrons to feed into oxidative phosphorylation, as we will see in the next chapter. The pyruvate formed in glycolysis has two major fates: (1) Pyruvate can be transported into a mitochondrion and be completely oxidized using the pathways of oxidative phosphorylation; (2) pyruvate can be converted to lactate as the next reaction illustrates.

$$pyruvate + NADH + H^+ \leftrightarrow lactate + NAD^+$$

This latter reaction is catalyzed by *lactate dehydrogenase*, abbreviated LDH. We described the LDH reaction in chapter 2 as an illustration of a redox reaction. Going from left to right, pyruvate is being reduced and NADH is being oxidized. The double-headed arrow reminds us that the reaction can also go from right to left. What actually happens—that is, the net direction—depends on the relative concentrations of the various constituents on the left and right sides of the equation.

When the pyruvate from glycolysis is converted to lactate, the overall process from glycogen (or glucose) to lactate is referred to as *anaerobic glycolysis*. The word "anaerobic" is used because ATP is formed without the need for any oxygen. When the pyruvate in glycolysis enters mitochondria and gets completely broken down to carbon dioxide and water, the process is often termed *aerobic glycolysis* because the final breakdown of pyruvate does require oxygen. Oxidation of pyruvate is a rich source of ATP; two pyruvate can be obtained from each glucose unit, and the complete oxidation of two pyruvate can produce 25 ATP. We will look at the oxidation of pyruvate in the next chapter.

The net production of ATP from glycolysis and PCr is known as **substrate-level phosphorylation,** in contrast to the production of ATP by oxidative phosphorylation. The enzymes catalyzing the reactions of glycolysis producing pyruvate from either glucose or glycogen have generally high activities in skeletal muscle. Thus pyruvate can be generated at a considerably faster rate than it can be completely oxidized in mitochondria. Therefore, when glycolysis is proceeding at a rapid rate in a vigorously contracting muscle, much of the pyruvate formed will be reduced to lactate because it is formed at a rate exceeding the capacity of mitochondria to oxidize it, even if there is plenty of oxygen available. For this reason, we say that anaerobic glycolysis has a moderate power to generate ATP. As we will see, the power of glycolysis in fast-twitch muscle fibers is larger than in slow-twitch muscle fibers because there is a richer concentration of enzymes of glycolysis in the former. The capacity of glycolysis to produce ATP is moderate, much larger than the PCr system, but much less than that of oxidative phosphorylation. The moderate capacity may be related to the availability of glycogen to feed into the glycolytic pathway. A more likely explanation relates to the fact that skeletal muscle can become fairly acidic when high rates of ATP hydrolysis are accompanied by very rapid rates of anaerobic glycolysis. Unlike oxidative phosphorylation, which may take 2 to 3 min to reach its peak rate at the onset of exercise, glycolysis can be quickly started at the beginning of exercise; and if the exercise is intense enough, the glycolytic pathway can reach a maximum rate in about 5 to 10 s. The quick start for glycolysis is not as rapid as the onset of ATP regeneration from PCr, but it makes an important contribution to ATP formation at the beginning of intense muscle work.

> **KEY POINT**
>
> Since ADP is a substrate, glycolysis will be turned on whenever the rate of ATP hydrolysis by a fiber abruptly increases. For low-intensity exercise, little lactate formation will take place, but for higher rates of exercise the concentration of lactate in muscle and blood will increase over time.

Phosphocreatine System

Earlier in the chapter we discussed the phosphagen PCr as a readily available, energy-rich molecule capable of regenerating ATP, as seen in the following equation.

$$ADP + PCr + H^+ \leftrightarrow ATP + Cr$$

Use of PCr to regenerate ATP is sometimes referred to by exercise physiologists as the anaerobic alactic system because it does not need oxygen and does not generate lactate. At the onset of very vigorous muscle activity, creatine kinase drives this reaction to the right, depleting PCr, maintaining ATP, and increasing creatine. The consumption of a proton (H^+) as PCr concentration decreases can be beneficial to the muscle during high-intensity exercise or games when high rates of ATP hydrolysis will acidify the muscle.

There are four genes for creatine kinase (CK) in mammals, coding for four protein monomers of about 40 kD each. Two of these monomers are designated B (for brain) and M (for muscle), and there are two mitochondrial subunit forms. Of the mitochondrial subunits, one is expressed primarily in heart and skeletal muscle and the other in brain and other tissues. The mitochondrial creatine kinase subunit is involved in the facilitation of oxidative phosphorylation, a topic we will discuss in the next

chapter. The nonmitochondrial form of creatine kinase in skeletal muscle is active as a dimer, containing two M subunits that we could represent as MM. This is found in the cytosol attached to contractile filaments, at the inner side of the sarcolemma, and at the outer face of the SR membrane, as well as free in the cytosol during exercise. These locations ensure that creatine kinase is appropriately positioned to sites where ATP is hydrolyzed and regenerated.

> **KEY POINT**
>
> Cardiac muscle contains creatine kinase as a dimer that includes both an M subunit and a B subunit. Damage to cardiac muscle can involve loss of the MB isozyme to the extracellular fluid, including blood. Measurement of the MB isozyme has been used to determine that damage to the heart has occurred. Skeletal muscle damage results in elevated levels of the MM subunit in blood.

Because creatine kinase activity is so high, it can maintain ATP levels remarkably well even during intense exercise. This ATP-buffering effect is substantial, and for this reason we can say that creatine kinase has a high power for regenerating ATP. The actual PCr concentration in resting human skeletal muscle based on muscle biopsy sample analysis is approximately 18 to 20 mmol per kilogram wet weight of muscle (or 23-26 mM). No matter how it is expressed, there is a limited supply of PCr, and therefore the CP system has a low capacity. During exercise PCr levels fall in proportion to the relative intensity of the exercise, and the free creatine concentration rises in parallel. For all-out efforts to fatigue, PCr levels can decrease by 90% or more. During recovery or rest periods, the reverse of the creatine kinase reaction dominates, and phosphate transfer to creatine from ATP produced by oxidative phosphorylation regenerates PCr. The half-time for recovery of PCr during rest periods is approximately 30 s. This can vary, since a fit person with a higher capacity for oxidative ATP formation should recover PCr stores at a faster rate.

Properties of Creatine

There is no doubt that PCr plays a very important role at the start of exercise and during sprint-type activities. Since the early 1900s, it has been known that feeding people extra creatine leads to retention of some of this creatine, presumably in the muscles. Indeed, Chanutin (1926) suggested that the retained creatine could have an effect on increasing body protein content. The issue of creatine as a supplement to improve exercise performance and muscle mass did not approach the significance it now has until the early 1990s. Researchers in Sweden and the United Kingdom initially published a number of articles suggesting that supplemental creatine could increase muscle PCr levels and improve performance in exercise activities. Additional research followed in many other laboratories and countries using a variety of human and animal feeding and exercise models. Today research on creatine may not be the "hot topic" it once was, but many serious exercise trainers consider supplementing with creatine to be a necessity.

To understand this interesting topic, it is necessary to learn something about creatine. Humans get much of their creatine in a two-step sequence of reactions that take place primarily in the liver, using the amino acids glycine and arginine. We also know that creatine can be obtained in significant quantities by those who consume creatine-containing tissues—that is, meat. Being a simple, water-soluble molecule, creatine is easily absorbed and in the blood would be indistinguishable from the creatine made by the liver. The average person needs only about 2 g of creatine a day to replace losses, an amount that can be obtained through biosynthesis or a combination of biosynthesis plus dietary sources. The usual supplemental regimen is 20 to 30 g per day for about six days, followed by maintenance doses of 3 to 5 g per day. Many studies have demonstrated that carbohydrate ingestion along with the creatine supplement enhances extra creatine retention. Retention of supplemental creatine is also increased with exercise training compared to inactivity.

An overall snapshot of creatine is shown in figure 3.9. Creatine in the blood, whether absorbed from food or synthesized by the liver, is generally in the

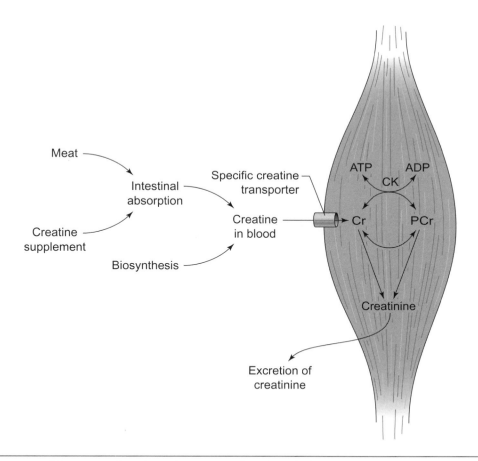

Figure 3.9 A summary of the sources of creatine and creatine uptake and metabolism in skeletal muscle. Creatine is synthesized in the body, obtained from dietary meat or through supplements. Creatine from all sources is taken up from the blood against a concentration gradient through a specific creatine transporter that also lets in Na^+ and Cl^- ions that move down their concentration gradients. In the cell, about 60% to 80% of creatine is phosphorylated from ATP to make creatine phosphate. About 2 g per day (1.7%) of cellular creatine or creatine phosphate undergoes nonenzymatic conversion to creatinine, which is excreted in the urine.

concentration range of 25 to 100 μM. The lower end would likely be seen in those who are vegetarians and the upper range in those who consume a diet with plenty of meat. With large supplemental doses, blood creatine concentrations can be increased by more than 10-fold. Creatine concentration inside muscle fibers at rest is generally in the 7 to 8 mM range, so transport of creatine into a muscle fiber is an uphill process, going against a significant concentration gradient. A specific *creatine transporter* allows creatine to enter a cell, driven by the simultaneous transport of sodium and chloride ions down their significant concentration gradients. Once inside the muscle fiber, much of the creatine is phosphorylated using ATP and the creatine kinase enzyme. Loss of creatine from cells occurs when either free creatine or CP undergoes nonenzymatic loss of a water molecule, producing a new molecule, *creatinine*. Creatinine is released from cells, filtered from the blood by the kidneys, and excreted in the urine. The daily loss of body creatine due to formation of creatinine is roughly 1.7% of the total body amount, or about 2 g per day.

KEY POINT

Since the preponderance of all creatine is found in muscle, the excretion of creatinine each day should reflect skeletal muscle mass. Indeed, the ratio of 1 g of creatinine excreted per 20 kg of muscle is commonly accepted.

The total concentration of creatine in a muscle fiber (TCr) is the sum of the concentrations of PCr and Cr. Mean values for creatine, PCr, and total creatine in muscle are shown in table 3.2. These values are based on chemical analysis of skeletal muscle biopsy samples in human subjects at rest, before and after a program of creatine supplementation. On the basis of data from table 3.2, supplementation with creatine leads to a 16% increase in total creatine and an 11%

increase in PCr. However, data from a variety of studies reveal that the amount of extra TCr retained can range from 10% to 30%. Although there are significant variations in TCr and the ratios of PCr to TCr in the literature, generally we can say the following:

1. The PCr/TCr ratio in rested muscle is generally in the range of 0.6 to 0.7.

2. The differences in total TCr and the ratio of PCr to TCr in human skeletal muscles are less pronounced compared to the values in animal muscles.

3. Low TCr and PCr values tend to be found in those eating a meat-free diet.

4. Concentrations of TCr and PCr in individual type II muscle fibers isolated from human skeletal muscle biopsy samples tend be about 20% higher than in type I muscle fibers.

Effect of Creatine Supplementation on Exercise

Research data show that creatine supplementation does increase PCr and total creatine concentrations in skeletal muscles of most subjects, but does this lead to an improvement in exercise and sport performance? Hundreds of published studies have looked at the effects of creatine supplementation on human exercise and sport performance, and some conclusions can be made on the basis of these reports. Creatine supplementation appears to be most effective in short-term, high-intensity exercise lasting up to 3 min in duration. Creatine appears to be especially helpful if the high-intensity activity is repeated with only a brief recovery period. We would expect results such as this because the half-time for restoration of PCr levels after an intense bout of exercise is approximately 30 s. Therefore, starting out with a higher PCr concentration before the exercise and a higher Cr concentration after the first exercise bout would seem to be advantageous to performance on the next bout. Many athletes report that they can train harder

Table 3.2 Creatine (Cr), Phosphocreatine (PCr), and Total Creatine (TCr) Concentrations in Muscle

Condition	Cr (mmol/kg dw)	PCr (mmol/kg dw)	TCr (mmol/kg dw)	PCr/TCr
Presupplementation	46.4 ± 8.1	76.3 ± 7.1	122.7 ± 7.8	0.62
Postsupplementation	57.1 ± 7.9	84.7 ± 6.2	141.8 ± 6.4	0.60

Values are based on data from 10 selected studies published between 1994 and 2003 and are reported as means ± SD.
The PCr/TCr ratio was calculated from the mean values for PCr and TCr.

in the gym or on the track with creatine because they recover faster after each set or repeat.

Many advocates of resistance training also believe that creatine increases body weight and strength gains. Such beliefs might suggest that creatine increases muscle protein synthesis rates either as an independent effect of creatine or because, as just noted, athletes can train harder. Michael Rennie, Jacques Poortmans, and their collaborators recently reported that creatine supplementation does not have any independent effect to increase the rate of synthesis or breakdown of skeletal muscle protein in the postabsorptive or fed states in human volunteers (Louis, Poortmans, Francaux, Berré, et al. 2003). A follow-up investigation in which subjects were studied after a resistance training workout again showed that creatine had no independent effect on skeletal muscle protein synthesis, whether in the fed or unfed state, following exercise (Louis, Poortmans, Francaux, Hultman, et al. 2003a). More recent studies by Tarnopolsky and his group demonstrated that creatine supplementation can upregulate the expression of a variety of skeletal muscle genes, especially those involved in intracellular signaling (Safdar et al. 2004).

There is another side to creatine supplementation, the long-term effects of which are not well known. Taking 10 times or more of the daily creatine need has been shown both to decrease the biosynthetic pathway for creatine in the liver and to decrease the expression of creatine transporters required for creatine uptake into cells (Snow and Murphy 2003).

▷ K E Y   P O I N T ◁

Many creatine supplementation enthusiasts may be unaware that only a fraction of the large supplemental doses are actually retained by the body. Our kidneys can filter excess levels of creatine from the blood, with the excess excreted in the urine.

Role of Adenylate Kinase and AMP Deaminase

When we consider the metabolism of muscle during exercise, two other reactions must be taken into account. The first is one we have seen previously. It is written here in the direction opposite to the way it was shown earlier in this chapter, but this does not matter because the reaction is freely reversible with an equilibrium constant near 1.

$$2 \text{ ADP} \leftrightarrow \text{ATP} + \text{AMP}$$

The enzyme for this reaction is named *adenylate kinase,* but you will also see it identified as adenylyl kinase. It has also been called *myokinase* (*myo* for muscle) because its activity is so high in skeletal muscle. The significance of this reaction is as follows. During hard muscle activity, the rate of ATP hydrolysis is high. Although muscle efficiently regenerates ATP from ADP by the three energy systems previously described, an increase in ADP concentration occurs. What the adenylate kinase does is to cause two ADP molecules to interact, transferring a phosphate from one to the other, creating one AMP and one ATP. This interaction keeps the ADP concentration from building up to the extent it would otherwise. Obviously, forming an ATP from two ADP is advantageous during exercise, but the AMP has an important role in stimulating glycolysis, as we will see in chapter 5.

In the second reaction, *AMP deaminase* (also called adenylate deaminase) irreversibly deaminates (removes the amino group from) the base adenine of AMP, producing ammonia, NH_3. Deaminated AMP becomes *IMP* (inosine monophosphate).

$$\text{AMP} + H_2O \rightarrow \text{IMP} + NH_3$$

Because NH_3 is a weak base, it takes a proton (H^+) from the medium, becoming ammonium ion (NH_4^+). The ammonium ion formed during the AMP deaminase reaction can leave the muscle cell and travel in the blood to the liver and kidney for further metabolism. Adenosine monophosphate deaminase is found in higher levels in the fast-twitch (type II) muscle fibers. These fibers are more active during intense exercise activity. The activity of AMP deaminase is low at rest but increases due to changes within a muscle fiber under vigorous exercise conditions such as the fall in muscle pH and rise in ADP concentration. In a rested muscle cell, most of the AMP deaminase is free in the sarcoplasm, but increasing contractile activity facilitates AMP deaminase binding to myosin in the thin filaments. This myosin binding increases AMP deaminase activity.

Adenylate kinase and AMP deaminase can work in concert during vigorous muscle activity. In effect, two ADP molecules are converted into an ATP and an AMP by the AMP kinase reaction. Then in the AMP deaminase reaction, the AMP is irreversibly converted into IMP. These related reactions play a major role in maintaining optimal energy status in the muscle fiber, especially during intense muscle work. The irreversible AMP deaminase reaction drives the reversible adenylate kinase reaction to the right, to diminish the increase in the concentration of ADP ([ADP]). This reaction also maintains a high [ATP]/[ADP] ratio, which is absolutely important to maximize the energy of ATP hydrolysis, so essential for driving the ATPase reactions associated with muscle contraction.

Since the AMP deaminase reaction is increased by a decrease in muscle pH, we would expect this reaction to be dominant in fibers with a high rate of ATP hydrolysis, a high capacity for glycolysis, and a lower capacity for oxidative phosphorylation. Not surprisingly, type II fibers are the primary sources of blood lactate and ammonia during exercise. The two reactions involving AMP are very important in muscle under certain exercise conditions, as evidenced by the fact that people with a genetic AMP deaminase deficiency have a poor ability to perform intense exercise and find that attempts at this produce muscle pain, cramping, and early fatigue.

The product, IMP, which accumulates in muscle during severe exercise, has several fates in the postexercise period. Most of the IMP is converted back to AMP during a two-step process, part of the *purine nucleotide cycle* that we will examine in chapter 7. Some of the IMP can undergo a loss of its phosphate group to form *inosine*. Inosine can lose its ribose group to form *hypoxanthine*. Without phosphate groups, inosine and hypoxanthine can leave the muscle fiber. Following hard exercise, both inosine and hypoxanthine can be found in blood draining active muscle.

Muscle Metabolism in Exercise

When we ask our muscles to do something for us, whether it is getting out of a chair, walking to the store, doing a set of resistance exercises, or running a 10 K race, we are placing a specific energy demand on the muscle. Activated by our nervous system, individual fibers contract at certain rates (or perhaps not at all), hydrolyzing ATP. Via the energy systems already described, the ATP used is regenerated as fast as possible. Nonetheless, there are changes within the muscle fiber that reflect its ability to regenerate ATP as well as its metabolic displacement from the rest condition. The concentrations of ADP and AMP generally increase; anaerobic glycolysis can produce lactate; PCr is used to make ATP; and we may actually lose some adenine nucleotide (ATP, ADP, and AMP) if IMP is formed. We call these molecules *metabolites*, and how their concentrations differ from those at rest can tell us a lot about the previous contractile

activity. For example, glycogen is an important fuel for anaerobic glycolysis. It is also a primary source of pyruvate that is oxidized in mitochondria. We would expect glycogen concentration to decrease in proportion to how much it is used and how it is used (anaerobic vs. aerobic glycolysis) because the anaerobic pathway uses glycogen at 10 times the rate of the aerobic pathway to get the same amount of ATP. Phosphate (Pi) will increase if there is a net decrease in PCr, and of course Cr will increase in proportion to the breakdown of total PCr.

> ### ▶ KEY POINT ◀
>
> Activation of muscle fibers during any type of physical activity is controlled by neurons that originate in the spinal cord. During low-level activity, the fibers that are active tend to be type I, with a high capacity for oxidative phosphorylation. As one increases the demand on a muscle, more muscle fibers are brought into play, including type II. At the highest levels of force demand, almost all muscle fibers are active. High levels of force can be maintained only for brief periods, as some large type II fibers with a high actomyosin ATPase activity and glycolytic capacity but a low ability for oxidative phosphorylation can fatigue easily and thus are able to participate only for brief bursts.

Researchers have various techniques to measure changes in muscle metabolites during exercise. One of these is to sample the muscle with a biopsy needle, which can extract a small (about 50 mg) piece of muscle. Chemical analysis of this muscle sample can give us a picture of the muscle at the time of the biopsy if the sample is immediately frozen in liquid nitrogen (–196 °C) upon being removed from the muscle. This is considered an invasive technique because the biopsy needle must be inserted into the muscle each time a sample is needed. An alternative technique based on the properties of specific atomic nuclei can be used. For example, the isotope of phosphorus with mass number 31 (^{31}P) can give rise to a specific signal depending on the type of molecule in which the phosphorus atom is found. Thus we can get a particular signal from each of the three phosphate groups in ATP, the phosphate in PCr, or the two forms of Pi (HPO_4^{2-} and $H_2PO_4^{1-}$). This technique, known as

^{31}P *NMR* (nuclear magnetic resonance) spectroscopy, has been used for more than 25 years to study human skeletal muscle. The term *magnetic resonance imaging* or *MRI* is often used now instead of NMR. This technique is noninvasive, and sampling of muscle can be done repeatedly. There are some drawbacks, not the least of which is the expensive equipment needed. Moreover, the person has to exercise inside the bore of a very large magnet, which limits the type of exercise that can be used. In addition, the sensitivity of the technique is limited, so metabolites such as ADP and AMP cannot be measured, although these can be estimated based on the assumption of equilibrium for the creatine kinase and adenylate kinase reactions, respectively.

The concentrations of metabolites in muscle are generally expressed in one of three ways. If metabolites from the frozen sample are extracted out and measured, we can express the quantity on the basis of the number of millimoles of particular metabolite per kilogram of wet tissue weight. If the muscle sample is first freeze-dried to remove all the water, we can express the concentration as millimoles per kilogram of dry weight of muscle. The water content of fresh

muscle is roughly 77% of the total weight. Thus a wet sample weighing 100 mg would be approximately 23 mg if completely dried. It is quite common now to express the concentrations of metabolites in muscle in mM units (i.e., millimoles of metabolite per liter of intracellular water). It is accepted that for every kilogram of muscle, 70% will be represented by intracellular water or 0.7 L. The difference between intracellular and total water is the extracellular water, that is, the fluid around cells and in capillaries, small arteries, and veins.

Table 3.3 outlines some representative values for human muscle metabolites, expressed in mM units for conditions representing rest; mild fatigue; fatigue from severe, repeated contractions; and that following a marathon run. These values do not reflect any specific person or study but give an indication of how metabolites change under different contractile conditions. Notice that for mild fatigue and that following a marathon race, concentrations of many metabolites are not very different from those at rest. In particular, notice that ATP concentration is well defended by the energy systems operating in muscle. The severe fatigue is not something many people experience, but there

Table 3.3 Hypothetical Metabolite Concentrations in Human Skeletal Muscle at Rest and After Different Kinds of Physical Activity

Metabolite	Cellular concentration (mM)			
	Rest	Mild fatigue[a]	Severe fatigue[b]	Prolonged fatigue[c]
ATP	8.2[d]	8.1	6.2	8.0
ADP (free)[e]	0.016	0.022	0.13	0.020
AMP (free)[e]	0.00008	0.00012	0.002	0.0003
TAN[f]	8.217	8.123	6.337	8.032
IMP	<0.02	0.09	1.8	0.1
Pi	3.0	12	28	14
PCr	32	22	5	23
Cr	10	20	37	19
Lactate	2.2	5.9	26.3	2.5
pH	7.1	7.0	6.3	7.1
Glycogen[g]	175	140	80	10

[a]Mild fatigue refers to a state associated with moderate-speed cycling or after 30 min of an easy game of pickup basketball.

[b]Severe fatigue could be induced by an intense training session for a 400 m running race—a metabolic state that can be generated only in the trained muscles of well-motivated subjects.

[c]Prolonged fatigue is best represented by the state of muscle after a major marathon race.

[d]Metabolite concentrations are expressed as millimoles of metabolite per liter of intracellular water. It is assumed that each kilogram of skeletal muscle contains 0.7 kg (L) of water.

[e]The word "free" is used for the concentrations of ADP and AMP, meaning the fraction of each that is not bound to proteins in the cell.

[f]TAN (total adenine nucleotide) concentration is the sum of the concentrations of ATP, ADP, and AMP.

[g]The number refers to the millimoles of glucose units in glycogen, assuming they were free in the cell.

are very large differences in PCr, Cr, Pi, IMP, lactate, and pH, suggesting the stress on the ATP-regenerating systems. For the marathon, the major difference is the huge drop in glycogen concentration, suggesting that this must be a very important fuel for this long race. In subsequent chapters we will look at exercise metabolism in greater detail.

> ▶ **K E Y   P O I N T** ◀

The major change in metabolite concentrations after a marathon run is in glycogen. The lack of glycogen to support aerobic glycolysis is a principal reason most runners are slower at the end of a marathon race. Of course, the quadriceps muscle group can become very sore, especially if there are significant downhill portions in the marathon.

BIOENERGETICS

So far our discussion has been very general in terms of energy. This is produced by the hydrolysis of ATP and PCr, glycolysis, and oxidative phosphorylation. This energy is used for the anabolic part of metabolism (synthesis of protein, RNA, cell constituents, etc.), contraction (skeletal, cardiac, and smooth muscles), transport processes, and control mechanisms. We have neither described the energy in any detail nor shown how it can be quantified. These are the aims of this section.

The Concept of Free Energy

Before we begin, however, let us review a few simple concepts using figure 3.10 as our model. A loose cable is strung between two poles and fastened at positions A and B in figure 3.10a and C and D in figure 3.10b. You can slide down the cable holding a handle attached to a ring that fits over the cable. Intuitively, you know that if you start from position A, you will slide along the cable, stopping not partway down, but at B. If you start from C, you will slide toward D but will not reach it, stopping, as shown, closer to D than to C. You could also start from D and slide along the cable; you know, though, on the basis of the relative heights of attachment of C and D, that you will slide along toward C, but not far. The examples of energy involving a mechanical apparatus are simple. Potential energy based on the

weight of your body and its elevation can be converted into kinetic energy in your sliding along the cable. The kinetic energy of the ring sliding over the cable causes the molecules in the cable to become warmer; that is, they gain heat, meaning that the molecules in the cable and even some in the air around the cable increase their motion. You could use your kinetic energy of movement sliding down the cable to knock over a stack of blocks. Of course, when you are suspended between C and D and not moving, you will not be able to transfer your energy to anything else. You may also realize that some of the potential energy based on your elevation cannot be converted entirely into kinetic energy of motion. There will be losses of energy in friction with the cable, generating heat and resistance by the air. The lessons we can learn from a simple analogy such as this help us to understand the following statements about chemical reactions in cells.

- All chemical reactions involve energy changes. If there is no energy change, nothing happens.
- Chemical reactions in living organisms are catalyzed by enzymes.
- Enzymes attempt to drive the reaction they catalyze toward equilibrium.
- As enzyme-catalyzed reactions proceed toward equilibrium, they release energy.
- The farther a reaction is from equilibrium, the more energy it can release.
- At equilibrium no energy is released.
- Some energy released in a chemical reaction can be used to do work; the remainder is unavailable.

In thermodynamics, we focus only on differences between initial and final states; in this sense, we talk about changes. The following thermodynamic terms are defined as applied to biological processes:

- **Enthalpy change (ΔH):** In any chemical reaction or sequence of reactions, the change in energy of the reactants when they are turned into products is called the enthalpy change. This can be measured as the total heat energy change, described by the abbreviation ΔH. When heat energy is given off or released in a reaction or process, we have a negative value for ΔH; such a reaction is an **exothermic reaction.** Reactions or processes with a positive value for ΔH are said to be *endothermic* and can occur only with an input of energy.

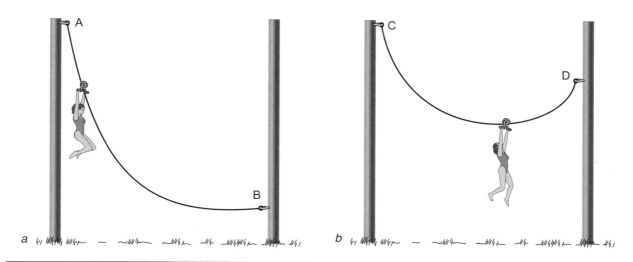

Figure 3.10 Example of energy involving a mechanical apparatus. A person sliding along a cable loosely hung between two posts from a higher to a lower position will stop (reach equilibrium) farther from the starting position the greater the difference in relative heights of the cable attachments. As in this mechanical example, chemical reactions proceed toward an equilibrium. The farther the reaction is from equilibrium, the more energy it can release. At equilibrium no energy is released.

• **Entropy change (ΔS):** Entropy is a measure of energy dispersal. The concept of entropy tells us that energy wants to move from where it is concentrated to where it is dispersed or spread out. Consider a hot frying pan after the heat is turned off. It is a concentrated source of heat energy, but we know that it will cool and the air around it will become warmer. A ball placed on the side of a steep hill will roll down the hill; this is a spontaneous process. The potential energy of the ball on the hill will be dispersed first into kinetic energy as it begins to move, then to heat energy as it proceeds to the bottom of the hill. If we place a rock below the ball, we simply prevent it from rolling, but we don't change the fact that the tendency is for the ball to roll down the hill if given a chance. This is entropy at work. Adenosine triphosphate is a concentrated form of energy as we have discussed. If given a chance, say when a myosin cross bridge interacts with an actin monomer, the chemical energy stored in ATP is dispersed and turned into a bit of work and mostly heat. Processes that can occur by themselves if given a chance have a positive value for entropy change (positive ΔS). The reverse of the examples just given cannot occur along the same path. That is, a cool frying pan does not become hot by accepting heat energy from the air around it, and a ball does not spontaneously roll up a hill.

• **Free energy change (ΔG):** Of the total energy released in a reaction or process, not all is available to do something useful. The free energy change, or ΔG, is the maximum energy available from a reaction or process that can be harnessed to do something useful. From a biological perspective, something useful would be a muscle contracting and lifting a load, moving ions across a membrane against their concentration gradient, or synthesizing a protein from amino acids. When free energy is released, the ΔG is negative; the reaction is an **exergonic reaction.** A reaction with a positive value for ΔG cannot occur by itself—as a ball cannot spontaneously roll up a hill—and is said to be an **endergonic reaction.**

The relationship between changes in enthalpy, entropy, and free energy for a reaction or process is given by the expression

$$\Delta G = \Delta H - T\Delta S \qquad (3.1)$$

This equation tells us that not all of the energy that is released in a reaction or process can be used to do something useful. The unavailable portion is given by the TΔS term. This term reflects loss of energy to the surroundings in the form of increased motion of molecules, perhaps in sound or light. We know from the law of the conservation of energy that we can neither create nor destroy energy. Thermodynamics tells us that we cannot capture 100% of the energy released in a reaction or process; some must be lost to the surroundings.

From the earlier parts of this chapter, you should understand that reactions or processes with a positive value for free energy change can take place only if there is a simultaneous reaction with a negative ΔG. As you have seen, the simultaneous reaction that drives most endergonic processes in living organisms is ATP hydrolysis. Thus a cross bridge in muscle can interact with an actin subunit in the thin filament and exert a force (a process with a +ΔG per se) only when an ATP is hydrolyzed. The key point is that the energy required must be less than the energy provided by ATP hydrolysis so that overall the combination has a negative value for ΔG.

Quantitative Values for Free Energy

In thermodynamics, one cannot measure exact values associated with an initial and final state of a reaction in terms of enthalpy, entropy, and free energy. However, we can measure energy changes over the course of a reaction as changes in enthalpy, changes in free energy, and changes in entropy. For the reversible chemical reaction given by the equation

$$A + B \leftrightarrow C + D$$

the free energy change in the direction from left to right is given by the expression

$$\Delta G = \Delta G^{\circ\prime} + RT \ln \frac{[C][D]}{[A][B]} \qquad (3.2)$$

where ΔG is the free energy change for the reaction.

ΔG°' is the **standard free energy change** for the reaction under what are considered to be standard conditions, when each of the reactants and products is at a concentration of 1.0 M. A prime is added to indicate a pH of 7.0; the traditional thermodynamic standard free energy is defined at a most nonbiological pH of zero. ΔG°' reflects the energy-generating potential of the reaction, and a value for ΔG°' can be determined from the equilibrium constant at a pH of 7.0 (K'_{eq}, discussed later).

R is the *gas constant*, which has a value of 8.314 joules per mole per degree K (1.987 cal per mole per degree K). Remember, 1 calorie (cal) is equivalent to 4.18 joules (J) or 1 kilocalorie (kcal) = 4.18 kilojoules (kJ). T is the absolute temperature expressed in degrees K. This can be determined by adding 273 to the temperature in Celsius units. [A] is the concentration of species A, [B] is the concentration of species B, and so forth.

In equation 3.2, ΔG°' and R are constants. At a particular temperature T, you should see that ΔG will depend only on the relative concentrations of A, B, C, and D. Suppose we start the reaction by adding A and B together. The reaction will quickly proceed to

the right; and as C and D accumulate, the net reaction (toward C and D) will gradually slow down. Eventually the forward (toward C and D) and backward (toward A and B) reactions will occur at the same rate, and we will have reached **equilibrium.** From an energy perspective and using the analogies in figure 3.10, we know the following:

1. The farther a reaction is from equilibrium, the larger the value of ΔG (as a negative value). In equation 3.2, when the reaction begins, [A] and [B] will be large, whereas [C] and [D] will be small. Accordingly, the natural logarithm (ln) term will be large and negative. Thus at the start of the reaction, the ΔG will be large and negative.

2. As the reaction proceeds toward equilibrium, the numerical values for [A] and [B] will decrease, whereas the values for [C] and [D] will increase. Accordingly ΔG will decrease but still be negative.

3. At equilibrium, there is no net reaction and no net energy change, so the value for ΔG is zero. Therefore, a reaction at equilibrium releases no free energy.

Rewriting equation 3.2 at equilibrium, where ΔG = 0, we get

$$0 = \Delta G^{\circ\prime} + RT \ln \frac{[C][D]}{[A][B]}$$

However, at equilibrium (and a pH of 7), the concentration ratio [C][D]/[A][B] is given by the equilibrium constant K'_{eq}. We use this fact and rearrange the equation:

$$\Delta G^{\circ\prime} = -RT \ln K'_{eq} \qquad (3.3)$$

This equation allows us to determine the value of the standard free energy change for a reaction at pH 7 if we know the temperature and the value of the equilibrium constant. We can say the following about the ΔG°' value for a reaction:

1. It is determined by the value for the equilibrium constant.

2. It reflects the energy-generating potential for the reaction.

3. It does not define the actual energy change for the reaction inside a cell, since this depends on the relative concentrations of reactants and products.

The actual free energy change for a reaction inside a tissue is not difficult to determine. Take a sample of tissue and quickly freeze it; this immediately stops all chemical reactions. Analyze chemically the amount of reactant(s) and product(s) for the reaction in question. From this, one can determine the **mass action ratio** (many people use the Greek symbol Γ for this) for the reaction. This is the ratio of product concentration(s) to reactant concentration(s) as they existed at the moment the tissue was frozen. One can determine the K'_{eq} for the reaction by allowing it to come to equilibrium in a test tube under conditions of temperature, pH, and ion concentration identical to those of the living tissue, and then analyzing the equilibrium concentrations of reactant(s) and product(s). An equilibrium constant having been determined, this can be plugged into equation 3.3 to determine $\Delta G^{\circ}{}'$. Now we can determine the actual free energy change for the reaction in question at the exact time it was frozen by employing a variation of equation 3.2.

$$\Delta G = \Delta G^{\circ}{}' + RT \ln \Gamma \qquad (3.4)$$

We can rearrange this equation, taking advantage of the relationship between $\Delta G^{\circ}{}'$ and K'_{eq}.

$$\Delta G = -RT \ln K'_{eq} / \Gamma \qquad (3.5)$$

This equation tells us that if K'_{eq} is $>\Gamma$, the free energy of the reaction (ΔG) will be negative; this means the reaction can proceed on its own, releasing energy. The larger the ratio K'_{eq}/Γ (that is, the more the reaction is displaced from equilibrium), the more free energy will be released by the reaction. If K'_{eq} is exactly the same as Γ, then ΔG is zero.

Free Energy in the Cell

We have emphasized that the free energy of ATP hydrolysis (or GTP or UTP) drives just about every reaction or process in our bodies. Therefore, it is essential to ensure that the free energy released during the hydrolysis of each ATP molecule is as high as possible. If the ATP hydrolysis reaction is allowed to come to equilibrium, the concentration of ADP is about 10 million times greater than that of ATP. In the cell, the complete opposite prevails; the concentration of ATP is hundreds of times larger than that of ADP under all but the most severe conditions of fatigue. The $\Delta G^{\circ}{}'$ for ATP hydrolysis is –30.5 kJ/mol (–7.3 kcal/mol), but the real ΔG is far larger in magnitude with a negative sign. We could determine the actual free energy for ATP in the cell by using equation 3.4 and using actual values for [ADP] [Pi]/[ATP] to get the mass action ratio (Γ).

$$\Delta G = -30.5 \text{ kJ/mol} + RT \ln [\text{ADP}] [\text{Pi}]/[\text{ATP}]$$

We exclude the role of water in this equation, focusing on the key metabolites whose concentrations can change in the cell.

If you use the concentrations for ATP, ADP, and Pi from table 3.3, use 37 °C (310 K) for temperature, and do the calculation for ΔG using equation 3.4, you will see that this number is well defended (large and negative) except under conditions of severe stress. The smaller release of free energy under severe fatigue conditions can affect the force generated by each cross bridge or the ability of the two ATPases to pump sodium, potassium, and calcium ions.

The strategy of the cell to maximize the ΔG for ATP hydrolysis is based on keeping the ratio of [ATP] to [ADP] as high as possible. This is assisted by the location of several key enzymes. For example, creatine kinase activity is high in the region of the myofibrils, the SR, and the sarcolemma so that as rapidly as ATP is hydrolyzed to ADP and Pi by the actomyosin, SR, and sodium-potassium ATPases, the ADP will be immediately converted back to ATP at the expense of PCr. In addition, adenylate kinase plays an important role in reducing ADP concentration by reacting two ADP molecules together to get an AMP and an ATP. Besides helping to keep [ADP] low, the AMP produced is a potent stimulator of glycolysis.

SUMMARY

The thousands of chemical reactions in our bodies that constitute our metabolism involve energy changes. Adenosine triphosphate, a nucleoside triphosphate, is the key mediator of metabolism because its hydrolysis generates a great deal of free energy that can be used to drive processes unable to occur by themselves. Adenosine triphosphate is generated when the energy stored in fuel molecules is released during catabolism. While ATP ensures that a host of energy-requiring reactions and processes in all cells can take place, its rate of use can be extremely high in skeletal muscle. Each tiny cross bridge on a myosin molecule in a thick filament in a muscle fiber can bind to actin in a thin filament, generating force. Each of these interactions uses an ATP, and millions of these interactions take place each second as a muscle fiber shortens. Adenosine triphosphate is also needed to regulate the concentration of calcium, which activates myosin and actin interactions, and to control the sodium and potassium ion concentrations that can change so much as the muscle fiber is activated by its neuron.

The ATP concentration in a muscle fiber is sufficient to drive only about 2 s of maximal work. Its concentration is maintained remarkably well by three key energy systems. Transfer of a phosphate group from PCr to ADP is a mechanism to renew ATP rapidly, but supplies of PCr are limited. Glycolysis, an anaerobic process in which glycogen or glucose is broken down to lactate, can provide ATP at a fairly rapid rate; but there is a limited capacity for this process. Oxidative phosphorylation, however, provides ATP at a low rate for a prolonged period of time. It is the dominant mechanism for producing ATP in almost all cells in our body and provides all the ATP during steady state exercise. Exercise at a high rate to fatigue can lead to severe metabolic displacement in muscle as the contracting fibers attempt to generate ATP to match its high rate of breakdown by the ATPases. Severe decreases in PCr and pH and large increases in ADP, lactate, ammonium, and phosphate ions characterize the fatigued state as a result of maximal exercise lasting 1 to 5 min.

Differences in energy states between reactants and products in a chemical reaction contribute to the overall energy changes of the reaction or process designated the enthalpy change (ΔH). Not all of the energy released in a chemical reaction can be captured to do something useful because there is an unavoidable dispersal of energy based on changes in entropy (ΔS) over the course of the reaction. What we can use is known as the free energy change (ΔG). Free energy release during hydrolysis of ATP is critical to cell function, and this is ensured as the concentration of ATP is kept high relative to its hydrolysis products ADP and Pi. The three energy systems play critical roles in this, aided under more severe contractile conditions by the enzymes adenylate kinase and AMP deaminase.

REVIEW QUESTIONS

1. Using the data in table 3.3, calculate the mass action (Γ) ratios for the hydrolysis of ATP for each of the four conditions shown. Be sure to use the free concentration of ADP.

2. Convert a metabolite concentration of 4.0 mmol per kilogram wet weight to the following:
   a. Micromoles per gram wet weight
   b. Millimoles per kilogram dry weight
   c. Millimoles per liter of intracellular water

3. Assuming that the adenylate kinase reaction is at equilibrium at rest in skeletal muscle, determine the equilibrium constant in the direction of ATP formation using the data in table 3.3.

4. Using equation 3.4 and the standard free energy of ATP hydrolysis for ATP as 7.3 kcal/mol, calculate the free energy for ATP hydrolysis in a skeletal muscle cell at rest at 37 °C using data for ATP, ADP (free), and Pi concentrations found in table 3.3.

5. If the free concentration of magnesium ions in a resting muscle cell is 2 mM, what is the approximate total concentration of magnesium in the cell?

6. Answer question 4 again, but use the values for severe fatigue and the same temperature.

7. Algebraically sum the ATPase reaction and the creatine kinase reaction to show that the overwhelming source of increased Pi in an actively working muscle fiber is from PCr.

CHAPTER
4

OXIDATIVE PHOSPHORYLATION

Humans are open thermodynamic systems. This means we take in matter and energy from our environment and release matter and energy to our environment. On the intake side, we take in oxygen and sometimes a bit of heat, food, and water from the world around us. We release heat, carbon dioxide, urine, and fecal waste back to the environment. Our source of energy to carry out all the actions associated with life is the food we eat. In this chapter, we go through the steps in which oxygen is involved in the generation of more than 90% of all the ATP we need. Because this chapter on oxidative phosphorylation represents our first detailed study of a metabolic pathway, we begin with a general overview of cellular energy metabolism. Oxidative phosphorylation is a complex process taking place in mitochondria, where we will start. The citric acid cycle is our first major pathway that generates electrons eventually transferred to the oxygen we breathe. The energy released is used to make ATP. Oxidative phosphorylation must be coupled to and regulated by our energy needs, so more detailed accounts of regulation of oxidative phosphorylation and how this is quantified are included. We finish with a basic section on oxidants and antioxidants.

OVERVIEW OF METABOLISM

The food we eat is broken down in the digestive tract to simplest components, absorbed into the blood, and distributed to various tissues. There, the simple, absorbed molecules may be broken down to yield energy in the form of ATP, stored as energy molecules for future needs (glycogen and triglyceride), or used to create large molecules (e.g., proteins) specific to our needs or to maintain a proper internal milieu and support metabolism (minerals and vitamins).

We have already covered briefly the ATP-generating systems important to muscle: phosphocreatine, glycolysis, and oxidative phosphorylation. The only fuel source for glycolysis is carbohydrate, but oxidative phosphorylation can use carbohydrate, fat, and amino acids. The amino acids that are oxidized are obtained from the normal breakdown of proteins or from excess proteins in the diet that are not immediately converted into fat or glucose. Most fat is stored in specialized cells known as fat cells or **adipocytes.** Small amounts of fat are also stored in other cells, such as muscle. When metabolism needs to be increased, nerve and hormonal signals cause triglyceride molecules to be broken down, and fatty acids are released to be used as fuel. The body carbohydrate stores are not large; most carbohydrate is stored in liver and muscle as glycogen—a polymer made of individual glucose molecules joined together. Approximately 80% of all the body's carbohydrate is found in muscle as glycogen. Although liver has a higher concentration of glycogen than muscle, the liver is so much smaller in mass compared to the skeletal muscles that it holds only 10% to 15% of total body carbohydrate. The remainder of the carbohydrate in the body is the glucose in blood and extracellular fluids. When needed, glycogen in liver is broken down to maintain a proper glucose concentration in the blood. In muscle, glycogen is broken down and fed into the glycolytic reactions. Glucose is normally the sole fuel for the brain, the tissue that the body treats as most important from an energy perspective.

Figure 4.1 illustrates a section of a muscle fiber, representing the tissue that is or has the potential to be the main energy consumer. For athletes in hard training, the consumption of fuel during training may almost double the energy expenditure of the body for the remainder of the day. Figure 4.1 illustrates the major energy pathways in the cell, which are located in the cytosol and mitochondria. The fuels illustrated for these pathways are glucose

from blood, either used directly in glycolysis or temporarily stored as glycogen, and fatty acids (FA), which can be taken up from the blood and used immediately in mitochondria or temporarily stored as *intramuscular triglyceride* (IMTG). Although our focus is on muscle, these same pathways operate in other cells as well.

In chapter 3 we defined oxidative phosphorylation as the formation of ATP from ADP and Pi in association with the transfer of electrons from fuel molecules to coenzymes to oxygen. Although probably the best term to use, oxidative phosphorylation is also known by other names. The biologist may call it cellular respiration, whereas the exercise physiologist may refer to it as oxidative metabolism. As shown in figure 4.1, oxidative phosphorylation

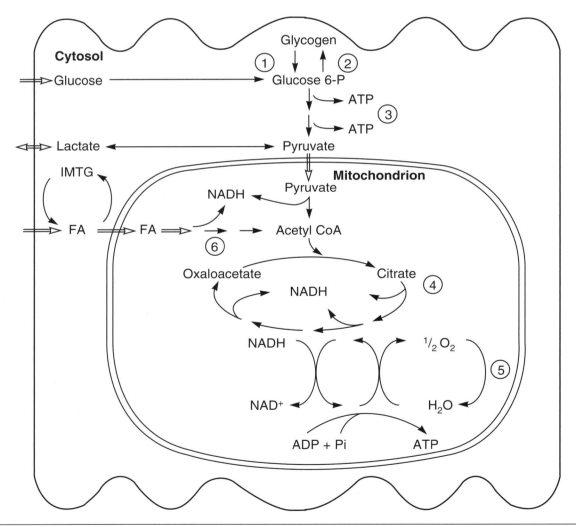

Figure 4.1 Overview of energy metabolism in a cross section of a muscle cell. Illustrated are the cytosol and mitochondrial compartments and the major pathways discussed in this chapter: (1) glycogen breakdown, (2) glycogen synthesis, (3) glycolysis, (4) citric acid cycle, (5) electron transport chain, and (6) beta-oxidation. A double-headed arrow means that a reaction or sequence of reactions is reversible. An open arrow means that a substance is crossing a membrane. IMTG represents intramuscular triglyceride, and FA represents fatty acids.

takes place in mitochondria. It encompasses the **citric acid cycle** (shown by the circle of arrows with oxaloacetate and citrate) and the **electron transport chain** (also known as the respiratory chain), which is shown below the citric acid cycle (CAC). This cycle starts with acetyl coenzyme A (CoA), which can be derived from any kind of fuel molecule. The citric acid cycle produces electrons in the form of the reduced coenzymes (mostly NADH). These electrons feed into the electron transport chain and are passed through a series of carriers to the oxygen we breathe. Addition of electrons to oxygen allows it to combine with two protons (H^+) to produce water. Free energy produced in the electron transport chain is used to phosphorylate ADP with Pi to make ATP.

As shown in figure 4.1, acetyl CoA is the substrate for the citric acid cycle. This can arise from the breakdown of fatty acids using **beta-oxidation** (number 6 in figure 4.1). Acetyl CoA is also formed in mitochondria from pyruvate, which is produced during glycolysis in the cytosol. In exercise, the proportion of acetyl CoA coming from pyruvate in the breakdown of glycogen and glucose increases with exercise intensity. We discuss this in chapter 5.

▶ K E Y   P O I N T ◀

When we exercise, we place demands for ATP use on our muscles. This must be matched by ATP synthesis. For most activities the bulk of the ATP is generated in mitochondria, when electrons on fuel molecules are transferred to coenzymes, through the electron transport chain, to oxygen. Free energy released during this process is captured via the formation of ATP from ADP and Pi. In this process, fuels and oxygen are consumed; the harder we exercise, the greater the requirement for fuels and oxygen.

MITOCHONDRIA

Mitochondria are often called the powerhouses of the cell because so much of our ATP is regenerated through oxidative phosphorylation in these organelles. This simple view of mitochondria has been updated, for we now know that they are involved in cell calcium ion control, cell signaling, and apoptosis (programmed cell death). Problems with and within mitochondria give rise to a number of pathological conditions, including tissue damage when blood flow is sharply reduced (ischemia) and then resumed (reperfusion), cardiomyopathy, diabetes, and many neurodegenerative diseases such as Parkinson's, Alzheimer's, and Lou Gehrig's (ALS) diseases (Duchen 2004). As shown in figure 4.2, many mitochondria have an elongated shape, although mitochondria can also form very complex structures, interacting with each other to form a reticulum and with other cell organelles such as the sarcoplasmic reticulum in skeletal muscle and endoplasmic reticulum in other tissues (Moyes and Hood 2003). In skeletal muscle, mitochondria are found just beneath the sarcolemma **(subsarcolemmal mitochondria)** and between myofibrils **(intermyofibrillar mitochondria)**.

Mitochondria have two membranes. The *outer membrane,* consisting of roughly equal proportions of proteins and lipid, contains *porins* (special pores) that allow entry to most ions and molecules up to a molecular weight of 1,000 daltons (D). The *inner membrane* is even higher in protein content (nearly 80%) and is impermeable to most ions and polar molecules unless they have specific transporters or carriers. The two membranes come together at intervals to form junctional complexes. Between the two membranes is the **intermembrane space.** Within the intermembrane space are some key proteins, including mitochondrial creatine kinase and cytochrome c. The inner membrane is formed into bulges called **cristae,** which greatly increase its surface area. The density of cristae in mitochondria is generally much higher than shown in figure 4.2, especially in tissues where the rate of oxidative phosphorylation is high, such as the heart. The inner membrane is loaded with enzymes and proteins for transferring electrons to oxygen, as well as the enzyme ATP synthase, which converts ADP and Pi to ATP. In figure 4.2, ATP synthase is shown as small knobs in the mitochondrial inner membrane.

In the center of a mitochondrion is the **matrix,** a viscous medium containing all the enzymes of the citric acid cycle (except succinate dehydrogenase), three enzymes of beta-oxidation of fatty acids, other enzymes, and *mitochondrial DNA.* This small, circular DNA molecule codes for 13 of the thousand or so mitochondrial proteins (found in the outer and inner membranes, the intermembrane space, and the matrix); the remainder are coded by nuclear genes. Following transcription and translation, proteins coded by nuclear genes are imported into the mitochondrion, specifically targeted because of presequences of amino acids. Mitochondrial DNA

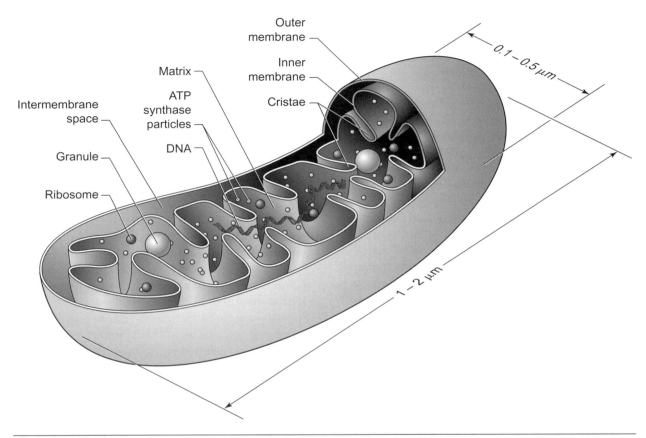

Figure 4.2   Cross section of a simple mitochondrion, showing the various components and approximate dimensions. Mitochondria have an outer membrane that is permeable to small molecules and an inner membrane that is folded into cristae studded with small knobs, identified as the ATP synthase. Between the outer and inner membrane is the intermembrane space. The interior space, known as the matrix, contains DNA, granules, and ribosomes. In the cell, mitochondria can assume quite complex shapes, but almost all have more densely packed cristae than shown here.

also codes for two rRNA (ribosomal ribonucleic acid) molecules and 22 tRNA (transfer ribonucleic acid) molecules—in sum, a total of 37 genes. In each mitochondrion, there are multiple copies of mitochondrial DNA. Researchers have discovered that mitochondrial DNA is more sensitive to damage by a variety of substances than is nuclear DNA. Lacking histone proteins and with fewer repair enzymes than found in nuclear DNA, mitochondrial DNA deteriorates as we age, compromising our ability to generate ATP. This is manifest in part by a reduced $\dot{V}O_2$max as we age, although many systems participate in this loss of aerobic function. As well, a variety of neurological and neurodegenerative disorders owe their origin to **mutations** in mitochondrial DNA or damaged mitochondrial proteins. Virtually all mitochondrial DNA is maternally contributed—less than 0.1% arises from sperm. Mitochondria also play a major role in programmed cell death, a process known as **apoptosis.**

The actual amount of mitochondria in various muscle types can be expressed by the percent volume of the total muscle cell occupied by mitochondria. This can range from about 1% in glycolytic fibers with a low oxidative capacity to 50% in cardiac muscle.

KEY POINT

The content of mitochondria in muscle parallels the need for oxidative phosphorylation by that muscle. It is exceptionally high in cardiac muscle and high in muscles involved in breathing. Signals generated by regular endurance training increase the expression of a variety of genes found in nuclear and mitochondrial DNA, leading to the formation of proteins that are imported into existing mitochondria and thus expanding the total mitochondrial volume to better handle the exercise stress.

Each milliliter of muscle mitochondrial volume has the remarkable ability to use up to 3 to 5 mL of oxygen per minute when maximally functioning (Moyes and Hood 2003). This is due primarily to the enormous ability to pack cristae inside the mitochondrion, creating crista areas up to 40 m^2 per milliliter of mitochondrial volume (Leary et al. 2003). Exercise training has been shown to increase mitochondrial volume by more than twofold, whereas bed rest, space travel, and immobilization can lead to a reduced volume of mitochondria (Hood 2001).

General Mechanism of Oxidative Phosphorylation

The chemical equations that follow illustrate the complete oxidation of two types of fuels: glucose, a representative carbohydrate, and palmitic acid, a very common fatty acid.

$$C_6H_{12}O_6 + 6\ O_2 \rightarrow 6\ CO_2 + 6\ H_2O$$

$$C_{16}H_{32}O_2 + 23\ O_2 \rightarrow 16\ CO_2 + 16\ H_2O$$

These balanced equations summarize the complete oxidation of these fuels as would occur in the cell by oxidative phosphorylation. The equations also describe the oxidation of these fuels if they were completely broken down by burning in a very hot flame.

The **respiratory quotient (RQ)** is the molar ratio of CO_2 produced divided by the O_2 consumed during fuel oxidation. Using the previous equations, RQ is 6 / 6 = 1.0 for glucose and 16 / 23 = 0.7 for palmitic acid. When the oxygen consumed and carbon dioxide produced are measured at the mouth of an animal or human as $\dot{V}O_2$ and $\dot{V}CO_2$, respectively, we use the term **respiratory exchange ratio** or **RER** (i.e., $\dot{V}CO_2/\dot{V}O_2$) instead of RQ.

In the cell, oxidation of glucose and fatty acids is tightly coupled to ADP phosphorylation, that is, phosphorylation of ADP with Pi to make ATP. Let's rewrite these two equations but now include the phosphorylation part; that is, we will include the number of moles of ATP that can be formed for each mole of fuel oxidized in the cell.

$$C_6H_{12}O_6 + 6\ O_2 + 32\ (ADP + Pi) \rightarrow 6\ CO_2 \\ + 32\ ATP + 38\ H_2O$$

$$C_{16}H_{32}O_2 + 23\ O_2 + 106\ (ADP + Pi) \rightarrow 16\ CO_2 \\ + 106\ ATP + 122\ H_2O$$

The number of water molecules generated is increased by 32 for glucose and 106 for palmitic acid in relation to the equations that showed only oxidation, because when ATP is formed from ADP and Pi, a water molecule results. In contrast, with ATP hydrolysis, a molecule of water is needed to hydrolyze the ATP.

In these latter two equations, the numbers of molecules of ATP that can be formed during the oxidation of a molecule of glucose or palmitate are considered the maximal amounts; they may be lower, as we will discuss later in this chapter. These numbers may also be lower than those you have seen in early textbooks, but revisions to the ATP yield from specific fuels are based on a better understanding of the overall mechanism of oxidative phosphorylation. The **P/O ratio** (sometimes called the ATP/O ratio) is the number of ATP formed for each atom of oxygen consumed. For palmitic acid, the maximum value for the P/O is

$$106 / (23 \times 2) = 2.3$$

For glucose, the maximum value for P/O is

$$32 / (6 \times 2) = 2.7$$

Comparing these two numbers reveals that, for the same amount of oxygen consumed, you get more ATP from glucose than you do from a fatty acid. Because of the recent revision in the number of ATP generated from glucose or palmitic acid oxidation, the P/O ratios just presented are lower than the numbers 3.0 and 2.8, respectively, for glucose and fatty acids as found in earlier biochemistry textbooks.

As we are defining it here, two major metabolic pathways are involved in oxidative phosphorylation. The citric acid cycle breaks down acetyl units derived from fuel molecules and generates the reduced coenzymes NADH and $FADH_2$ as well as CO_2. In the second pathway, the electron transport chain, the free energy released when electrons are transferred from reduced coenzymes (NADH and $FADH_2$) to oxygen, is channeled into the phosphorylation of ADP with Pi to make ATP; that is, it drives the reaction

$$ADP + Pi \rightarrow ATP + H_2O$$

The mechanism behind oxidative phosphorylation parallels the way electricity is generated by falling water. A dam creates potential energy by raising water to a high level. When the water falls down through special channels, the kinetic energy of the falling water rotates turbine blades in a magnetic field, producing an electric current. During electron transfer from NADH and $FADH_2$ to oxygen, free energy is released. At three complexes of the electron transport chain, the free energy release from electron transport is employed to pump protons (H⁺) from the matrix side of the inner membrane of the mitochondria to the outside or cytosolic side, that is, the intermembrane space. As a result, an **electrochemical gradient** is created in which the cytosolic side of the inner membrane is more positive in charge (the *electro* part of the gradient) and has a higher concentration of H⁺ (the *chemical* part of the gradient). When protons (H⁺) return down the gradient

through a special protein complex, the free energy released is used to make ATP from ADP and Pi. In other words, return of protons down their gradient is harnessed into driving ADP phosphorylation much as the energy of falling water is harnessed to make electricity. Figure 4.3 summarizes this process.

The term *proton pumping* refers to the forced movement of H⁺ ions across the inner membrane using the free energy released when electrons are passed through the inner membrane protein complexes. The outer (cytosolic) side of the inner

Figure 4.3 An overview of oxidative phosphorylation in mitochondria. Electrons on substrates (SH_2) are transferred to NAD^+, making NADH. The citric acid cycle is the principal pathway for this. Two electrons ($2e^-$) on NADH are then transferred through protein complexes in the inner mitochondrial membrane (I, III, IV) to oxygen, reducing it to H_2O. Energy released during the electron transfer is captured as protons are pumped from the matrix to the intermembrane (cytosolic) side of the inner membrane, creating an electrochemical gradient with a lower pH (higher [H^+]) and a more positive charge compared to the matrix. Protons flowing back down this gradient through a special complex V, the ATP synthase, are coupled to ADP phosphorylation to make ATP. Because of the electrochemical gradient, mitochondria can also take up Ca^{2+} ions using a calcium uniport. The source of the Ca^{2+} in muscle is primarily the sarcoplasmic reticulum. Removal of matrix Ca^{2+} occurs through a sodium-calcium antiport, which again takes advantage of the electrochemical gradient.

KEY POINT

We describe the regions on either side of the inner mitochondrial membrane as the matrix and cytosolic sides. Yet as figure 4.2 illustrates, what we call the cytosolic side is in reality the intermembrane space. However, the term *cytosolic side* is used in this text because the outer membrane is so porous that the ion composition of the inner membrane space is essentially cytosolic in nature.

membrane (the intermembrane space) has a lower pH (higher [H^+]) by about 1.0 pH unit. This is the chemical part of the electrochemical gradient. In addition, the intermembrane space is more positively charged by about 0.14 to 0.18 volts (140-180 mV); this represents the electrical gradient, often referred to as the *transmembrane electric potential,* given by the symbol $\Delta\psi$ (ψ is the Greek letter psi). Thus protons are pumped against both a concentration and a charge gradient, which means that this is an example of active transport. The transmembrane electrochemical potential gradient is also described as a **proton motive force (pmf)**. Some books and articles use the symbol $\Delta\mu_{H^+}$ rather than pmf.

$$\text{pmf } (\Delta\mu_{H^+}) = \Delta\psi + \Delta pH$$

Calcium and Mitochondria

The electrochemical gradient across the inner membrane favors the uptake of calcium into the matrix from the cytosol whenever the cytosolic calcium concentration is elevated through release of calcium from the sarcoplasmic reticulum of muscle or the endoplasmic reticulum in other cells. The movement of Ca^{2+} ions is driven by both electrical and concentration gradients into the matrix through a special protein called a calcium uniport. Remember that the matrix is more negatively charged than the cytosol and typically has a low calcium ion concentration. A membrane protein **uniport** allows one-way transport of a molecule or ion down its concentration gradient; this is an example of facilitated diffusion. Mitochondrial calcium uptake is facilitated in skeletal muscle because the calcium release areas on sarcoplasmic reticulum can be very close to mitochondria. When calcium release channels on sarcoplasmic reticulum are suddenly opened in response to depolarization

of the T-tubule of skeletal or cardiac muscle, the Ca^{2+} concentration in the region next to the release channel is transiently elevated by more than 100-fold. If a calcium uniport is in the immediate area, calcium can flow into the mitochondrial matrix. A **sodium-calcium antiport** removes the calcium from the matrix, exchanging three Na^+ ions for one Ca^{2+}. This exchange is driven by the fact that the matrix is more negatively charged than the cytosol. An **antiport** transports two molecules or ions simultaneously across a membrane in opposite directions. Matrix sodium ions are subsequently exchanged for intermembrane space protons via a sodium-hydrogen antiport (Na^+-H^+ antiport), which uses the proton electrochemical gradient to make this exchange. Figure 4.3 shows mitochondrial calcium uptake and removal through the calcium uniport and sodium-calcium antiport.

KEY POINT

Calcium can be a dangerous substance to the cell because it can initiate cell death or apoptosis. Accordingly, its cytosolic concentration is generally kept very low and carefully regulated. On the other hand, calcium has many positive effects for the cell since changes in its concentration can signal alterations in many reactions and processes, such as initiating muscle contraction and oxidative phosphorylation. One example is the ability of mitochondrial calcium to stimulate some key reactions that regulate oxidative phosphorylation.

THE CITRIC ACID CYCLE

The citric acid cycle, abbreviated CAC, is also known as the **TCA (tricarboxylic acid) cycle** or **Krebs cycle.** You should be aware of all three names, because you will encounter all of them. The citric acid cycle has been described as the "central metabolic hub of the cell" (Berg et al. 2002, p. 466).

Overview

The prime function of the citric acid cycle is to completely oxidize acetyl groups in a way that will result in ATP formation. These acetyl groups are formed from all the oxidizable fuels in the body. The citric acid cycle removes electrons from acetyl groups and attaches them to nicotinamide adenine dinucleotide (NAD⁺) and flavin adenine dinucleotide (FAD), forming NADH and $FADH_2$, respectively. In the electron transport chain, the electrons on the reduced coenzymes NADH and $FADH_2$ will be transferred through a series of carriers to oxygen. In two reactions in the citric acid cycle, the carbon atoms in the acetyl group are released as carbon dioxide. Each kind of fuel is converted to acetyl groups, attached to **coenzyme A (CoA)** (see figure 4.4). Our definition of oxidative phosphorylation is broader than in most biochemistry books in that we include the role of the citric acid cycle, the pathway where most of the electrons on fuels are transferred to coenzymes.

Coenzyme A is derived from **pantothenic acid,** a B vitamin. It acts as a handle to attach to a number of acyl groups, some of which we will see later. Coenzyme A has a terminal SH (sulfhydryl) group to which the acetyl group is attached, forming an energy-rich thioester bond. Coenzyme A and acetyl CoA are often written as CoASH and $CH_3COSCoA$, respectively. However, for simplicity, we use CoA and acetyl CoA.

Oxidation of acetyl CoA accounts for about two-thirds of the ATP formation and oxygen consumption in mammals. Acetyl groups enter the citric acid

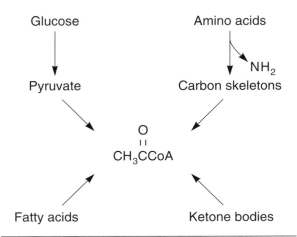

Figure 4.4 Oxidation of fuels. Glucose, fatty acids, ketone bodies, and amino acids are broken down into two-carbon acetyl groups attached to coenzyme A (CoA). The resulting acetyl CoA feeds into the citric acid cycle. The amino groups of amino acids, which cannot be oxidized, are removed and excreted in the urine.

cycle where their two carbon atoms appear as CO_2, while the hydrogens and their associated electrons are removed. Figure 4.5 summarizes and simplifies this process. This picture also reinforces the concept that removal of electrons from fuels occurs using hydrogen. The :H⁻ + H⁺ represents the hydride ion

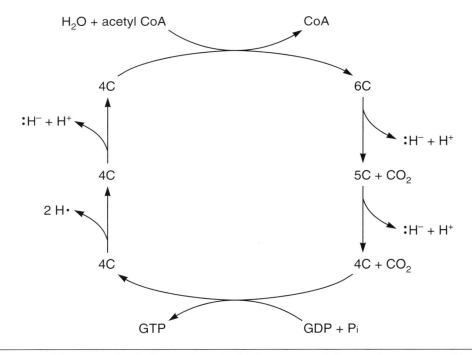

Figure 4.5 An overview of the citric acid cycle showing the path of carbon atoms and electrons that are recovered in association with hydrogen. The :H⁻ is recovered in the form of NADH, while the two H · are recovered in the form of $FADH_2$. The two-carbon acetyl unit feeds into the cycle attached to coenzyme A (CoA). The carbon atoms are recovered as carbon dioxide (CO_2). During one step in the cycle, enough free energy is released to phosphorylate guanosine diphosphate (GDP) to make guanosine triphosphate (GTP).

and proton, respectively, that are removed from many fuel substrates. As we saw in chapter 2, electrons accompany the hydrogen removed from substrates. The hydride with its two electrons, shown as dots, is attached to NAD^+, forming NADH; the negative charge on the hydride ion balances the positive charge with NAD^+. The two H· represent two hydrogen atoms removed from succinate, although succinate is not identified in the figure. The two hydrogen atoms, each containing one electron, become attached to FAD, forming $FADH_2$. Figure 4.5 shows that one turn of the citric acid cycle consumes one acetyl group and produces four pairs of electrons, one guanosine triphosphate (GTP), and two CO_2. Two water molecules are also consumed, although in the simplified diagram in figure 4.5 only one H_2O is shown. The GTP produced is an example of substrate-level phosphorylation, that is, formation of an energy-rich phosphate without the use of oxidative phosphorylation. The ATP produced in glycolysis also occurs via substrate-level phosphorylation.

KEY POINT

The citric acid cycle consumes acetyl groups. One limit to oxidative phosphorylation is the rate at which acetyl groups are provided to the citric acid cycle. The maximum rate of oxidative phosphorylation is markedly reduced if available carbohydrate stores are depleted. This suggests that the ability to produce acetyl CoA from fat can be a limiting factor in exercise.

Reactions of the Citric Acid Cycle

Figure 4.6 shows the complete citric acid cycle including the structures of intermediates. Since many readers are not comfortable with organic chemistry, figure 4.7 summarizes the citric acid cycle without the use of chemical structures. As is the custom for this book, double-headed arrows are used to show reactions that, if isolated, would reach equilibrium. Arrows showing one direction only define reactions in which there is a large free energy change; if studied in isolation these reactions would use up virtually all substrate. The overall direction of the tricarboxylic acid cycle is to consume acetyl groups.

Hydrolysis of acetyl CoA to acetate and CoA has a $\Delta G^{\circ\prime}$ of −32 kJ/mol (−7.7 kcal/mol). Thus the reaction combining acetyl CoA, oxaloacetate, and

water to form citrate, catalyzed by *citrate synthase*, is virtually irreversible. In the next step, the tertiary alcohol group on citrate is converted to a secondary alcohol group by two steps in the reaction catalyzed by the enzyme *aconitase*. The first step is a dehydration reaction in which the tertiary OH group is removed, producing cis-aconitate, which remains attached to the aconitase enzyme. A hydration reaction follows that results in the formation of isocitrate. Notice that the only difference between citrate and isocitrate is the position and type of OH group—tertiary alcohol on citrate and secondary alcohol on isocitrate.

Isocitrate undergoes oxidative decarboxylation, catalyzed by *isocitrate dehydrogenase,* to form α-ketoglutarate. This is the first of four oxidation–reduction reactions in the citric acid cycle. First, oxidation generates NADH and H^+; then decarboxylation forms CO_2. At this point in the cycle, one of the two carbon atoms on the acetyl group is removed as carbon dioxide.

In the next step, an α-ketoglutarate (some call this 2-oxoglutarate) undergoes oxidative decarboxylation to succinyl CoA. This step begins with decarboxylation, followed by oxidation, generating NADH and H^+. In this process, the second of the two acetyl carbon atoms is lost as CO_2. The enzyme *α-ketoglutarate dehydrogenase* (2-oxoglutarate dehydrogenase) catalyzes the same kind of reaction as pyruvate dehydrogenase, a reaction we will encounter soon, and contains the same kinds of subunits and coenzymes. Both enzymes contain three types of polypeptide subunits and five coenzymes. The coenzymes are NAD^+ and CoA, which are loosely bound, and thiamine pyrophosphate, **lipoic acid,** and FAD, which are tightly bound and not seen in the simple way the reaction is presented in figures 4.6 and 4.7. Recall from table 2.1 (p. 24) that **thiamine pyrophosphate (TPP)** is derived from the B vitamin, thiamine. During the α-ketoglutarate dehydrogenase reaction, enough free energy is released to generate the energy-rich succinyl CoA. Like the reactions catalyzed by citrate synthase and isocitrate dehydrogenase, the α-ketoglutarate dehydrogenase reaction is virtually irreversible, providing a one-way direction to the citric acid cycle.

In the next step, succinyl CoA is broken down to succinate and CoA in a reaction catalyzed by *succinyl CoA synthetase*. When this occurs, the free energy released when energy-rich succinyl CoA is hydrolyzed drives the substrate-level phosphorylation of GDP to make GTP. The succinyl CoA synthetase reaction is freely reversible, and the enzyme's name denotes the backward reaction, in keeping with the naming of

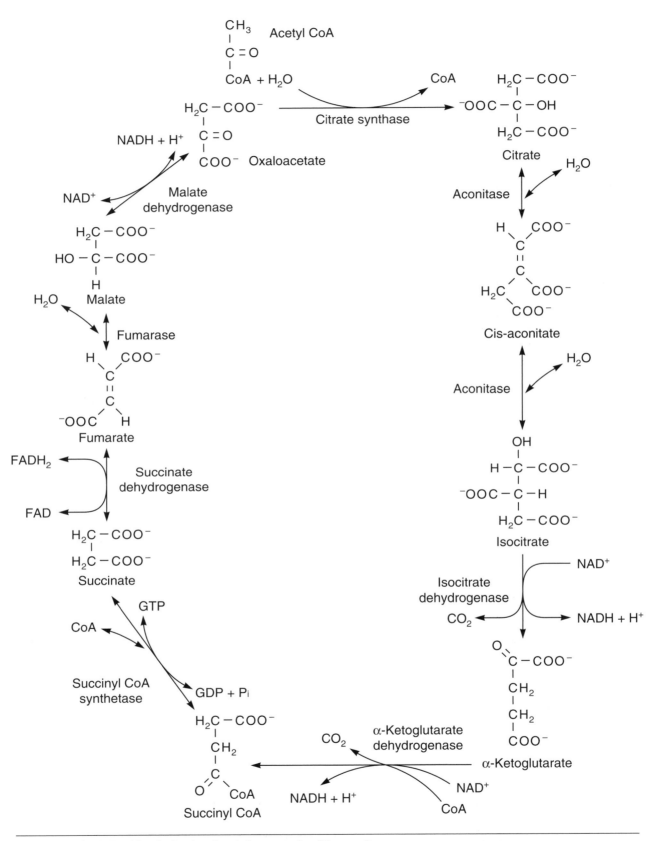

Figure 4.6 The citric acid cycle showing chemical structure for all intermediates.

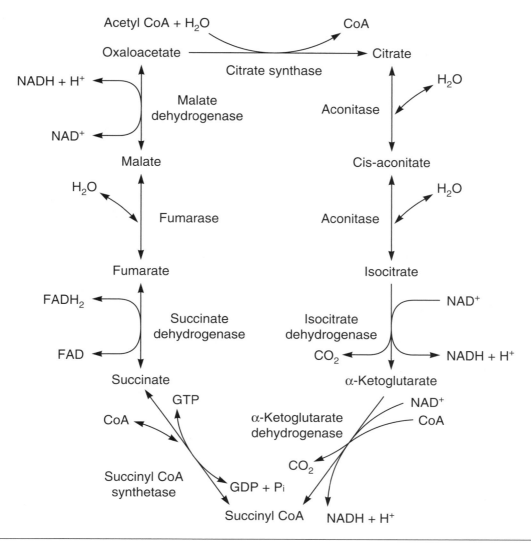

Figure 4.7   The citric acid cycle shown without chemical structures.

similar reactions in biochemistry. The GTP produced may be used (a) for peptide bond formation during the process of translation; (b) to phosphorylate ADP to make ATP using the enzyme nucleoside diphosphate kinase, described previously; or (c) in certain types of cell signaling processes.

Once succinate is formed, the remaining three reactions of the citric acid cycle regenerate oxaloacetate (one of the starting substances of the citric acid cycle, along with acetyl CoA) and generate electrons for the electron transport chain. *Succinate dehydrogenase* (SDH) is an oxidation–reduction enzyme that contains a tightly bound FAD. Unlike the other enzymes of the citric acid cycle that are found in the mitochondrial matrix, SDH is a component of the inner mitochondrial membrane. In the SDH reaction,

electrons are transferred from succinate to FAD and then to **coenzyme Q.** We will see later in this chapter that SDH and coenzyme Q are part of the electron transport chain. The product from the SDH reaction, fumarate, is hydrated (water molecule added) to malate by the enzyme *fumarase*. Malate contains a secondary alcohol group that is oxidized in the *malate dehydrogenase* reaction, generating NADH + H$^+$ and oxaloacetate, a starting substrate for a new round of the cycle.

The last three enzymes of the citric acid cycle carry out a three-step reaction sequence in which a methylene group (CH_2) in the first molecule (succinate) is converted to a carbonyl group (C=O) in the last molecule, oxaloacetate. This is accomplished by a dehydrogenation of succinate, generating FADH$_2$ and fumarate, a hydration reaction to make malate, followed by another dehydrogenation forming NADH and oxaloacetate. This sequence of dehydrogenation, hydration, and dehydrogenation is common in life; we will next see this in the sequence of reactions breaking down fatty acid molecules, a process known as beta-oxidation.

If we add algebraically all the reactions of the citric acid cycle we get the following:

$$\text{acetyl CoA} + 3 \text{ NAD}^+ + \text{FAD} + \text{GDP} + \text{Pi} + 2 \text{ H}_2\text{O} \rightarrow$$
$$2 \text{ CO}_2 + \text{GTP} + 3 \text{ NADH} + 3 \text{ H}^+ + \text{FADH}_2 + \text{CoA}$$

As shown in this summary equation, the citric acid cycle does not involve the net production or consumption of oxaloacetate or any other constituent of the cycle, and the only things consumed are an acetyl group and two water molecules. The summary equation also shows that oxygen is not directly involved in the citric acid cycle. As shown next, oxygen plays an obligatory secondary role in that without it, the citric acid cycle would quickly cease because there would be insufficient NAD⁺ and FAD for the cycle to continue. Evidence is accumulating to show that the enzymes of the citric acid cycle are physically located in the matrix to enhance the flux of this pathway by transferring the product of one reaction directly to the next enzyme. The word **metabolon** has been used to denote a multienzyme complex in which products and substrates are physically channeled to enzymes, as opposed to cycle intermediates diffusing randomly in the matrix.

▶ K E Y   P O I N T ◀

For the citric acid cycle to operate at a high rate, there must be sufficient oxaloacetate to accept acetyl groups from acetyl CoA in the first reaction of the cycle. If oxaloacetate is used for any other purpose, the maximal power of the citric acid cycle may be reduced.

The reduced coenzymes produced in the citric acid cycle (NADH and FADH₂) are oxidized in the electron transport chain and their electrons transferred to oxygen. We can show this electron transfer as follows:

$$3 \text{ NADH} + 3 \text{ H}^+ + \text{FADH}_2 + 2 \text{ O}_2 \rightarrow 3 \text{ NAD}^+ + \text{FAD} + 4 \text{ H}_2\text{O}$$

Associated with the transfer of electrons from the reduced coenzymes to oxygen is the tightly coupled ADP phosphorylation reaction, producing ATP. On the basis of conservative estimates, the transfer of electrons from three NADH to oxygen will yield 7.5 ATP. The transfer of electrons from one FADH₂ to O_2 will generate 1.5 ATP. In counting the energy-rich phosphates, one must include the GTP formed during the succinyl CoA synthetase reaction. In summary, the complete oxidation of one acetyl group is associated with the formation of 10 ATP. Figure 4.8 illustrates the close coupling of the citric acid cycle, the electron transfer chain, and ADP phosphorylation.

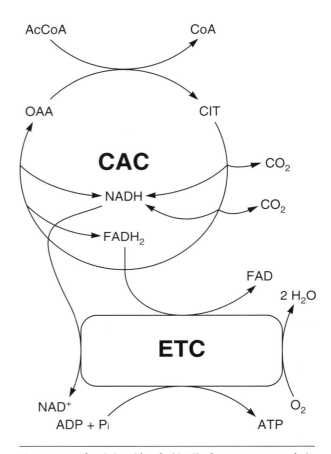

Figure 4.8 The citric acid cycle (CAC), electron transport chain (ETC), and ADP phosphorylation to ATP. All these are tightly coupled in mitochondria. OAA is oxaloacetate and AcCoA is acetyl CoA.

ELECTRON TRANSFERS

From a functional perspective, the electron transport chain (ETC), commonly called the respiratory chain, consists of four protein-lipid complexes located in the inner membrane of the mitochondrion. The role of these complexes is to participate in the transfer of electrons to oxygen. It is worth emphasizing again that we talk about electrons being transferred to oxygen, and this is true. However, as we have seen repeatedly, the electrons are most commonly transferred in association with hydrogen atoms or hydride (H^-) ions. In three of the four complexes that make up the electron transport chain, the free energy released during electron transport is associated with proton pumping from the matrix to the cytosolic side of the inner membrane. As already indicated, the term *pumping* means that the protons are transported across the inner membrane against both an electrical and a concentration gradient. Figure 4.9 shows the four complexes (each identified with a Roman numeral and a name) involved in electron transport and provides a perspective on the energy released during electron transfer, based on the concept that moving down means free energy is released. The standard free energy axis in figure 4.9 helps to put the free energy release into numbers we discussed in the previous chapter. Electrons transferred from NADH through complex I to coenzyme Q (shown as Q) result in proton pumping. So also the free energy released when electrons are transferred from Q to cytochrome c, utilizing complex III, results in enough free energy released to pump protons. When electrons

on cytochrome c are transferred through complex IV to oxygen, more energy is captured in pumping protons. On the other hand, the free energy release in transferring electrons from succinate through complex II to coenzyme Q is too little to be captured as protons transported from the matrix to the intermembrane side on the inner membrane.

Two electrons are transferred via NAD^+, requiring dehydrogenases from fuel substrates (many in the citric acid cycle), through a series of electron carriers to oxygen. The sequences of electron flow are nothing more than a series of oxidation–reduction reactions, each of which can be shown as

reduced A + oxidized B → oxidized A + reduced B

As we will show at the end of this chapter, energy release during electron transfer can be measured in units of volts. When reduced substrate A in the preceding equation transfers electrons to oxidized substrate B, this is a downhill-type reaction that can be measured in volt units and also displayed as energy units as seen in figure 4.9. B has a greater affinity for electrons than does A. We can show this more clearly as the reduction of B, when it accepts two electrons associated with hydrogen in the equation.

$$AH_2 + B \rightarrow A + 2e^- + 2 H^+ + B \rightarrow A + BH_2$$

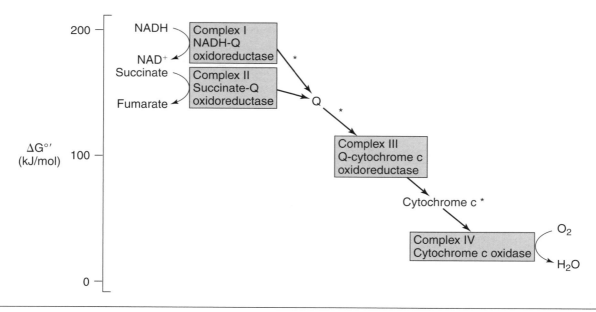

Figure 4.9 The path of electrons from NADH to oxygen uses complexes I, III, and IV. Electron transfer from succinate to oxygen goes through complex II with $FADH_2$, complex III, and complex IV. A perspective on the free energy released during the electron transfers can be seen by the relative position of the complexes in the chain, given as drops in standard free energy ($\Delta G^{\circ\prime}$). Asterisks indicate where the free energy released can be captured by the pumping of protons across the inner mitochondrial membrane. The free energy release in electron transfer from succinate to Q, utilizing complex II, is insufficient to be captured as proton pumping.

This is exactly what takes place in the electron transfer or respiratory chain. As shown in figure 4.9, electrons on NADH are transferred to Q using complex I. Q has a higher affinity for electrons than does NADH, so this reaction is associated with a release of energy. Similarly, cytochrome c has a higher affinity for electrons than Q, and so, utilizing complex III, electrons on reduced Q are transferred to cytochrome c. Oxygen has a very high affinity for electrons, so electrons on cytochrome c are passed to oxygen in a reaction catalyzed by cytochrome c oxidase. When oxygen accepts electrons and hydrogen, it becomes water. In each of the three electron transfer reactions, the energy released is captured through proton pumping that creates the electrochemical gradient across the inner mitochondrial membrane. Return of these protons through the ATP synthase is coupled to ADP phosphorylation with Pi to make ATP.

Figure 4.10 illustrates electron transfer in the inner membrane of a mitochondrion, showing the spatial relationships between the four electron transfer protein complexes and the two other intermediates of the electron transfer (respiratory) chain, coenzyme Q (Q) and cytochrome c (Cyt c). There are two sources of electrons, NADH and succinate through FADH$_2$. Electrons from NADH are transferred to oxygen using complexes I, III, and IV. During the transfer of two electrons from NADH to oxygen, the free energy release is captured as a total of 10 protons are pumped across the inner membrane, creating the electrochemical gradient. Four protons are pumped at complexes I and IV and two at complex III. The second source of electrons in figure 4.10 is succinate. Here, electrons are first transferred to an FAD in complex II, then through Q, to complex III, cytochrome c, and finally to oxygen using complex IV. The free energy released when electrons on succinate are passed to FAD in complex II and then to coenzyme Q is insufficient to pump protons across the inner membrane. Thus a total of six electrons are pumped per pair of electrons passed from succinate to oxygen. In the following sections, we will look at each complex in more detail.

> ### ▶ K E Y   P O I N T ◀
>
> It has been common to describe a pair of electrons as **reducing equivalents.** Reducing equivalents could be the electrons that are transferred from NADH or FADH$_2$. Since we talk about oxidation and reduction in terms of losing and gaining electrons, respectively, the term *reducing equivalents* is appropriate.

Complex I: NADH-Coenzyme Q Oxidoreductase

Complex I, which is also known as NADH dehydrogenase, is a huge complex consisting of at least 42 different polypeptide subunits, seven of which are coded by genes in mitochondrial DNA; the remainder are coded by genes in nuclear DNA. The role of this largest complex is to transfer a pair of

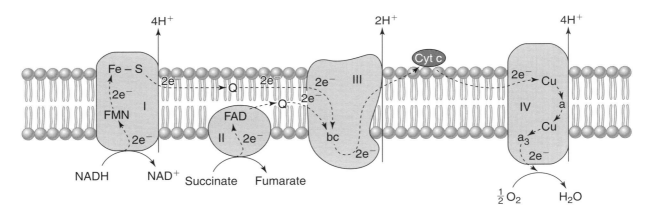

Figure 4.10 The flow of electrons in the inner mitochondrial membrane showing spatially how the four complexes of the electron transfer (respiratory) chain are located with respect to each other and how the two other electron carriers, coenzyme Q (Q) and cytochrome c (Cyt c), are arranged. Electrons are transferred from NADH through complexes I, III, and IV to oxygen. Electrons on succinate are transferred from FAD to Q and then through complexes III and IV to oxygen. The path of electron transfer is shown with the dotted lines. The number of protons pumped at each complex is illustrated. Complex I has an intermediate flavin mononucleotide (FMN). Complex III contains two other cytochromes, b and c$_1$, as intermediates.

electrons from NADH in the matrix to the oxidized form of coenzyme Q (abbreviated Q; also known as **ubiquinone**) in the inner membrane, reducing it to QH_2 (**ubiquinol**). We have already seen a variety of substrates (lactate, isocitrate, α-ketoglutarate, and malate) whose oxidation is associated with the reduction of NAD^+ generating NADH. We will see more of these substrates and their reduction by NAD-dependent dehydrogenases later. Complex I also contains another cofactor, *flavin mononucleotide* (FMN), that is related structurally to FAD and is formed from the B vitamin riboflavin (see figure 2.1, chapter 2). Proteins containing tightly bound cofactors such as FMN and FAD are known as *flavoproteins*. There are also clusters of iron complexed with sulfur in this complex. Researchers who study the electron transport chain learn details about the individual complexes by blocking certain steps with inhibitors. The plant toxin rotenone specifically blocks complex I. The remainder of the electron transport chain can be activated through the addition of succinate, the specific substrate for complex II.

Reduction of coenzyme Q (ubiquinone) by NADH to make NAD^+ and QH_2 (ubiquinol) is catalyzed by complex I. Details of this reaction are shown in figure 4.11. Overall the reaction involving complex I causes the fully oxidized coenzyme Q (ubiquinone) to accept two electrons and two protons to form the fully reduced coenzyme QH_2 (ubiquinol). The reaction proceeds in two steps. The first step involves the transfer of a single electron to coenzyme Q, forming a short-lived species known as ubisemiquinone, which is a free radical. Normally a second electron is quickly added to the **ubisemiquinone,** along with two protons, to produce the fully

reduced ubiquinol form. However, in resting cells, the overall flow of electrons from NADH to oxygen through complexes I, III, and IV is slower, so the lifetime of the free radical ubisemiquinone is fairly long. As we will see later, ubisemiquinone can pass its one free electron to oxygen to produce **superoxide,** a free radical of molecular oxygen containing one extra electron. It is shown as O_2^-. Coenzyme Q is a very small molecule compared to proteins. It contains a large hydrophobic side chain, represented by R in figure 4.11. Thus coenzyme Q is at home in the hydrophobic interior of the mitochondrial inner membrane where it shuttles electrons back and forth. A considerable amount of free energy is released when two electrons are transferred from NADH to coenzyme Q using complex I. This is captured via pumping of four protons across the inner mitochondrial membrane.

Complex II: Succinate-Coenzyme Q Oxidoreductase

Succinate-coenzyme Q oxidoreductase is commonly known as succinate dehydrogenase (SDH); we have seen this as a component of the citric acid cycle. The role of SDH is to transfer electrons on succinate to coenzyme Q (ubiquinone), forming fumarate and coenzyme QH_2 (ubiquinol). The actual enzyme complex contains four polypeptide subunits, none of which are coded by mitochondrial DNA. Flavin adenine dinucleotide, a tightly bound cofactor in SDH, accepts electrons from succinate to become the reduced $FADH_2$ form. Subsequently, the electrons are transferred to coenzyme Q, reducing it to coenzyme QH_2. Although we show the citric acid cycle enzyme

Figure 4.11 Reduction of coenzyme Q, also widely known as ubiquinone. The reduction is accomplished in two stages. In the first stage, a single electron is added, creating a charged, free radical species known as ubisemiquinone. Subsequently, a second electron and two protons are added to make the final product coenzyme QH_2, also known as ubiquinol.

SDH as producing $FADH_2$ as a product, strictly speaking it is QH_2 that is the reduced product of this reaction. We acknowledge the true final product QH_2 but show $FADH_2$ in keeping with a tradition passed down through a succession of biochemistry textbooks.

Other flavoprotein enzymes are also part of the inner mitochondrial membrane and, using the tightly bound cofactor FAD, transfer electrons from substrates to coenzyme Q. Strictly speaking, these other FAD-containing enzymes are also part of complex II. Unlike what happens with complexes I, III, and IV, the free energy change associated with complex II is not large; therefore, no protons are pumped across the inner membrane during electron transfer in this type of dehydrogenation.

Complex III: Coenzyme Q-Cytochrome c Oxidoreductase

Complex III or coenzyme Q-cytochrome c oxidoreductase transfers electrons from reduced coenzyme Q (QH_2) to cytochrome c. Complex III is a dimer, with each monomer containing 11 different polypeptide subunits, all but one coded by nuclear DNA. The **cytochromes** are a class of heme electron transport proteins located in or on the inner membrane of the mitochondrion. In the center of the heme group is an iron ion that can exist in an oxidized (Fe^{3+}) or reduced (Fe^{2+}) state. In complex III, electrons on reduced coenzyme Q (i.e., QH_2) are transferred to cytochrome c, changing the iron from Fe^{3+} to Fe^{2+}. Since reduction of oxidized iron involves accepting only one electron, two cytochrome c molecules must be reduced to accept the electrons from each QH_2. Although complex III involves electron transfer from coenzyme Q to cytochrome c, there are two other types of cytochromes (known as cytochrome b and cytochrome c_1) that are intermediates in the stage from Q to cytochrome c. Electron transfer from ubiquinol (QH_2) to cytochrome c using complex III is not a simple process. More comprehensive sources provide details on the actual path of electron transport and proton pumping (described as the Q *cycle*) (Trumpower 1990). Cytochrome c is a small protein that is not a part of any of the four electron transfer complexes but is attached to the intermembrane side of the inner mitochondrial membrane. In its location, it can accept electrons from QH_2 and transfer them to oxygen using complex IV. During electron transfer from QH_2 to cytochrome c, enough free energy is released so that two protons are pumped across the inner mitochondrial membrane. In the following summary equation, note that only the electrons are transferred from QH_2 to cytochrome c. As a result, two protons are left over. Complex III can be specifically inhibited by the antibiotic antimycin A.

$$QH_2 + 2 \text{ cytochrome c-Fe}^{3+} \rightarrow Q + 2 \text{ H}^+$$
$$+ 2 \text{ cytochrome c-Fe}^{2+}$$

Complex IV: Cytochrome c Oxidase

Cytochrome c oxidase contains 13 different polypeptide subunits coded by 10 nuclear and three mitochondrial genes. Two of these are cytochrome a and cytochrome a_3; there are also two copper ions in another polypeptide chain. This complex accepts electrons from reduced cytochrome c and passes them to an oxygen molecule, reducing it (after combination with four protons) to two water molecules. The sequence is as follows. A single electron on reduced cytochrome c is transferred to a protein-bound Cu^{2+}, reducing it to Cu^+. The electron is then transferred to cytochrome a, then to another protein-bound copper ion, to cytochrome a_3, and finally to oxygen. Because the oxygen molecule (O_2) contains two atoms of oxygen, we need four electrons to reduce it. We cannot break up the oxygen molecule as suggested earlier by the use of $\frac{1}{2}\ O_2$; this form of expression is for convenience. We can summarize in the following way the cytochrome oxidase reaction as it really occurs:

$$4 \text{ cytochrome c-Fe}^{2+} + O_2 + 4 \text{ H}^+ \rightarrow 4 \text{ cytochrome}$$
$$\text{c-Fe}^{3+} + 2 \text{ H}_2O$$

The free energy released during electron transfer in cytochrome oxidase (complex IV) results in the pumping of four protons across the inner membrane. Complex IV can be strongly inhibited by cyanide, which accounts for the deadly nature of this substance.

Summary of the Electron Transfers

Two electrons in the form of a negatively charged hydrogen ion (the hydride ion, $:H^-$) are transferred from most substrates to NAD^+, creating NADH and a free proton (H^+). These reactions are catalyzed by NAD-dependent dehydrogenase enzymes, three of which we have encountered in the citric acid cycle. The electrons on NADH are transferred to oxygen utilizing three large protein complexes in the inner mitochondrial membrane: complexes I, III, and IV. Two other intermediates play key roles in this electron transfer. Coenzyme Q, often called ubiquinone, is a relatively small, hydrophobic molecule in the inner mitochondrial membrane. Coenzyme Q can accept two electrons and two protons from complex I to become the reduced form coenzyme QH_2 or ubiquinol. Electrons on ubiquinol are transferred to a small heme-containing protein, cytochrome c, on the cytosolic side of the inner mitochondrial membrane. Reduction of cytochrome c is catalyzed by complex III. Finally, electrons on reduced cytochrome c are transferred to oxygen utilizing complex IV. During the transfer of electrons from NADH to oxygen, a great deal of free energy is released. This is captured by the creation of a proton and charge gradient across the inner membrane through a process of proton pumping at complexes I, III, and IV. For each pair of electrons transferred from NADH to oxygen, 10 protons are pumped across the inner membrane. The stoichiometry for this could be described as $10H^+/2e^-$.

Other reduced substrates (e.g., succinate) are oxidized by another class of dehydrogenase enzymes containing FAD. Two electrons in the form of two hydrogen atoms are transferred to FAD, reducing it to $FADH_2$. The electrons on $FADH_2$ are transferred to coenzyme Q, complex II, and then to oxygen using complexes III and IV. Less free energy is released in electron transfers from succinate to oxygen, and a total of only six protons are pumped across the inner membrane. The proton pumping to electron transfer stoichiometry for this would be $6H^+/2e^-$.

> ## ▶ K E Y   P O I N T ◀
>
> The free energy released during electron transfers from NADH to oxygen is very large. Under standard conditions it is –220 kJ/mol (–52.5 kcal/mol). Less free energy is released from electron transfer from substrates such as succinate that use FAD-dependent dehydrogenases. Under standard conditions, –200 kJ/mol (–47.8 kcal/mol) free energy is released.

COUPLED PHOSPHORYLATION

We intuitively know that water flows downhill but must be pumped uphill. Therefore, the term *proton pumping* should be clear. The pump is driven by the energy released when the electrons flow from one complex to another that is more easily reduced. The cytosolic (intermembrane) side of the membrane has a lower pH (a higher H^+ concentration) and would have a higher positive electrical charge. Thus, protons must be pumped across the inner membrane against both an electrical and a chemical gradient creating a proton motive force, based on a difference in electric potential ($\Delta\psi$) and a difference in pH (ΔpH). However, just as when water is raised above a dam, the electrochemical gradient can be exploited as the energy released when protons flow back into the matrix is allowed to drive the phosphorylation of ADP, making ATP.

The concept that electron transport is linked to ATP synthesis by way of proton pumping originated with Peter Mitchell. His concept is known as the **chemiosmotic hypothesis,** for which Mitchell received a Nobel prize. As already mentioned, 10 protons (H^+) are pumped per pair of electrons transferred from NADH to oxygen. Similarly, for each pair of electrons transferred from $FADH_2$ to Q to oxygen, six protons are pumped.

ATP Synthase

The **ATP synthase,** or complex V, couples energy released during proton flow down the gradient into the matrix to phosphorylation of ADP with Pi to make ATP. This concept is similar to that of using energy from the flow of water to make electricity as previously described.

A schematic of the ATP synthase is shown in figure 4.12. Adenosine triphosphate synthase consists of two parts. The F_o subunit, embedded within the inner membrane, acts as a pore to allow protons to pass into the matrix. F_o consists of a disc of 10 *c* subunits plus an *a* and *b* subunit. Tightly associated with F_o is the F_1 subunit, which bulges into the matrix. It is composed of a central stalk (γ subunit) that can rotate and an immobile head complex composed of three α and three β subunits involved in ADP and Pi binding and ATP formation. The current model, as shown in figure 4.12, has the protons passing through F_o. Energy released during this passage creates a rotational motion in F_o that is transmitted to the F_1 part. The rotational energy in F_1 is thought to drive ADP phosphorylation to make ATP. In essence, the ATP synthase is acting like a rotatory engine, utilizing the rotational energy created by the flow of protons down their gradient to combine ADP and Pi to make ATP. The ATP is released to the matrix. The symbols N (for negative) and P (for positive) are often used to identify the matrix and intermembrane spaces, respectively, reminding us that the matrix is more negatively charged than the cytosolic side.

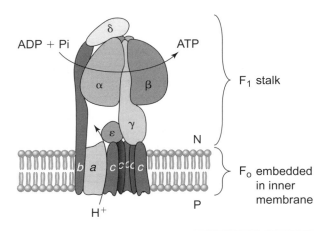

Figure 4.12 The synthesis of ATP accomplished by complex V, the ATP synthase. The ATP synthase consists of two major components. The F_o part is composed of a disc of c subunits and an a and a b subunit that make up the core within the inner membrane. Attached to this is the F_1 complex, which extends into the matrix side (N for negative) of the inner membrane and consists of a number of other polypeptide subunits, given by Greek letters. Protons passing through F_o create a rotation of the c subunits of F_o and the γ subunit of F_1. The rotational energy of the γ stalk of F_1, inside the fixed α and β subunits, drives ADP phosphorylation to make ATP. Three protons are thought to pass through the membrane from the cytosolic (P for positive) to the matrix side for each molecule of ATP manufactured by the complex.

Adapted from http://www.biologie.uni-osnabrueck.de/biophysick/Feniouk/Home.html. Copyright © Dr. Boris Feniouk.

Three protons are thought to pass through the inner membrane from the cytosolic to the matrix side for each molecule of ATP manufactured by the complex. We have already confirmed that the proton pumping to electron transport stoichiometry is $10H^+/2e^-$ for electron transfer from NADH to oxygen and $6H^+/2e^-$ for electron transport from succinate through FAD to oxygen. Putting together this information, we could say that the ATP/O ratio for NADH is 3.3 ($10H^+/3$ protons for each ATP) and for $FADH_2$ is 2 ($6H^+/3$ protons for each ATP). Indeed, these are the numbers that have been used for many years as the P/O ratios for NADH and $FADH_2$. However, we will see that this is a gross oversimplification, and it is important to be mindful of this fact.

Complex V, ATP synthase, is also known as the F_o-F_1 ATPase because it can run backward. In this case, the free energy released from ATP hydrolysis (the ATPase part) drives proton translocation from the matrix to the cytosolic side—the opposite of what occurs normally during the synthesis of ATP. Such a reverse movement of protons driven by ATP hydrolysis would increase the electrochemical potential across the inner membrane.

Uncoupled Oxidative Phosphorylation

Normally, electron transport from substrates to coenzymes to oxygen is tightly coupled to ADP phosphorylation, making ATP. However, protons can also leak across the inner membrane from the cytosolic side to the matrix side without accompanying ATP formation. This is known as *uncoupled respiration* because oxygen would be consumed as the electrons are accepted from substrates, but ATP would not be made. Such leakage appears to be most prominent in the liver and more important in small mammals, where it can account for as much as 35% to 45% of mitochondrial oxygen consumption. Uncoupling of oxidation from phosphorylation allows the proton flow to be dissipated as heat only, helping small mammals maintain body temperature. This uncoupled respiration can also be caused by specific chemicals that make the inner membrane permeable to protons.

One type of adipose tissue is rich in mitochondria, giving it a brownish color. This **brown adipose tissue (BAT)** has **uncoupling protein** 1 (UCP1) in its mitochondrial inner membranes. The protein allows the electrochemical gradient across the membrane to be dissipated by letting protons flow down their gradient into the matrix without being coupled to ATP formation using the ATP synthase. This is shown in figure 4.13. In this way, the free energy of the electrochemical gradient is immediately released as heat, warming the animal. Adult humans do not have BAT, but they do have uncoupling protein 2 (UCP2) in many tissues and UCP3 in skeletal muscle (Rousset et al. 2004). As you may suspect, uncoupled respiration actually decreases the ATP/O (P/O) ratio.

▶ KEY POINT ◀

The human body contains UCP2 and UCP3, but there is still some disagreement about their roles. Because these mitochondrial proteins uncouple electron transport from ATP formation, it is suggested that people who are especially endowed with these proteins would have a higher resting metabolic rate and thus would be less likely to develop obesity. Data supporting such an assertion are still not compelling. On the other hand, there seems to be little doubt about the theory that uncoupling proteins can reduce the rate of formation of the free radical superoxide in mitochondria of resting cells. This may be the most important role for these proteins.

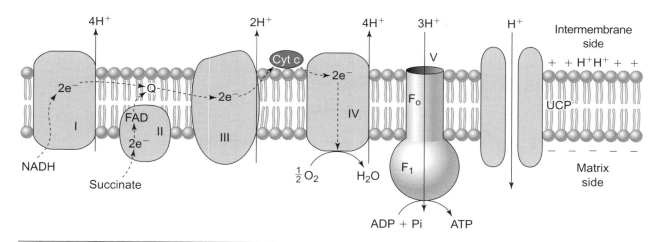

Figure 4.13 The four mitochondrial inner membrane protein complexes involved in electron transfer from NADH and succinate to oxygen. Energy is captured at complexes I, III, and IV as protons are transported across an electrical and concentration gradient. Free energy released on return of the protons through the F_o and F_1 components of complex V, the ATP synthase, is coupled to ATP formation from ADP and Pi. Protons can also return from the intermembrane to the matrix side of the inner membrane through uncoupling proteins (UCP). Such flow dissipates the electrochemical gradient across the inner membrane, but does not result in ATP formation.

Mitochondrial Transport of ATP, ADP, and Pi

Most synthesis of ATP occurs in the mitochondria, but most ATP is hydrolyzed in the cytosol. Thus, we must have a way of getting ADP and Pi into the mitochondrion and ATP out of the mitochondrion. Adenosine diphosphate and Pi must enter the mitochondrial matrix by crossing the inner membrane, whereas ATP must exit the matrix in the opposite direction. Recall that the inner membrane is impermeable to most substances and that polar or charged molecules can cross only if they are transported (translocated) by a specialized carrier protein. Figure 4.14 illustrates how ADP and Pi enter the matrix and how ATP crosses to the cytosolic or intermembrane side. Remember that we ignore the outer mitochondrial membrane because it is so permeable to small molecules.

The **ATP-ADP antiport** (also commonly called the adenine nucleotide translocase [ANT] or ATP-ADP translocase) transports ADP and ATP. Similarly, the **phosphate transporter** allows Pi in the form of dihydrogen phosphate ($H_2PO_4^-$) to be exchanged for hydroxide ion (OH^-). This is an example of an antiport mechanism. The driving force for the ATP-ADP antiport comes from the charge difference between the ADP and ATP. Thus the movement of ATP^{4-} from the more negative matrix region to the more positive region on the cytosolic side of the inner membrane allows the translocation to proceed with a free energy release. The movement of ATP^{4-} out of the matrix, coupled to movement of an ADP^{3-} into it, reduces

the charge gradient by 1 but does not alter the proton concentration gradient. The phosphate transporter is electrically neutral, but acts to dissipate a proton on the cytosolic side of the membrane because OH^- can combine with a proton (H^+) to form water. Thus it is driven by discharge of part of the chemical imbalance of H^+ across the membrane. In summary, moving the constituents to take an ATP molecule into the matrix (e.g., ADP and Pi) and transferring the ATP out of the matrix costs the equivalent of one of the protons pumped during electron transport from reduced coenzymes to oxygen.

Figure 4.14 also shows transport across the inner mitochondrial membrane of calcium, as already discussed, and pyruvate, which we will discuss later in this chapter. There are many other substances that must be transported across the inner mitochondrial membrane, each requiring a specific carrier and each driven by energy release based on charge or chemical differences across the membrane.

In trying to account for the ATP produced during transfer of a pair of electrons (reducing equivalents), we are assuming that 10 protons are translocated across the inner membrane when NADH is the source of electrons and that six protons are pumped when the source of electrons is $FADH_2$. We also assume that only three protons are needed to flow down their gradient through the ATP synthase to provide enough energy to phosphorylate one ADP with Pi, making ATP. Further, transporting one ADP and one Pi into the mitochondrial matrix, across the inner membrane, is equivalent to using up one proton. Therefore, we should be

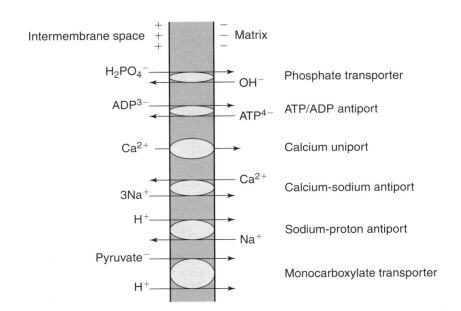

Figure 4.14 Transport of charged substances across the inner membrane of the mitochondrion, based on two principles. First, there is an electrochemical gradient (proton motive force) across the membrane, based on the fact that there is a higher concentration of protons (ΔpH) and a more positive charge ($\Delta\psi$) in the intermembrane space or cytosolic side than the matrix side. Discharge of this gradient provides the energy to drive movement of charged particles. Second, all transport involves specific transporters or carriers that recognize certain charged species. In the examples shown, action of each transporter leads to a reduction in either ΔpH or $\Delta\psi$ or both.

able to generate 2.5 ATP from each pair of electrons transported in the respiratory chain from NADH, and 1.5 ATP for each pair of electrons transported from $FADH_2$. Nonintegral P/O ratios (e.g., 2.5 and 1.5) may be intuitively disturbing to some. As already mentioned, many earlier sources cite P/O ratios of 3 and 2 for NADH and $FADH_2$, but this is changing as the real stoichiometry of proton pumping and ATP synthesis is increasingly appreciated. We have been careful to refer to 2.5 and 1.5 as maximum P/O ratios for NADH and $FADH_2$, respectively, because there is also leakage of protons across the inner mitochondrial membrane through uncoupling proteins—and we also discharge some of the proton electrochemical gradient in moving molecules and ions (besides ATP, ADP, and Pi), as figure 4.14 indicates. Therefore, true P/O ratios will be difficult to determine given species and individual differences, but the values will be less than the 2.5 and 1.5 we are using.

REGULATION OF OXIDATIVE PHOSPHORYLATION

Oxidative phosphorylation involves the oxidation of fuels to make ATP. From a whole-body perspective, we need systems to digest, absorb, and store the fuels that must come from our diets. We need systems to deliver oxygen from the air to the mitochondria where it can accept electrons from fuels and use the energy released to generate ATP. This is the whole-body picture. For the most part, however, oxidative phosphorylation is controlled at the level of the cell. It is here that ATP is hydrolyzed to generate the energy to drive endergonic reactions, and it is in each cell that this ATP is regenerated. Because oxidative phosphorylation generates most of the energy for the cell in the form of ATP, its rate should be precisely connected to the rate of ATP hydrolysis.

The citric acid cycle is one component of oxidative phosphorylation. As shown in figure 4.8, the citric acid cycle and electron transport chain are tightly linked together because the citric acid cycle is the major producer of reduced coenzymes needed to funnel electrons into the respiratory chain. Therefore, what regulates the citric acid cycle can influence the electron transport chain and vice versa.

If we neglect transfer of electrons from $FADH_2$ to oxygen, a single equation can represent oxidative phosphorylation, in which four electrons reduce a complete oxygen molecule and ADP is phosphorylated with Pi to make ATP:

$$5\ ADP + 5\ Pi + 2\ NADH + 2\ H^+ + O_2 \rightarrow 2\ NAD^+ + 5\ ATP + 7\ H_2O$$

Two of the seven molecules of water are generated by reduction of a molecule of oxygen, and five come about when five ATP are formed.

> ### ▶ KEY POINT ◀
>
> FAD is reduced to FADH₂ and electrons are transferred to oxygen in only a few reactions. Reduction of NAD⁺ to form NADH occurs in many reactions—three are in the citric acid cycle. For this reason, we simplify our thinking by focusing on NADH and not FADH₂.

A simple way to understand the regulation of oxidative phosphorylation is to consider which of the substrates on the left side of the equation (i.e., ADP, Pi, NADH, and O_2) actually limit the overall process. Our discussion focuses on muscle because it has an enormous range of metabolic rate, from complete rest to the vigorous contractions of sprinting. Figure 4.15 aids in this discussion. First we must ask, where do the four potentially

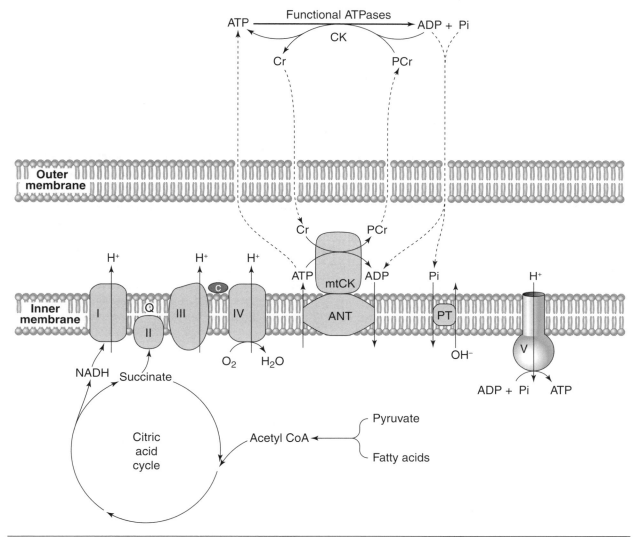

Figure 4.15 ATP hydrolysis in the cytosol and ATP formation by oxidative phosphorylation in mitochondria. If the rate of ATP production by oxidative phosphorylation is to closely match ATP hydrolysis by the various ATPases, a continuous supply of ADP and Pi must be available to complex V, the ATP synthase. Adenosine diphosphate and Pi can diffuse (dotted arrows) from the site of ATPases; and when formed, ATP can diffuse back. Alternatively, Cr and creatine phosphate (PCr) can act as energy transfer agents, diffusing toward the mitochondrion and back to the ATPases, respectively. The location of creatine kinase (CK) at sites of ATP hydrolysis and mitochondrial CK (mtCK) in the intermembrane space, adjacent to the adenine nucleotide translocase, promotes efficient PCr and Cr diffusion. Acetyl CoA, generated principally from pyruvate and fatty acids, is a source of electrons using the citric acid cycle. Electrons are transferred to oxygen using the electron transfer chain (complexes I, II, III, and IV); and the free energy released is coupled to proton (H⁺) translocation across the inner membrane, creating an electrochemical gradient within the intermembrane space. When protons flow down their gradient through the ATP synthase (V), the energy released is used to phosphorylate ADP to make ATP. ANT is adenine nucleotide transporter and PT is the phosphate transporter.

limiting substrates for oxidative phosphorylation come from?

Adenosine diphosphate and Pi are formed mainly in the cytosol when ATP is hydrolyzed to drive endergonic reactions. The increase in Pi is directly proportional to the decline in the concentration of phosphocreatine (PCr); that is, PCr decreases and free creatine (Cr) and Pi increase.

The NADH comes from the citric acid cycle, beta-oxidation of fatty acids in the matrix of the mitochondrion, and the cytosol when pyruvate formed during glycolysis is oxidized in the mitochondrion. NADH is a potential limiting factor because adequate NADH for oxidative phosphorylation requires available substrates for the dehydrogenase reactions that generate it, as well as sufficient activity of the dehydrogenase enzymes to drive the NADH-forming reactions at a sufficient rate.

Oxygen is taken into the lungs when you breathe, diffuses to hemoglobin molecules in red blood cells (i.e., erythrocytes) in the capillaries in the lung, is pumped throughout the body from the heart, and is unloaded from hemoglobin molecules in the capillaries that reach all parts of the body. Oxygen is delivered by way of diffusion from the small capillaries, across the cell membrane and the cytosol, and then to mitochondria.

Figure 4.15 shows that ATP is hydrolyzed to ADP and Pi by the functional ATPases we discussed in chapter 3, whereas ATP is primarily regenerated via oxidative phosphorylation in the mitochondrion. For tightly coupled ATP hydrolysis and ATP regeneration to happen, ADP and Pi must cross from the cytosol into the matrix, and ATP must cross from the matrix to the cytosol to be used again. Transport of ADP, Pi, and ATP across the inner mitochondrial membrane is aided by specific carriers that we have already described. It has been generally assumed that ADP diffuses from sites of ATPase activity to mitochondria and that ATP diffuses back to the ATPase sites. However, it is widely believed that in skeletal and cardiac muscle, much of the cytoplasmic transport of ADP and ATP occurs via Cr and PCr, respectively, as is shown in figure 4.15.

Adenosine triphosphate is hydrolyzed in the cytosol of muscle at three major sites: (a) where myosin interacts with actin (the actin-activated ATPase discussed in chapter 3), (b) when calcium ions are pumped back into the sarcoplasmic reticulum (the sarcoplasmic reticulum-Ca^{2+} ATPase), and (c) when sodium ions are pumped out of the cell and potassium ions are pumped back in (the sodium-

potassium ATPase). At the site of these ATPases are the CK enzyme and PCr. Creatine kinase catalyzes the phosphorylation of ADP to make ATP, utilizing PCr and producing Cr. At this level, the net direction of the CK reaction is toward ATP formation. Next, Cr can diffuse through pores in the outer membrane to the inner membrane, where it is phosphorylated by an ATP, producing PCr and ADP. A mitochondrial creatine kinase (mtCK) catalyzes this reaction. The resulting PCr can then diffuse back to the site of the ATPases to rephosphorylate ADP. This process is called the *phosphocreatine shuttle* (or creatine phosphate shuttle) and is illustrated in figure 4.15.

Use of this shuttle does not prevent ADP from diffusing to the mitochondrion or ATP from diffusing back as shown in figure 4.15. It just means that Cr and PCr can carry out the same process. Since diffusion depends on a concentration gradient, the fact that there are significantly larger changes in PCr/Cr concentrations than in ATP/ADP concentrations during muscle work makes the former molecules more likely candidates in this energy transport process. Moreover, both Cr and PCr are smaller molecules than ADP and ATP, making diffusion easier, and in addition the normal cellular concentrations of the former are larger than those of ADP and ATP. The fact that the active form of mtCK (an octamer with eight subunits) is located on the intermembrane space side of the inner membrane, adjacent to the ATP-ADP antiport (adenine nucleotide translocase), lends further support to the phosphocreatine shuttle mechanism.

> ### ▶ KEY POINT ◀
>
> For simplicity, we talk about limiting factors as if we have a chain with links that are not equal. It is important to point out that what limits oxidative phosphorylation in one metabolic state may not be the weak link in another condition.

Figure 4.15 shows the role of the major players in oxidative phosphorylation (NADH, ADP, Pi, and O_2). Now let us take a more detailed look at what limits the rate of oxidative phosphorylation.

Regulation of the Citric Acid Cycle

It has been estimated that in the transition from rest to very intense exercise, the flux through the citric acid cycle may increase 100-fold in contracting muscles of well-trained humans. Because the citric acid cycle is so tightly coupled to the electron transport chain, anything that limits the activity of electron transport to oxygen and ADP phosphorylation will stop the citric acid cycle. In other words, the citric acid cycle is the primary source of electrons (three-fourths as NADH, the remainder as succinate) for the electron transport chain. If the flow of electrons from NADH to oxygen in the electron transfer chain is slowed or blocked, NADH will accumulate and NAD^+ will be sharply reduced. Because NAD^+ is a substrate for three of the dehydrogenase reactions in the citric acid cycle, its concentration is critical for cycle operation. There are other considerations, since three of the citric acid cycle enzymes catalyze reactions that are essentially irreversible. If uncontrolled, these reactions could convert all available substrate into product, thus compromising mitochondrial metabolism.

Table 4.1 summarizes the major loci of control for the citric cycle. At the level of the whole cycle, acetyl units attached to CoA are absolutely essential for the cycle to operate. The acetyl units come from fuel molecules such as carbohydrate, following breakdown to pyruvate, fatty acids, ketone bodies, and carbon skeletons of amino acids. How fast these fuel molecules can be converted into acetyl CoA is critical to regulation of the citric acid cycle. As we will see later, the inability to generate acetyl CoA at a sufficient rate is a major mechanism accounting for the "hitting the wall" phenomenon that affects many endurance athletes near the end of a prolonged event.

Citrate synthase (CS) should be controlled because if it were not, it could consume the available acetyl CoA and oxaloacetate. Citrate synthase is inhibited by the allosteric effector NADH, which binds to an allosteric site on CS, increasing the K_m of CS for its substrate acetyl CoA (see the discussion on allosteric enzymes in chapter 2). In addition, citrate is a competitive inhibitor for oxaloacetate at the active site of CS. This means that a rise in citrate can competitively inhibit the binding of oxaloacetate (see chapter 2).

Isocitrate dehydrogenase (ICDH) is likewise inhibited by NADH at a negative allosteric site. Thus,

at rest, when NADH concentration is high, ICDH is inhibited. In addition, ICDH is activated by Ca^{2+} ions. Flow of calcium into the matrix through its uniport increases the matrix $[Ca^{2+}]$, as we have already discussed. Calcium ions lower the K_m for the substrate isocitrate, thus increasing enzyme activity for the same isocitrate concentration. The more active a muscle fiber is, the higher and more sustained is the rise in $[Ca^{2+}]$ in both the cytosol and matrix, and the more ICDH is activated.

α-Ketoglutarate dehydrogenase (α-KGDH) is inhibited allosterically by NADH (just like CS and ICDH). Moreover, like ICDH, α-KGDH is activated by a rise in the concentration of Ca^{2+}, which lowers the K_m for its substrate α-ketoglutarate. In addition, succinyl CoA is a competitive inhibitor for CoA. This means that a rise in the product succinyl CoA will competitively inhibit the binding of the normal substrate CoA. Thus, α-KGDH cannot tie up the available CoA in the matrix.

Control of the ICDH and α-KGDH by Ca^{2+} in muscle is such an elegant mechanism because the Ca^{2+} that initiates contraction, stimulating ATP breakdown, also signals the need for ATP synthesis. Moreover, the increase in mitochondrial Ca^{2+} that activates the citric acid cycle also activates the breakdown of pyruvate to acetyl CoA catalyzed by pyruvate dehydrogenase, as well as enhancing complex V activity in the inner membrane, the ATP synthase. Such coordinated regulation of the citric acid cycle, pyruvate dehydrogenase, and ATP synthesis reminds us of the extremely tight relationship between the primary mitochondrial reactions generating NADH and the electron transport system that consumes electrons.

Regulation of Pyruvate Oxidation

We cover carbohydrate metabolism in the next chapter. However, the control of pyruvate oxidation is closely linked with the regulation of oxidative phosphorylation. As shown in figure 4.1, pyruvate is formed during the glycolytic reactions. It may be

Table 4.1   Regulation of the Citric Acid Cycle

Site of regulation	Substrate(s)	Negative effectors	Positive effectors
Cycle as a whole	Acetyl CoA, NAD^+	–	–
Citrate synthase	Acetyl CoA, oxaloacetate	NADH, citrate	–
Isocitrate dehydrogenase	Isocitrate, NAD^+	NADH	Ca^{2+}
α-Ketoglutarate dehydrogenase	α-Ketoglutarate, NAD^+, CoA	NADH, succinyl CoA	Ca^{2+}

reduced to lactate in the cytosol or may enter mitochondria to be oxidized to acetyl CoA. The latter then enters the citric acid cycle. Pyruvate is an important source of acetyl CoA, and as we will see shortly, pyruvate availability and its oxidation to acetyl CoA can significantly influence peak rates of oxidative phosphorylation. The reaction for pyruvate oxidation, catalyzed by pyruvate dehydrogenase (PDH) in the mitochondrial matrix, is as follows:

$$\text{pyruvate} + NAD^+ + CoA \rightarrow \text{acetyl CoA} + NADH$$
$$+ H^+ + CO_2$$

The PDH reaction as shown in the equation is catalyzed by an enzyme complex with multiple copies of three enzyme subunits, identified as E1, E2, and E3. Although the reaction seems simple, in reality much takes place that is not shown and is beyond the scope of this book. Three coenzymes are involved in this reaction that are not visible in the equation presented. *Thiamine pyrophosphate* is involved in the decarboxylation of pyruvate, and *lipoic acid* and FAD are involved in the oxidation part of the reaction in which NADH is formed as a final product. As mentioned earlier, the PDH reaction is similar to that of α-KGDH in that both use the same five coenzymes (NAD+, CoA, TPP, lipoic acid, and FAD), although the subunits of the enzyme complexes are different. We will use PDH to represent pyruvate dehydrogenase, but some sources use the abbreviation PDC for pyruvate dehydrogenase complex.

The PDH reaction must be carefully regulated because the irreversible conversion of pyruvate to acetyl CoA means that a potential precursor to make glucose is lost. That is, pyruvate can be converted to glucose in the liver, but acetyl CoA cannot. Because the brain needs glucose and is treated biochemically as the most important tissue in the body, the PDH reaction must be regulated to spare pyruvate from being irreversibly lost. Regulation of PDH occurs via phosphorylation and dephosphorylation and allosteric mechanisms, discussed in chapter 2. A summary of the control of PDH is shown in figure 4.16. To prevent the unnecessary oxidation of pyruvate to acetyl CoA when other fuels such as fat can provide the acetyl CoA, PDHa is phosphorylated by *PDH kinase* into the inactive form, PDHb. Removal of the phosphate group, catalyzed by *PDH phosphatase*, leads to the activated form, PDHa. The relative activities of the two enzymes, PDH kinase and PDH phosphatase, determine the extent of phosphorylation and hence activity of PDH. Both the phosphate and kinase are part of the overall PDH complex. There are different

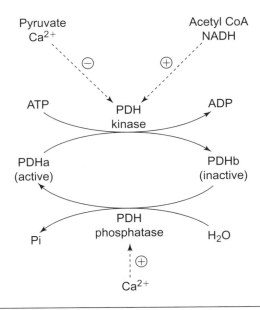

Figure 4.16 The activity of pyruvate dehydrogenase (PDH), regulated mainly by its state of phosphorylation. When the enzyme is phosphorylated by PDH kinase, it becomes inactive (PDHb). When the phosphate group is removed by PDH phosphatase, it is active (PDHa). The balance between the active (PDHa) and inactive (PDHb) forms determines the rate of formation of pyruvate to acetyl CoA. Pyruvate dehydrogenase kinase is activated allosterically by NADH and acetyl CoA. It is inhibited by Ca²⁺ ions and pyruvate. Pyruvate dehydrogenase phosphatase is powerfully activated by Ca²⁺ ions. Dotted arrows with + or – are used to show activation and inhibition, respectively.

isoenzyme forms for pyruvate dehydrogenase kinase (PDKs 1-4) and pyruvate dehydrogenase phosphatase (PDPs 1 and 2). PDK4 and PDP1 are the predominant isoforms in cardiac and skeletal muscle. PDK4 is the isozyme in skeletal muscle that is most sensitive to changes in exercise and carbohydrate content of the diet (Horowitz et al. 2005).

In resting skeletal muscle, where fat oxidation provides the bulk of fuel oxidized to make ATP, the PDH is predominantly in the inactive (PDHb) form. Under these conditions, the ratios NADH/NAD+, ATP/ADP, and acetyl CoA/CoA are elevated, and intramitochondrial calcium concentration is low. At the onset of exercise, when there is a demand for acetyl CoA for the citric acid cycle, dephosphorylation of PDH is stimulated and the kinase is inhibited as the intramitochondrial concentration of Ca²⁺ increases and the concentrations of acetyl CoA and NADH decline. Research on the activation of PDH at the onset of exercise shows that changes in its activity can take place quickly (in <1 min) and that activation is graded to the intensity of the exercise. It is suggested that the rise in intramitochondrial [Ca²⁺] provides the immediate increase in PDH activation, and that the relative changes in NADH/NAD$^+$, ATP/ADP, and acetyl CoA/CoA are responsible for fine-tuning the activation of PDH to match the cell's need for ATP. Indeed, following a seven-week endurance training program, subjects had lower levels of activation of PDH, accompanied by lower pyruvate concentrations and increased ratios of acetyl CoA/CoA and ATP/ADP when performing an identical exercise task as before training (LeBlanc et al. 2004). Such results point to a training-induced adaptation that can spare the use of carbohydrate during exercise at the same absolute workload.

One of the interesting research topics in the past few years relates to whether the relative activity of PDH can regulate oxidative phosphorylation at the onset of exercise. One theory holds that, at the transition from rest to exercise, a lag in PDH activation slows down the availability of acetyl CoA for the citric acid cycle and thereby slows $\dot{V}O_2$ (oxygen uptake) response. This theory has been tested by prior administration of a drug, *dichloroacetate* (DCA), into a vein. Dichloroacetate acts as an analog to pyruvate, which as you can see from figure 4.16 inhibits the activity of PDH kinase. With DCA administered before exercise, PDH is primarily in the active PDHa form. Although many studies have employed DCA to enhance activation of PDH before, rather than during, exercise, the effects have generally been minor; in most cases DCA has not significantly affected the rate of increase of $\dot{V}O_2$ during the first few minutes of exercise. Thus it appears that the PDH reaction does not have a significant regulatory effect on oxygen consumption at the beginning of exercise.

It has been shown that a single prolonged bout of endurance exercise leads to a rapid increase in the expression of the PDK4 gene in the exercised muscle. It is believed that this leads to a sharp increase in PDK4 enzyme to better turn off pyruvate oxidation in

➤ K E Y   P O I N T ◄

Although results from exercise studies in humans do not favor a major effect for PDH activation in terms of regulating oxygen kinetics, some studies show that rapid activation of pyruvate dehydrogenase at the onset of exercise plays a role in determining the extent of lactate buildup. This likely occurs because the glycolytic pathway can produce pyruvate at a much greater rate than it can be oxidized in the mitochondrion even if PDH is maximally activated.

the postexercise period. This would lead to a decrease in glucose oxidation in the exercised muscle, so that the glucose taken up by the fatigued muscle would be directed to the restoration of glycogen used up by the exercise (Booth and Neuter 2005).

Regulation of Oxygen Delivery

Since we will be spending considerable time on metabolic responses to exercise and how exercise training modifies metabolism during exercise, it is worthwhile to summarize some basic material on the delivery of oxygen from the air to mitochondria. The overall rate of oxidative phosphorylation by the body is given by the Fick equation:

$$\dot{V}O_2 = \dot{Q} \times (a\text{-}\bar{v})O_2 \text{ difference}$$

$\dot{Q}$ is the *cardiac output*, a measure of the amount of blood pumped (liters) from the left ventricle into the aorta each minute. This is based on the product of **stroke volume** and heart rate. Stroke volume is the volume of blood ejected during each beat. Both stroke volume and heart rate can increase with exercise; the changes in heart rate during exercise are more pronounced than that of stroke volume.

The *(a-$\bar{v}$)O$_2$ difference* is the average difference in oxygen content between the arterial blood and the mixed venous blood, usually expressed as milliliters of oxygen contained in 100 mL of arterial and venous blood. This increases during exercise. Table 4.2 shows representative values for $\dot{V}O_2$, heart rate, stroke volume, $\dot{Q}$, and (a-$\bar{v}$)O$_2$ difference for a young male before and after a comprehensive three-month program designed to improve his endurance capacity. For comparison purposes, cardiovascular values are given for rest, during submaximal exercise at 60% of maximal oxygen uptake, and during maximal exercise.

As the values in table 4.2 reveal, the body responds to an increase in exercise intensity by increasing heart rate, stroke volume, and the (a-$\bar{v}$)O$_2$ difference. With a relatively short-term exercise training program, an adaptation takes place in terms of stroke volume, which seems to be a major factor in dictating the maximal aerobic capacity ($\dot{V}O_2$max).

It is also important to understand how oxygen is transferred from the air to each individual mitochondrion where it will accept electrons to become water in the reaction catalyzed by cytochrome c oxidase (complex IV). Although cytochrome oxidase has a high affinity for oxygen, it is still an open question whether the actual concentration of oxygen at the

Table 4.2   Representative Values for Cardiovascular Function

Cardiovascular function	Rest		Submaximal exercise		Maximal exercise	
	Before training	After training	Before training	After training	Before training	After training
$\dot{V}O_2$ (L/min)	0.30	0.30	1.90	2.20	3.15	3.65
Heart rate (beats/min)	70	60	145	145	185	182
Stroke volume (mL)	75	87	92	107	94	110
Cardiac output (L/min)	5.25	5.20	13.34	15.46	17.40	20.00
$(a-\bar{v})O_2$ difference	6.0	6.0	14	14	18	18

level of cytochrome oxidase limits overall $\dot{V}O_2$ or whether there is some other limiting factor.

We know that the fraction of oxygen in air is 0.209, and thus oxygen should contribute 20.9% to the total pressure of air. At sea level, with an air pressure of 760 millimeters of mercury (760 mmHg), the fraction of total pressure of the air attributable to oxygen (PO_2) can be expressed as 0.209×760 mmHg or 159 mmHg. This is the *partial pressure* of oxygen in air. This value would be less at altitudes above sea level where air pressure is less, but the fraction of total pressure attributed to oxygen would remain the same.

> ### ▶ K E Y   P O I N T ◀
>
> It is common in physiology to express the partial pressure of oxygen in units of millimeters of mercury. Some people use *torr* instead of mmHg, where 1 torr is 1 mmHg. Expression of pressure using kilopascals (kPa) is also common. The relationship between the two is that a pressure of 760 mmHg is equivalent to 101.3 kPa.

Figure 4.17 summarizes the path of oxygen from the air to the mitochondria. When we breathe, oxygen is drawn into tiny alveoli sacs in our lungs by the action of respiratory muscles. In alveoli, the partial pressure of oxygen is less than in air, being diluted by water vapor and carbon dioxide returning in venous blood. Based on sea level values, the alveolar oxygen concentration, or P_AO_2, is now about 102 mmHg. As blood in capillaries passes alveoli in the lung, oxygen diffuses to capillaries where it binds to the protein *hemoglobin* inside red blood cells (erythrocytes). Hemoglobin contains four polypeptide subunits, each subunit including a heme group with a central Fe^{2+}

ion to which a single oxygen molecule can bind. Each hemoglobin molecule is thus capable of maximally transferring four oxygen molecules. Numerically, each gram of hemoglobin can transport 1.34 mL of oxygen. The more hemoglobin in the blood, the more oxygen can be transported. For a male with a blood hemoglo-

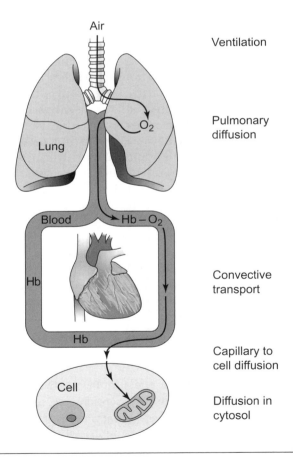

Figure 4.17   The path of oxygen from the air to mitochondria in cells. Air is drawn into the lungs by ventilation. Oxygen in the air diffuses from alveoli in the lung to the blood, where it is conveyed by convective transport bound to hemoglobin. At the cell level, oxygen is released from hemoglobin and diffuses from capillaries to cells, then diffuses in the cytosol to mitochondria.

bin concentration of 150 g per liter of blood (15 g/100 mL), the oxygen content of the blood is 201 mL/L (150 g/L $\times$ 1.34 mL/g). For a young woman with 130 g of hemoglobin per liter of blood, the oxygen content in blood is 130g/L $\times$ 1.34 mL/g or 174 mL of oxygen per liter. These blood oxygen concentrations are based on hemoglobin's being 100% saturated with oxygen. Normally in blood leaving the lungs, the saturation of hemoglobin with oxygen is about 98%. Although oxygen can dissolve in blood plasma, the amount that can be transported this way is tiny, so we will omit this aspect from consideration.

▶ K E Y   P O I N T ◀

Because more hemoglobin in the blood means a greater oxygen transport capacity, it is not surprising that athletes will go to great lengths to increase their hemoglobin concentration. Such practices as blood doping (adding extra red blood cells to the blood) or injection of erythropoietin, a factor that stimulates the formation of red blood cells, are unfortunately too common in endurance sporting activities.

Oxygen-rich blood leaves the lungs and is transported back to the heart, where it is pumped from the left ventricle into the aorta, to smaller arteries and arterioles and then capillaries that surround all cells in the body, in a process described as convective transport. The partial pressure of oxygen in the arterial blood leaving the heart (P_AO_2) is about 100 mmHg, although the actual content of oxygen can vary depending upon the concentration of hemoglobin in the blood. At the capillary level, where blood and individual cells interact, the partial pressure of oxygen is much lower, and oxygen can leave the hemoglobin because the ability of hemoglobin to bind oxygen is critically dependent on the partial pressure of oxygen. Released from hemoglobin, oxygen diffuses down a concentration gradient, through the erythrocyte cell membrane, the endothelial cells of the capillary wall, the interstitial space surrounding cells, and the cell membrane of the cell and into the cell cytoplasm. During maximal exercise, the PO_2 gradient between the capillary and the cytosol of the adjacent cell is considered to be very high ($\sim$30 mmHg), but the PO_2 gradient from the cell membrane to mitochondria is very shallow, approximately 2 mmHg (Richardson

et al. 1999). Cardiac and skeletal muscle cells have the oxygen-binding protein *myoglobin*. The content of myoglobin parallels the mitochondrial content of the cell and is higher in the muscles from endurance-trained individuals (Duteil et al. 2004). Because oxygen is not very soluble in aqueous solutions, myoglobin can facilitate oxygen diffusion in the cytosol. The extent of oxygen bound to myoglobin can be used to determine the oxygen content in cells. In cell cytosol, the partial pressure of oxygen can be in the range of 3 to 18 mmHg.

Our discussion to this point has been from a physiological perspective. Now let us switch to how a biochemist views the situation. Molecules interact with each other on a molecule-to-molecule basis (mole to mole), so the biochemist is interested in the concentration of oxygen expressed in molar units. In typical cells, the concentration of oxygen is in the 2 to 13 μM range (PO_2 3-18 mmHg). Oxygen is a substrate for the enzyme cytochrome c oxidase, whose affinity for oxygen is quite high. The concentration of oxygen needed to generate 50% of the maximal rate of cell respiration (K_m) is in the range of 1.5 to 1.7 μM for isolated liver and cardiac muscle mitochondria (Costa et al. 1997). This is well below the typical oxygen concentration in the cell cytosol, suggesting that oxygen is unlikely to limit oxidative phosphorylation under normal circumstances. However, the concentration of oxygen needed to support maximal cytochrome c oxidase activity (often described as the critical oxygen concentration) is estimated to be in the 2 to 5 μM range (Boveris et al. 1999). Alternatively, we could say that the critical mitochondrial partial pressure to support maximal respiration ($P_{mito}O_2$) is in the range of 3 to 7 mmHg. Although there is an apparent overlap between the critical oxygen concentration and the normal cytosolic oxygen concentration range in cells (i.e., 3-18 mmHg), it is far from clear whether there is oxygen-limited oxidative phosphorylation under exercise conditions when demand for oxygen may limit its supply.

▶ K E Y   P O I N T ◀

Mitochondria are wonderfully endowed for oxidative phosphorylation. It is estimated that each milliliter of mitochondrial volume has a maximum respiration rate of 3 to 5 mL of oxygen per minute and possesses about 40 m^2 of crista surface area (Moyes and Hood 2003).

Regulation of Oxidative Phosphorylation in Rested Muscle

In a muscle at rest the rate of energy expenditure is quite low, based mainly on maintaining protein synthesis and normal cell function. Now consider the substrates in the simple equation to describe oxidative phosphorylation (shown earlier in this chapter). Which of these is likely to limit the rate of oxidative phosphorylation?

$$5 \text{ ADP} + 5 \text{ Pi} + 2 \text{ NADH} + 2 \text{ H}^+ + O_2 \rightarrow 2 \text{ NAD}^+ + 5 \text{ ATP} + 7 \text{ H}_2O$$

As just discussed, oxygen is readily available in rested muscle; therefore it cannot be limiting. The concentration of Pi is also high enough to sustain a modestly high rate of oxidative phosphorylation (approximately 3 mM; see table 3.3) and is therefore not limiting. There is typically sufficient reducing power in the form of NADH in rested muscle. This leaves the availability of ADP to the respiratory chain as the weak link limiting oxidative phosphorylation including the citric acid cycle. Of course, the availability of ADP depends on the rate of ATP hydrolysis in the cytosol plus the entry of ADP into the mitochondria via the ADP-ATP antiport. On the basis of studies using isolated mitochondria, some authorities use the term **state 4 respiration** to describe the situation of mitochondria in a rested cell where the rate of oxidative phosphorylation is limited by availability of ADP. Actual consumption of oxygen occurs under state 4 conditions, but much of this is due to uncoupled oxidative phosphorylation as we have discussed previously.

Many studies support the fact that availability of ADP limits oxidative phosphorylation for a variety of situations in muscle as well as other tissues. For example, if mitochondria are isolated and placed in a well-oxygenated medium, the rate of oxygen consumption (used as an index of the rate of oxidative phosphorylation) is low. If either ADP or Cr is added, the rate of oxygen utilization greatly increases. We have already discussed how addition of ADP stimulates oxidative phosphorylation. However, addition of Cr also effectively increases oxidative phosphorylation by stimulating the mitochondrial creatine kinase enzyme, generating ADP and PCr (see figure 4.15). The ADP then enters the matrix using the ADP-ATP antiport to stimulate oxidative phosphorylation. When the oxygen utilization of isolated mitochondria is sharply increased by addition of an excess of ADP or Cr (the latter generates ADP), we say it is **state 3 respiration.**

Overall, the rate of oxidative metabolism for a muscle at rest, and for many other tissues, fits neatly into the concept that a limitation is due primarily to ADP within the mitochondria. This view of the regulation of oxidative phosphorylation is called the kinetic model or **acceptor control** model.

Regulation of Oxidative Phosphorylation at the Onset of Exercise

When muscle undergoes a transition from rest to moderate exercise or when the intensity of moderate exercise is modestly increased, the rate of ATP hydrolysis undergoes an abrupt step increase to match the new exercise intensity. There will be an increase in the rate of oxidative phosphorylation, measured at the mouth as $\dot{V}O_2$, but this will follow an exponential time course with a half-time of approximately 20 s (see figure 4.18). This is described as the $\dot{V}O_2$ kinetics or the kinetic response of $\dot{V}O_2$ and simply refers to the shape of the curve when breath-to-breath $\dot{V}O_2$ measures are plotted against time. Why is there a lag in oxygen utilization, a measure of the rate of oxidative phosphorylation, before it reaches a new level corresponding to the increased exercise intensity? Two hypotheses have been put forth. The "metabolic inertia" hypothesis suggests that there is a lag in electron transport and ATP synthesis because of substrate limitations other than oxygen (Grassi 2001). This could arise from a lack of acetyl CoA for the citric acid cycle or from the partial inhibition of key enzymes in

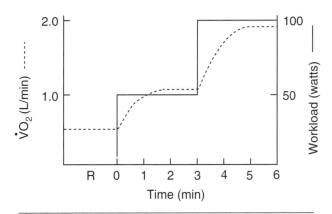

Figure 4.18   The relationship between the rate of oxidative phosphorylation, measured as the $\dot{V}O_2$, and the rate of ATP hydrolysis that tracks exercise workload for a hypothetical subject. $\dot{V}O_2$ is determined while a subject sits resting on a cycle ergometer (R). The clock starts and the subject immediately begins pedaling at a workload of 50 W (watts). After 3 min, the workload is increased to 100 W. There is a lag of approximately 2 min before the $\dot{V}O_2$ reaches a new steady state for both workloads.

the cycle because mitochondrial Ca^{2+} is insufficient to fully activate them. The alternate view is that there is an oxygen limitation—an inability to supply oxygen to the electron transport chain due to a limitation in oxygen transport (Hughson et al. 2001).

The simple kinetic or acceptor control concept explains quite well what keeps oxidative phosphorylation at a low rate in rested muscle. It does not adequately account for the responses that accompany changes in exercise intensity, because it is difficult to demonstrate a change in the concentration of any one of the potential limiting substrates that corresponds to the increase in oxidative phosphorylation. Use of ^{31}P nuclear magnetic resonance spectroscopy (NMR—see chapter 3) can provide concentrations of ATP, PCr, and Pi. From this, we can make calculations to determine ADP concentration. Values for intracellular PO_2 during exercise can be determined for exercising humans using the magnetic resonance signal from myoglobin, which is based on the extent of oxygen binding to myoglobin (Richardson et al. 2002). From these and other techniques, including studies with lower (hypoxia) or elevated (hyperoxia) inspired oxygen concentrations, we learn that muscle is effective in maintaining adequate oxidative ATP formation under most situations. Despite different intracellular concentrations of oxygen, ATP formation by substrate oxidation can be maintained through the manipulation of concentrations of energy-rich phosphates, especially PCr and ATP, substrates as in NADH and controlling molecules such as ADP, AMP, and Pi (Hughson 2005).

Exercise biochemists have been more accepting of the idea that regulation is based on a combination of factors. For submaximal exercise, across a range of intensities, intracellular PO_2 remains consistently in the 4 to 5 mmHg range during the breathing of room air (normoxia) (Richardson et al. 2002). With a lower oxygen content in the air, exercise $\dot{V}O_2$ can still be maintained although intracellular PO_2 is lower. With hyperoxia, in which inspired oxygen content far exceeds the normal 21%, intracellular PO_2 is elevated beyond that in normoxia, but $\dot{V}O_2$ for the same exercise intensity remains the same. How is this achieved? During hypoxic exercise, oxidative phosphorylation can be maintained with lower concentrations of PCr and ATP and elevated levels of ADP, Pi, and NADH. The increase in substrates for oxidative phosphorylation (ADP, Pi, and NADH) would help to maintain ATP formation to balance the reduced cellular oxygen concentration. With elevated cellular oxygen levels, oxidative phosphorylation could be maintained with

higher PCr and ATP and lower ADP, Pi, and NADH concentrations (Hughson 2005).

Biochemists have used the terms **phosphorylation potential** (that is, [ATP]/[ADP] $\times$ [Pi]) and the *mitochondrial redox state* (the ratio [NADH]/[NAD$^+$]) to define the energetic and metabolic potential of a cell. Combined with the cellular oxygen content as a measure of electron-accepting potential, these variables describe the effects of energy demand and potential for ATP supply by oxidative phosphorylation. For example, the phosphorylation potential ([ATP]/[ADP] [Pi]) will decrease when any increase in exercise intensity occurs. While ATP concentration remains remarkably constant, it will decrease a little. Couple this to a roughly parallel increase in ADP and a much larger increase in Pi due to a decline in PCr, and you can see how the phosphorylation potential can vary more widely than any of its constituents. Note also that the phosphorylation potential reflects PCr concentration because the increase in Pi reflects the decline in PCr. The mitochondrial **redox potential** ([NADH]/[NAD$^+$]) changes when the rate of electron transfer from NADH to oxygen is not matched by the rate of formation of NADH through dehydrogenase enzymes. Recall that the activities of three of these enzymes (pyruvate, isocitrate, and α-ketoglutarate dehydrogenases) respond to a complex pattern of substrate and allosteric effectors, including an increase in matrix [Ca^{2+}]. Finally, the oxygen availability to the respiratory chains in each mitochondrion ($P_{mito}O_2$) depends on a complicated combination of gas exchange, blood flow, and diffusion in getting from the outside air, through the lungs, into the arterial blood flow, and to the cytosol of each muscle fiber. Thus the exponential response of $\dot{V}O_2$ measured at the mouth at the onset of exercise, or during a step increase in exercise intensity, is the result of a complex mix of whole-body and cellular events providing the substrates for ATP synthesis.

Any step increase in exercise intensity will cause phosphorylation potential to decline. The extent of the decrease will reflect the increase in exercise intensity. This will stimulate the entry of ADP and Pi into the mitochondrion and electron transport from NADH to oxygen. As NADH is oxidized to NAD$^+$, the inhibitory effect of NADH on the three irreversible citric acid cycle enzymes is reduced and the citric acid cycle will speed up. In addition, the gradual rise in matrix calcium as a result of an increased cytosolic concentration in the more active fibers will further stimulate isocitrate and α-ketoglutarate dehydrogenase and activate PDH. Because of the initial mismatch between oxygen utilization by the cytochrome c oxidase complex and oxygen delivery from the air to the fiber, the oxygen tension within the fiber will decline. However, the rate of oxidative phosphorylation can be increased by a combination of a decrease in phosphorylation potential, an increase in mitochondrial redox potential, and a gradual increase in oxygen transport to the muscle mitochondria. The key point is that adjustments in phosphorylation and mitochondrial redox potentials can help to maintain oxidative metabolism in the face of declining oxygen availability to the respiratory chain (i.e., $P_{mito}O_2$). Any mismatch between ATP demand and ATP supplied by oxidative phosphorylation must be compensated with ATP provided by PCr and glycolysis. For small step increases in exercise intensity, the former will predominate. This has been demonstrated in a variety of models, showing that any slowing in the delivery of oxygen-rich blood to exercising muscle is compensated by a steeper decline in PCr. Of course this means that phosphorylation potential would be lower compared to that in the exercise condition with intact blood flow.

Oxidative Phosphorylation During Other Exercise Conditions

Much of the work and exercise we do is performed at a fairly constant rate of energy expenditure. For low to moderate sustained exercise, a steady rate of oxidative phosphorylation can supply virtually all the ATP needs. We call this steady state exercise because the rate of oxidative phosphorylation is precisely matched to the ATP demands by a combination of phosphorylation potential and redox potential adjusted to the steady state content of oxygen in the muscle fibers. The phosphorylation potential would be inversely related to the intensity of exercise, while the redox potential should parallel exercise intensity.

So far, we have looked at control of oxidative phosphorylation only when the step increases in muscle activity are relatively small. The situation is in part similar, but more complicated, if we consider transitions to maximum exercise. For a huge and sudden increase in muscle activity, phosphorylation potential will decline even more. Although PCr can partially buffer significant declines in ATP and increases in ADP, both of these will change to a greater extent than in the situation discussed previously. Further, PCr concentration will decline faster and more precipitously, and Pi will rise in parallel. Oxidative phosphorylation will be stimulated even more than before, with larger changes in mitochondrial redox and matrix calcium

concentration. However, we are talking about exercise intensities far beyond what could possibly be supported by even maximal rates of oxidative phosphorylation. Moreover, these supramaximal exercise intensities can be sustained for time periods much less than the time it takes to reach peak rates of oxidative phosphorylation. For these situations, the huge changes in cytoplasmic phosphorylation potential (and other factors we discuss in the next chapter) rapidly turn on glycolysis. With its moderately high power for ATP formation, glycolysis plays a prominent role in providing the ATP needs.

The oxygen available to muscle mitochondria can limit oxidative phosphorylation under certain circumstances. For example, at higher altitude, where the oxygen content of the air is low, the time course in the response of $\dot{V}O_2$ (reflecting active muscle oxygen consumption) to a modest step increase in muscle activity is delayed. This would suggest that the oxygen content available to mitochondria is lower for any absolute work rate when a person exercises at higher altitude. Therefore, compared to the same exercise intensity at sea level, we would expect a lower phosphorylation potential and higher redox potential to offset the lower oxygen content of altitude. Oxygen available to the electron transport chain could also be limiting during isometric contractions when intramuscular pressure builds up sufficiently to reduce or completely cut off the blood flow. In situations of reduced mitochondrial oxygen availability due to compression on blood vessel walls, glycolysis becomes extremely important as a source of ATP.

Athletes engaged in prolonged exercise such as marathons or triathlons can experience a situation in which performance falls off near the end. This has been called "hitting the wall," but its cause is easy to understand. As we will see, oxidation of carbohydrates and fats provides the ATP to support submaximal exercise. The higher the exercise intensity, the more carbohydrate is oxidized and the less the reliance on fat. Near the end of a marathon, for example, carbohydrate stores can become severely reduced. This means that the source of acetyl groups for the citric acid cycle must come increasingly from the beta-oxidation of fatty acids (discussed in chapter 6). However, numerous studies have revealed that provision of acetyl CoA to the citric acid cycle from beta-oxidation of fatty acids alone cannot match that when pyruvate or a mixture of pyruvate and fatty acids is used. This means that the primary supply of reducing equivalents (electrons on NADH) to the electron transport chain is compromised. Accordingly, the athlete must reduce

A keenly debated topic is what limits the maximum rate of oxidative phosphorylation by an individual. Is it limited by oxygen transport from the air to cytochrome c oxidase (i.e., transport limited)? Alternatively, is the limitation based on the production of reducing equivalents for the electron transport chain (i.e., a metabolic limitation)? A program of endurance training generally increases the activities of citric acid cycle enzymes and electron transport chain protein components about three to five times more than it increases the ability to transport oxygen as measured by $\dot{V}O_2$max. This is taken to mean that in fit individuals, oxygen transport places the upper limit on the maximum rate of oxidative phosphorylation. For people who are unfit, it is more likely that they cannot generate electrons at a rate sufficient to match their ability to deliver oxygen to the mitochondria of exercising muscle (Mourtzakis et al. 2004).

running, cycling, or skiing pace to a level where the rate of ATP demand by the exercise can be met by the new, lower level of oxidative phosphorylation using fatty acids as the primary fuel.

QUANTIFICATION OF REDOX REACTIONS

We have discussed oxidation and reduction reactions throughout this chapter, but have not put into numbers how easy it is for some molecules to be oxidized or reduced. Figure 4.9 was the closest we came to defining relative oxidation and reduction abilities based on a standard free energy scale. From this we could derive the interpretation that NADH is easily oxidized and oxygen is the easiest to reduce.

General Principles of Biological Redox Reactions

Biological oxidation–reduction (redox) reactions are catalyzed by enzymes. In a redox reaction, one substrate is reduced and one substrate oxidized. The names of the enzymes catalyzing biological redox reactions, besides describing a substrate, end with such words as dehydrogenase (pyruvate dehydroge-

nase), oxidoreductase (NADH-Q oxidoreductase), oxidase, peroxidase, oxygenase, or reductase. Regardless of the name, the reactions catalyzed are redox reactions, and one substrate is oxidized and one reduced as shown next.

$$A_{red} + B_{ox} \leftrightarrow A_{ox} + B_{red}$$

Not all redox reactions are reversible or will reach equilibrium. As we discussed in chapter 3, the energy change in the reaction dictates whether it is an equilibrium or a nonequilibrium reaction.

Redox reactions involve addition or removal of electrons. Students can be reminded of what addition or removal of electrons means by using acronyms such as *LEO* (losing electrons oxidation) *says GER* (gaining electrons reduction) or *OIL* (oxidation is losing) *RIG* (reduction is gaining). In the general example just presented, A_{red} becomes oxidized as it loses electrons, whereas B_{ox} is reduced by gaining electrons. What is oxidized is the reducing agent or reductant, and what gets reduced is the oxidizing agent or oxidant. From a biological perspective, the term *pro-oxidant* refers to the oxidizing agent or oxidant, and *antioxidant* refers to the reductant or reducing agent. Table 4.3 explains these terms.

Table 4.3 Common Terms and Actions in Biological Redox Reactions

Process	What happens	Abbreviation	Term for what loses or gains electrons
Oxidation	Loss of electron(s)	LEO	Reducing agent Reductant Antioxidant
Reduction	Gain of electron(s)	GER	Oxidizing agent Oxidant Pro-oxidant

Redox Potentials

In oxidative phosphorylation, the electron transfer potential of NADH or $FADH_2$ to oxygen is converted into phosphoryl transfer potential of ATP, which we can measure as free energy change. Under defined conditions in which concentrations of reactants and products are kept at 1 M concentration and pH is 7.0, we use standard free energy change ($\Delta G°'$). For electron transfer reactions, we use E°' as the **standard reduction potential,** under defined conditions in which the reactants and products are at 1 M concentration. The value of E°' reflects the ability of a substance to act as an electron acceptor, that is, to be reduced, under standard conditions. Unlike $\Delta G°'$, where a negative sign preceding a large number reflects a reaction that is energetically favorable, the actual value of $E°'$ provides a direct measure of the ability to accept electrons under standard conditions. That is, a positive value shows strong ability as an oxidizing agent (to gain electrons). On the other hand, a negative value for $E°$ shows a weak ability to be reduced (i.e., as an oxidizing agent or oxidant) but a strong ability to be oxidized, since oxidation and reduction are opposite events. Therefore, if a substance has a negative value for $E°'$ it will be a poor oxidizing agent, but on the other hand it will be a good reducing agent (have the ability to be oxidized or lose electrons).

Standard reduction potentials are measured in volts or millivolts (mV), based on the reduction of hydrogen ions (protons) to form hydrogen gas as shown in the following equation:

$$2\ H^+ + 2e^- \rightarrow H_2$$

In this reaction, the hydrogen gas is at a pressure of 1 atmosphere (760 mmHg), and the $[H^+]$ is 1.0 M (pH = 0). Under these conditions, the standard reduction potential ($E°$) is arbitrarily set at 0 V. When corrected to the biological pH of 7.0, the resulting $E°'$ becomes –0.42 V (–420 mV). Table 4.4 provides representative values for $E°'$ for a variety of common substances. The table illustrates the reduction of the substance, but it shows only half a reaction. As we have discussed, redox reactions involve something that is being oxidized and something that is being reduced, but this table focuses only on the relative ability of a single substance to be reduced.

From the table, you can see that the easiest to reduce is oxygen and the hardest to reduce is acetyl CoA (when it is converted to pyruvate). However, we know that the reverse reaction will take place readily; that is, when pyruvate is converted (oxidized) to acetyl CoA. With some of the other examples, you can see that they are written in the direction opposite to what occurs in the cell, but be reminded that the only reason is that we are writing them as reductions. The reduction reactions for some common biological antioxidants are also shown. The reaction involving lipoic acid, a commonly purchased supplemental antioxidant, is written in the opposite way it would work in the cell in an antioxidant function because an antioxidant donates electrons to something else. GSH, representing reduced **glutathione,** is the common name for a tripeptide that plays an enormous role as an antioxidant in the cell. When it

Table 4.4 Standard Reduction Potentials for Common Biological Reactions

Reduction half-reaction	$E°'$ (volts)
Acetyl CoA + CO_2 + H^+ + 2e⁻ → pyruvate + CoA	–0.48
$2H^+$ + 2e⁻ → H_2	–0.42
α-Ketoglutarate + CO_2 + $2H^+$ + 2e⁻ → isocitrate	–0.38
NAD^+ + $2H^+$ + 2e⁻ → NADH + H^+	–0.32
Lipoic acid + $2H^+$ + 2e⁻ → dihydrolipoic acid	–0.29
GSSG + 2e⁻ + $2H^+$ → 2GSH	–0.23
FAD + 2e⁻ + $2H^+$ → $FADH_2$	–0.22
Pyruvate + 2e⁻ + $2H^+$ → lactate	–0.18
Oxaloacetate + 2e⁻ + $2H^+$ → malate	–0.17
Fumarate + 2e⁻ + $2H^+$ → succinate	0.03
Ubiquinone + 2e⁻ + $2H^+$ → ubiquinol	0.04
Dehydroascorbic acid + 2e⁻ + $2H^+$ → ascorbic acid	0.08
Cytochrome c Fe^{3+} + e⁻ + $2H^+$ → cytochrome c Fe^{2+}	0.23
½ O_2 + 2e⁻ + $2H^+$ → H_2O	0.82

functions as an antioxidant, it donates electrons and becomes the oxidized form, shown as GSSG. Similarly, ascorbic acid (vitamin C), a common water-soluble biological antioxidant, donates electrons to become the oxidized or dehydroascorbic acid form.

As already mentioned, table 4.3 shows only half reactions, but in the cell one half reaction would be coupled to another half reaction to generate the real reaction that takes place in vivo. When we combine two half reactions, such that one substance is reduced and another is oxidized, we describe the overall reaction as a redox reaction. The overall change in redox potential for the combined reaction is given by the following:

$$\Delta E°' = E°'_a - E°'_b$$

Let us see how this works. In the citric acid cycle, malate can be oxidized to oxaloacetate by transferring electrons to NAD^+, making it NADH. In this example, NAD^+ would behave just as shown in its half reaction from table 4.3. However, we would be doing the opposite to the reduction of oxaloacetate to make malate, so we would need to write this half reaction the opposite way. We would determine the overall standard redox ($\Delta E°'$) for this combined reaction by algebraically combining the two half reactions to make the overall net reaction. Notice that when the first reaction is written the opposite way, we reverse the value of the $E°'$ by changing the sign from minus to plus.

malate → oxaloacetate + 2 H$^+$ + 2e$^-$

E°' = 0.17 V

NAD$^+$ 2 H$^+$ + 2e$^-$ → NADH + H$^+$

E°' = –0.32 V

The net reaction is

malate + NAD$^+$ → oxaloacetate + NADH + H$^+$

ΔE°' = –0.15 V

If we think of the overall process in oxidative phosphorylation, electrons from substrates such as malate, isocitrate, α-ketoglutarate, and lactate are transferred to NAD$^+$, making it the reduced form, NADH. These electrons on NADH are then passed through the electron transport chain to oxygen, reducing it to water. We can show the overall ΔE°' for this by writing the two half reactions. We take the half reaction for NAD$^+$ reduction and reverse it to make it NADH oxidation. Then we take the E°' value and change its sign. Then we take the half reaction for oxygen reduction. Combining these two reactions, we get the overall reaction in which electrons on NADH reduce oxygen.

NADH + H$^+$ → NAD$^+$ + 2H$^+$ + 2e$^-$

E°' = 0.32 V

½ O$_2$ + 2 H$^+$ + 2e$^-$ → H$_2$O

E°' = 0.82 V

NADH + H$^+$ + ½ O$_2$ → NAD$^+$ + H$_2$O

ΔE°' = 1.14 V

This shows how we can describe the propensity to accept electrons measured in volts. The larger the value in volts, the greater the propensity to accept electrons. Thus the reduction of oxygen is the most powerful reduction process on the list. We could also say that oxygen is the strongest oxidizing agent (oxidant) in the table. Moreover, when electrons are transferred from NADH to oxygen, the overall reaction is even more spontaneous, measured in volts. The key point is that electrons move spontaneously to the compounds with the more positive redox potential.

We can put the ΔE°' values into more familiar terms by converting this redox drive measured in volts into energy units measured in kilojoules per mole. We do this using the equation

ΔG°' = –nF ΔE°'

where n is the number of electrons transferred, F is the **Faraday constant** with a value of 96.5 kJ per volt per mole, ΔE°' is the algebraic sum of the two E°' for the two half reactions, and ΔG°' is the standard free energy change.

Working this equation out for the transfer of electrons from NADH to oxygen, we get

ΔG°' = –2 × 96.5 kJ/V × 1.14 V or –220 kJ/mol

Considering that the standard free energy for hydrolysis of ATP is –30.5 kJ/mol, it is easy to understand how much free energy is released during electron transfer from substrates to oxygen. We can also learn that, when two half reactions are combined to create a redox reaction, a positive value for ΔE°' will signify that the reaction occurs with a negative value for standard free energy change (ΔG°'). Such reactions are energetically favorable (i.e., downhill reactions). The larger the value of ΔE°', the less likely that reaction is to be an equilibrium (reversible) reaction.

The discussion so far has dealt with standard values, those expressed using the designation ΔE°', where everything is at 1 M concentration. In reality, concentrations of reactants and products in redox reactions in the cell are far less than 1 M, so we use the term ΔE, not ΔE°'. This argument is similar to that made in the previous chapter regarding actual free energy changes for reactions in vivo and thus the use of ΔG rather than ΔG°'. Using the general redox reaction that follows as a model, we could determine the actual ΔE for the general reaction by using the equation that follows the reaction:

$$A_{red} + B_{ox} \leftrightarrow A_{ox} + B_{red}$$

$$\Delta E = \Delta E°' - \frac{RT}{nF} \ln \frac{[A_{ox}] [B_{red}]}{[A_{red}] [B_{ox}]}$$

This equation is known as the **Nernst equation.** R is the gas constant we saw in the previous chapter, with a value of 8.314 J/mol × K; F is the Faraday constant with a value of 96,500 J per volt per mole; n is the number of electrons transferred in the redox reaction; and T is the absolute temperature (K). Since R is expressed in joules, not kilojoules, the Faraday constant is also expressed in joules. If we plug in the values for the two constants R (8.314 × 10^{-3} kJ per mole per degree K) and F (which we defined before as 96.5 kJ per volt per mole), use T = 310 K (i.e., 37 °C), and call the ratio of the oxidized and reduced products over reactants Q, we can simplify the equation to the following:

ΔE = ΔE°' – 0.0267/n ln Q

OXIDANTS AND ANTIOXIDANTS

In our discussion on electron transfers, we noted that during electron transfers from NADH to coenzyme Q (ubiquinone) using complex I, a ubisemiquinone intermediate is formed by the acceptance of a single electron (see figure 4.11). This intermediate can accept another electron and two H^+ to become coenzyme QH_2 (ubiquinol), or it can transfer the single electron to oxygen, forming the superoxide anion (O_2^-). Normally, at least 95% of the O_2 consumed by the body accepts four electrons to make two molecules of water, catalyzed by cytochrome c oxidase; the remainder is reduced by one electron, forming superoxide (figure 4.19).

Figure 4.19 The stages of reduction of oxygen in cells. Most oxygen molecules are reduced to water as they accept four electrons and four protons in a reaction catalyzed by cytochrome c oxidase. One-electron reduction of oxygen forms superoxide anion, which can undergo a dismutation reaction, accepting another electron and two protons to form hydrogen peroxide. Reduction of hydrogen peroxide through acceptance of another electron forms the hydroxyl radical and the hydroxide anion. Addition of another electron creates water. The bold dot identifies the substance as a free radical with an unpaired electron.

Superoxide is known as a **free radical** because it contains an unpaired electron. We typically show its structure as $O_2^-\cdot$, where the dot represents the unpaired electron. As a free radical, superoxide is quite reactive; and not all the reactions it can undergo are helpful to the cell. For example, superoxide can be a reducing agent by donating its unpaired electron to ions such as Fe^{3+} (ferric ion), forming an oxygen molecule and the reduced Fe^{2+} (ferrous ion). It can also act as an oxidizing agent, accepting another electron and two H^+, forming hydrogen peroxide.

We refer to highly reactive oxygen compounds like superoxide as **reactive oxygen species** or ROS. They may or may not be free radicals.

Superoxide can also be formed at complex III, when electrons on coenzyme QH_2 or ubiquinol are transferred to cytochrome c. Again, the electrons are transferred one at a time from ubiquinol to cytochrome c, such that an intermediate ubisemiquinone is present. This intermediate can also undergo a one-electron transfer to oxygen to form superoxide. New research tells us that the proportion of oxygen that is converted to superoxide at either complex I or III is highest when the rate of electron transport from substrates to oxygen is lowest. Under these state 4 conditions of electron transport, when ADP availability limits oxidative phosphorylation, proton motive force (the electrochemical gradient across the inner mitochondrial membrane) is high and the lifetime of the ubisemiquinone intermediate is longest. This provides a greater opportunity for one-electron reduction of oxygen, instead of the normal four-electron reduction that makes water. One of the helpful roles of uncoupling proteins is to allow electron transport without coupled phosphorylation, thus reducing the proton motive force, shortening the lifetime of ubisemiquinone, and reducing superoxide formation (see figure 4.13). During elevated or maximal rates of electron transport (state 3 conditions), less than 1% of cellular oxygen is reduced to form superoxide because the lifetime of the ubisemiquinone is shortest and there is less time for oxygen to accept just one electron.

Other Reactive Oxygen and Nitrogen Species

Superoxide is formed constantly in the body. Its presence allows other ROS to be created. Two superoxide radicals can undergo a spontaneous **dismutation reaction,** in which one of the superoxides donates its unpaired electron to the other. The former becomes an oxygen molecule, while the latter adds two protons and becomes *hydrogen peroxide*, H_2O_2. The spontaneous dismutation of superoxide is too slow to be of much value to the cell. However, both mitochondria and cytosol of cells are endowed with a **superoxide dismutase** (SOD), which quickly catalyzes the removal of superoxide (figure 4.20). The cytosolic form of SOD contains a copper and a zinc ion at the active site and is known as Cu-Zn SOD. The mitochondrial SOD contains a manganese ion at the active site and is known as Mn SOD. Two isomers of SOD are needed because complex I forms superoxide on the matrix side of the intermembrane whereas complex III forms and releases superoxide in the intermembrane space where it can be removed by Cu-Zn SOD. Superoxide is also produced during the enzymatic conversion of xanthine to hypoxanthine and of hypoxanthine to uric acid catalyzed by **xanthine oxidase** (figure 4.20). Superoxide is also

formed by cells of the immune system and in endothelial cells of the blood vessels in a reaction catalyzed by **NADPH oxidase** (figure 4.20). Formation of superoxide by such white blood cells as **neutrophils** and tissue-scavenging **macrophages** is essential to their roles in protecting us. Not only is superoxide an important weapon for these immune cells, but superoxide can be converted to hydrogen peroxide,

Superoxide dismutase reaction

$$2O_2^- \bullet + 2H^+ \longrightarrow H_2O_2 + O_2$$

Catalase reaction

$$2H_2O_2 \longrightarrow 2H_2O + O_2$$

Reduction of ferric to ferrous ion by superoxide

$$O_2^- \bullet + Fe^{3+} \longrightarrow O_2 + Fe^{2+}$$

Fenton reaction

$$Fe^{2+} + H_2O_2 \longrightarrow OH\bullet + OH^- + Fe^{3+}$$

Glutathione peroxidase breaks down hydrogen peroxide using glutathione

$$2GSH + H_2O_2 \longrightarrow GSSG + 2H_2O$$

Glutathione removes organic peroxides (ROOH)

$$2GSH + ROOH \longrightarrow GSSG + ROH + H_2O$$

Myeloperoxidase in leukocytes produces cytotoxic hypochlorite

$$H_2O_2 + Cl^- \longrightarrow ClO^- + H_2O$$

Xanthine oxidase oxidizes hypoxanthine and xanthine, producing superoxide

$$Hypoxanthine + O_2 + H_2O \longrightarrow Xanthine + O_2^- \bullet$$
$$Xanthine + O_2 + H_2O \longrightarrow Uric\ acid + O_2^- \bullet$$

NADPH oxidase generates superoxide

$$NADPH + 2O_2 \longrightarrow 2O_2^- \bullet + NADP^+ + H^+$$

Nitric oxide synthase (NOS) produces signaling molecule nitric oxide (NO)

$$Arginine + O_2 + NADPH \longrightarrow NO + Citrulline + NADP^+$$

Nitric oxide and superoxide easily react together to form peroxynitrite ion

$$NO + O_2^- \bullet \longrightarrow ONO_2^-$$

Figure 4.20 Important enzyme-catalyzed reactions that play a major role in producing or removing reactive species in cells. Two GSH or reduced glutathione molecules donate a pair of electrons and two protons to become glutathione disulphide or GSSG. Recipients of electrons and protons from two GSH molecules are organic peroxides (shown as ROOH), that upon reduction become alcohols (ROH).

which can react with a chloride ion in a reaction catalyzed by **myeloperoxidase** to form *hypochlorite* (figure 4.20). The latter compound, a component of bleach, is a particularly effective weapon by activated immune cells. The accelerated formation of superoxide by white blood cells in response to their activation is known as the **respiratory burst.**

Hydrogen peroxide is uncharged, so unlike the superoxide anion, it can cross cell membranes. Hydrogen peroxide is a strong oxidizing agent (easily reduced) and can cause damage to cell constituents. Perhaps its most damaging quality is that it can be split into the very dangerous **hydroxyl radical** and hydroxide ion when it accepts an electron from ferrous or cuprous (Cu^+) ions in a reaction known as the **Fenton reaction** (figure 4.20). Hydroxyl radicals are so reactive that their lifetime can be measured in billionths of second. Hydrogen peroxide is removed in reactions catalyzed by two enzymes (figure 4.20). **Catalase,** found in peroxisomes and in mitochondria in heart muscle cells, converts hydrogen peroxide to water. **Glutathione peroxidase** (GPX) is a selenium-containing enzyme that plays very important roles in the cell. It can rid mitochondria of hydrogen peroxide, and it is involved in removal of organic peroxides (figure 4.20). It uses glutathione, a tripeptide formed in all cells that can exist in a reduced GSH or oxidized form (GSSG). The sulfhydryl (–SH) group of GSH is very important as an electron donor (reducing agent), clearing cells of peroxides. In the process, glutathione is oxidized to the disulfide form, GSSG. The active, GSH form of glutathione is recovered by reduction of GSSG using electrons on NADPH, a molecule we will encounter later.

Many cells also produce a highly reactive nitrogen compound, **nitric oxide (NO)**, that is an important signaling molecule. Nitric oxide is formed by an enzyme known as **nitric oxide synthase** (NOS), using the amino acid arginine as the source of nitrogen (figure 4.20). There are at least four forms of NOS. In endothelial cells, formation of NO by endothelial NOS (eNOS) leads to relaxation of smooth muscle in artery walls and increased blood flow. In cells of the immune system, an inducible form of NOS (iNOS) produces NO that can combine with other substances such as superoxide to generate more potent responses to invading microorganisms. There is a mitochondrial form of NOS (mtNOS), similar to neuronal (nNOS) and eNOS in that it is activated by calcium ions. The matrix NO can combine with superoxide in a diffusion-controlled reaction to make the more dangerous substance peroxynitrite (ONO_2^-). Another NOS enzyme (mtNOS) was recently discovered in mitochondria. A rationale for production of NO in mitochondria is hard to reconcile given that NO is a competitive inhibitor to oxygen for cytochrome c oxidase In effect, NO in mitochondria should diminish ATP production by oxidative phosphorylation. It now appears that the concentration of NO produced in a calcium-activated mitochondrion is insufficient to do more than partially inhibit cytochrome c oxidase. Its role as a competitive inhibitor is to raise the K_m for oxygen at the active site of cytochrome c oxidase, an effect that can be overcome when oxygen concentration is elevated. Further, it is hypothesized that the mild inhibitory effect of NO in mitochondria facilitates oxygen diffusion to mitochondria farther distanced from the cell membrane, through which oxygen first passes. In this way, NO acts to distribute ATP formation more homogenously in all parts of the cell. Because of the significance of reactive nitrogen-containing species such as NO and peroxynitrite, we often include the word *nitrogen* with ROS to provide the more inclusive term *reactive oxygen and nitrogen species* (RONS).

Protection From Reactive Oxygen and Nitrogen Species

Our bodies produce a number of endogenous oxidants that we have categorized as RONS. Added to this are a host of exogenous substances such as pollution, ozone, automobile exhaust, solvents, pesticides, and cigarette smoke. Foreign molecules that may pose a hazard to us are generally categorized as **xenobiotics.** Some xenobiotics are free radicals or can produce RONS. There are three principal approaches to maintaining control over RONS. First, reducing the formation of or exposure to RONS and xenobiotics is an essential step. Second, our endogenous scavenging systems, including antioxidant enzymes and other antioxidant molecules, can scavenge those RONS already present. Finally, there are mechanisms in cells that can upregulate antioxidant defenses in the face of persistent problems (Halliwell and Gutteridge 1999). A failure to maintain a balance between the formation of oxidants (e.g., RONS) and their removal by our various antioxidant systems produces a state known as **oxidative stress.** Powerful oxidants such as those just discussed can damage proteins, DNA, and the unsaturated fatty acid molecules in membranes. It is no wonder, then, that a variety of diseases such as cancers, heart disease, neurodegenerative diseases, and type 2 diabetes, as well as aging, all have a relationship to oxidative stress.

Figure 4.20 summarizes a number of important reactions that produce and remove ROS, such as SOD, catalase, and GPX. In addition to specific antioxidant enzymes, the body contains a number of nonprotein antioxidant molecules. For example, the essential nutrients vitamin C, vitamin E, and carotenoids can by themselves scavenge free radicals. Minerals such as selenium, zinc, manganese, and copper are constituents of antioxidant enzymes. Finally, there are a host of nonnutrient but important antioxidants in a wide variety of foods of plant origin. The general name **phytochemical** is applied to those molecules that can function in our bodies as antioxidants. Fruits, vegetables, and grains, as well as common products derived from these such as tea and wine, are the major sources. Flavonoids, polyphenols, and lycopene are common examples of phytochemicals.

Exercise and Oxidative Stress

It is extremely difficult to measure directly free radical formation in a human being. Rather, we use indirect measures to point to increased or decreased RONS formation, and from this make inferences. From a theoretical perspective, during exercise we should expect a greater formation of superoxide, and therefore hydrogen peroxide and hydroxyl radicals, simply because of an increased flux through the electron transport chain with the increased need for ATP. Even though the proportion of oxygen that undergoes one- instead of four-electron reduction should decrease with increased electron transport chain activity, the fact that the overall flux may increase 10-fold or more above resting levels when we exercise should lead us to suspect an increase in superoxide formation. Bailey and colleagues (2003) were the first to actually measure increased formation of free radicals in venous blood during a single-leg exercise task, using an expensive technique known as electron paramagnetic resonance spectroscopy. In this study, the researchers noted that the outflow of free radical species in the venous blood increased as exercise intensity increased.

The active muscle is not the only source of increased RONS formation during exercise. It has been observed that ROS production by white blood cells increases with exercise. However, the effect of ROS production by white blood cells is blunted the more trained the person is for the particular exercise activity (Mooren et al. 2004), likely as a consequence of increased antioxidant enzyme activity in white blood cells with training (Elosua et al. 2003). Some types of exercise, such as intense isometric or eccentric contractions, can create inflammation in muscle. Part of the inflammatory response is caused by neutrophils (white blood cell subset) accumulating in the muscle. As mentioned earlier, these cells produce superoxide that can cause damage within the muscle and also produce other reactive species (Nguyen and Tidball 2003). Obesity, a major public health problem, is characterized by elevated markers of oxidative stress. When people who are obese perform either aerobic or resistance exercise, they produce higher levels of lipid peroxidation products than normal-weight individuals (Vincent et al. 2004). Such a response suggests either an enhanced production or a reduced ability to scavenge free radicals, or both, in the obese state.

Rested muscle produces RONS. When muscle becomes more active, formation of RONS increases. Having some RONS in muscle confers a benefit, but this benefit is lost if there are too few or if the muscle is in a state of oxidative stress (Reid 2001). Several recommendations can be made regarding minimizing the damaging effects of reactive species on muscle. Unaccustomed severe exercise leads to muscle soreness, inflammation, and damage. This can be reduced by repeated exposure to the activity, that is, training. Aging is associated with a decline in muscle performance, related in part to free radicals. Antioxidants have been shown to reduce the influence of both aging and free radicals (Cesari et al. 2004). Sound training along with a diet rich in antioxidants will optimize muscle function (Watson et al. 2005).

KEY POINT

The role of exercise training on the enzymes that remove RONS is unclear. One would expect mitochondrial enzymes such as Mn-SOD and GPX to be increased with endurance training since enzymes of the citric acid cycle and electron transport chain proteins can easily double with a serious endurance training program. However, for every study that shows a training-induced improvement in SOD and GPX, there is another showing no effect. We do know, however, that repeated doses of exercise make succeeding ones easier, reducing muscle soreness and indices of inflammation.

SUMMARY

Oxidative phosphorylation is the synthesis of ATP from ADP and Pi in association with the transfer of electrons from fuel molecules to coenzymes to oxygen. Oxidative phosphorylation, responsible for generating the preponderance of ATP in our bodies, takes place in mitochondria, utilizing the citric acid cycle and the electron transport (respiratory) chain. The citric acid cycle is the primary source of reduced coenzymes. The electron transport system, represented by four protein-lipid complexes in the inner mitochondrial membrane, oxidizes the reduced coenzymes and transports the electrons to oxygen. During the process of electron transport through these complexes, free energy is released, which is utilized to transport protons across the mitochondrial inner membrane against a concentration and electrical gradient. Protons are allowed to flow down their electrical and chemical gradient through a specialized inner mitochondrial protein complex known as ATP synthase. Free energy released during proton flow is harnessed to ADP phosphorylation, making ATP. The exact stoichiometry of ATP produced per atom of oxygen consumed (P/O ratio) during electron transport in humans is less than 2.5. Although electron transport is normally tightly coupled to ADP phosphorylation, leakage of protons across the inner membrane (i.e., uncoupled oxidation) takes place using inner membrane uncoupling proteins.

The citric acid cycle (Krebs or TCA cycle) is the pathway that removes the last carbon atoms in the form of CO_2 from all the body's fuels, while electrons associated with the hydrogen atoms of these fuels are used to reduce the coenzymes NAD^+ and FAD. The citric acid cycle is a circular pathway catalyzed by eight enzymes, all of which are located in the mitochondrial matrix except SDH. Acetyl groups containing two carbon atoms enter the pathway attached to CoA and join with oxaloacetate to form citrate. Because one turn of the cycle yields two CO_2, the citric acid cycle neither produces nor consumes oxaloacetate or any other intermediate of the cycle.

Since most of the ATP is consumed in the cytosol of cells, whereas ATP is produced from ADP and Pi in mitochondria, special transporters move ATP from the matrix of the mitochondria while simultaneously bringing ADP in. The rate of ATP hydrolysis in the cytoplasm is closely matched to the citric acid cycle, the rate of electron transport, and ADP phosphorylation in mitochondria. Calcium ion regulates the activity of pyruvate dehydrogenase that converts pyruvate produced in the cytosol to acetyl CoA. Calcium also regulates two enzymes in the citric acid cycle. Oxidative phosphorylation in skeletal muscle, generating ATP, is tightly regulated to rates of ATP hydrolysis in the cytosol; and calcium control of both processes is essential. Rather than focusing on changes in the concentrations of the individual substrates for oxidative phosphorylation, exercise biochemists have adopted two ratios that better account for the coupling of ATP demand to ATP provision. Changes in muscle tissue oxidation rates better match alterations in the energy demand based on the phosphorylation potential ([ATP]/[ADP] [Pi]) and mitochondrial redox potential ([NADH]/[NAD$^+$]) generated by the regulated dehydrogenases in mitochondria. The oxygen tension at the level of the mitochondria ($P_{mito}O_2$) can limit oxidative phosphorylation. This can occur during exercise at altitude or when blood flow to exercising muscle is partially or completely occluded. The rate of NADH formation in the citric acid cycle may limit the overall rate of oxidative phosphorylation in skeletal muscle if carbohydrate available to exercising muscle is depleted, forcing the muscle to use fat as the exclusive fuel. For an athlete, this means reducing the pace of the activity.

The basis for oxidative phosphorylation is the coupling of oxidation of fuel molecules to reduction of oxygen. The ability of substances to be reduced can be measured against the reduction of hydrogen ions and given a specific value described as the reduction potential. Reduction potentials can be converted into free energy values. Partial reduction of oxygen in the electron transport chain can give rise to superoxide, a free radical with one unpaired electron. Superoxide is converted to hydrogen peroxide by the enzyme superoxide dismutase. Hydrogen peroxide can be converted to the dangerous hydroxyl radical. Nitric oxide, an important signaling molecule, can combine with superoxide to form highly reactive peroxynitrite. Reactive oxygen and nitrogen species are collectively known as RONS. Although a number of enzyme-catalyzed reactions help to control levels of RONS, there are circumstances, such as exercise, during which the antioxidant enzymes, vitamins, and dietary phytochemicals are unable to maintain a redox balance. Such an imbalance is described as oxidative stress. Although exercise is associated with increased formation of RONS, this can be blunted by exercise training and antioxidant nutrients and nonnutrients.

R E V I E W   Q U E S T I O N S

1. Determine the number of ATP molecules generated from the complete oxidation of a pyruvate, based on the P/O ratios used in this book. In earlier times, P/O ratios were 3 for NADH and 2 for $FADH_2$. If you used these older numbers, how many ATP would be formed from the complete oxidation of pyruvate?

2. How many electrons are needed to reduce four molecules of oxygen? How many protons would be translocated (pumped) across the inner membrane during the reduction of four oxygen molecules?

3. A subject exercising at 110 W, breathing room air (21% O_2), has a $\dot{V}O_2$ of 2.0 L per minute. What would the $\dot{V}O_2$ be if the subject were performing at the exact same workload but breathing a gas mixture with only 14% O_2?

4. You have isolated mitochondria and are testing various substrates by measuring the oxygen consumption with a special electrode. You add the inhibitor rotenone. What must be your substrate? Why?

5. In figure 4.18, the $\dot{V}O_2$ is approximately 0.4 L per minute when the subject is merely sitting on the cycle ergometer doing no exercise. Why?

6. Approximately how many kilocalories (or kilojoules) of energy are expended when a subject exercises for 2 h at an average $\dot{V}O_2$ of 3.0 L per minute?

7. You determine the $\dot{V}O_2$max of a group of sedentary students and cross country runners while they are running at constant speed on a treadmill. The treadmill grade is increased every 2 min until the subjects are no longer able to continue. For the untrained subjects, the $\dot{V}O_2$ increases progressively up to the point at which the subjects can no longer continue. For the cross country runners, the $\dot{V}O_2$ levels off about 2 min before they are forced to stop running. Explain why there are these differences.

8. Calculate the standard free energy when electrons on $FADH_2$ are transported to oxygen.

9. Among cytochrome c reduced form, cytochrome c oxidized form, pyruvate, and oxygen, which can dehydroascorbic acid reduce?

10. Determine the $\Delta E°'$ for the enzymatic oxidation of lactate to pyruvate.

11. Using the Nernst equation, determine the ΔE for the oxidation of lactate to pyruvate at 25 °C at the time when the concentrations of the substances in the reaction are (in millimoles) as follows: lactate, 10; pyruvate, 0.2; NAD^+, 2; NADH, 0.05. Ignore the $[H^+]$ in your calculation.

CHAPTER

5

CARBOHYDRATE METABOLISM

Most active people are aware of the importance of carbohydrates to physical performance. Many others are likely aware that a diet with a greater proportion of carbohydrates to fats is healthier than one with an equal or lower proportion. Few people, however, know much about the chemical reactions in the body involving carbohydrates or how the body, especially skeletal muscle, adjusts its fuel utilization to spare the use of carbohydrates unless stores are high or the need for ATP production is acute. Carbohydrate is brain food, and our carbohydrate chemistry is set to favor scarce carbohydrate stores for brain use. Carbohydrates have received a lot of bad publicity, as exemplified by the focus on diets low in carbohydrates. Obesity, one of the fastest-growing health problems in the world, is associated with a diminished ability to handle carbohydrates. In the extreme, this can develop into type 2 diabetes, a condition that endangers the health of an increasing number of people.

We begin this chapter by looking at the various carbohydrates and the ways in which glucose entry into cells is regulated. The chapter includes a detailed examination of the glycolytic pathway. We also look at the metabolism of glycogen and how the synthesis and breakdown of liver and muscle glycogen are tied in to the priorities of carbohydrate storage and utilization. Lactate, a product of intense exercise, is a major focus, followed by mechanisms whereby electrons generated in the cytosol can feed into the electron transport chain. There are important routes to making carbohydrate when dietary sources are

inadequate. Gluconeogenesis, or the synthesis of glucose from noncarbohydrate sources, is vitally important. The **pentose phosphate pathway** is often overlooked because it is not active in muscle; we will consider why it is important. This chapter ends with a detailed look at signaling pathways that control cellular metabolism. Material is organized into the basic pathways, followed by the mechanisms that regulate these pathways. Each section ends with a discussion of the role of exercise.

CARBOHYDRATES

We classify carbohydrates as monosaccharides, disaccharides, and polysaccharides. Examples of **monosaccharides** are glucose, fructose, and galactose; these are simple sugars called hexoses containing six carbon atoms—*hex* for six carbon atoms and *ose* meaning sugar. We will encounter two other monosaccharides that are pentoses. Ribose and deoxyribose, each containing five carbon atoms, are found in ribonucleotides and deoxyribonucleotides, respectively. Figure 5.1 shows the structures for the predominant forms of α-D-glucose, α-D-galactose, and α-D-fructose. The D refers to the configuration about carbon atom 5 in each molecule. Only the D forms of the monosaccharides are acceptable to glucose-metabolizing enzymes in animals. In the remainder of this chapter, the D form of each monosaccharide is assumed and thus the letter will be omitted. Recall that only L amino acids can be

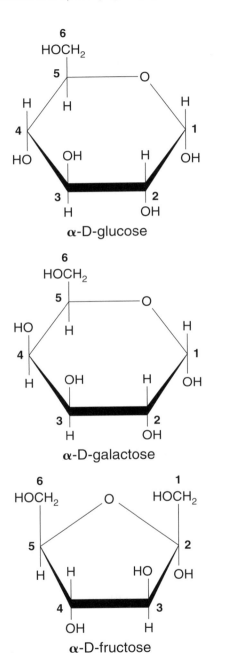

used by animals to make proteins. The α in front of the name refers to the configuration about carbon atom 1 in glucose and galactose and carbon 2 in fructose. The other configuration about these carbons is β, in which the hydroxyl group (OH on carbon 1

Figure 5.1 Structural formulas for the monosaccharides α-D-glucose, α-D-galactose, and α-D-fructose. The numbers identify the carbon atoms, and the D refers to the absolute configuration about carbon 5. The symbol α refers to the position of the OH groups on carbon atom 1 of D-glucose and D-galactose and carbon 2 on D-fructose. These molecules are all isomers with the formula $C_6H_{12}O_6$. Note that D-glucose and D-galactose differ only in the configuration about carbon atom 4.

for glucose and galactose and carbon 2 for fructose) would be pointing up. Sugars such as glucose and galactose in which carbon atom 1 is free (not attached to another molecule) are capable of reducing certain metal ions, producing a colored reaction product. They are said to be *reducing sugars*. Fructose is not a reducing sugar.

Disaccharides are formed when two monosaccharides join together. The common disaccharide sucrose is composed of the monosaccharides glucose and fructose, while lactose or milk sugar contains glucose and galactose. Maltose, produced during the digestion of dietary starch, contains two glucose molecules joined together. Specific digestive enzymes located on the surface of intestinal cells hydrolyze disaccharides into their monosaccharide constituents. Thus, sucrase, lactase, and maltase digest sucrose, lactose, and maltose, respectively. In sucrose, carbon atom 1 of glucose is joined to the fructose, so sucrose is a nonreducing sugar. In maltose or lactose, there is a free carbon atom 1, so both are reducing sugars.

Glycogen and starch are polysaccharides, but only starch is significant as a dietary source of carbohydrate. From a nutritional perspective, starch is a complex carbohydrate. When it is completely digested, its products are glucose molecules. Glycogen is a highly branched polysaccharide, composed only of glucose molecules joined together.

After a meal containing a variety of foods, the main carbohydrate digestion products are glucose, some fructose, and galactose from milk sugar. These substances are absorbed into the blood and transported to the liver, where galactose is converted to glucose. Fructose can be utilized as a substrate for glycolysis in muscle, liver, and adipose tissue. It enters the pathway at several locations; but in the end, its products are no different from those of glucose.

CELLULAR UPTAKE OF GLUCOSE

The normal blood glucose concentration, referred to as **euglycemia,** is approximately 5 to 6 mM, equivalent to 90 to 108 mg of glucose per deciliter (100 ml) of blood. Following a meal, blood glucose is elevated above normal and can increase to 9 mM (160 mg/dL) or more. Uncontrolled insulin-dependent **diabetes mellitus** (type 1 diabetes) can result in a glucose concentration over 20 mM (360 mg/dL). People with established type 2 diabetes can have resting blood glucose concentrations in the range 7 to 9 mM or 126 to 162 mg/dL (Burtis and Ashwood 2001). We call elevated blood glucose concentrations **hyperglycemia.** Blood glucose concentration well below normal (about 2.5 mM or less, or less than 45 mg/dL) is called **hypoglycemia.** You can become hypoglycemic if you do not eat for a long period of time or if you exercise for hours without ingesting carbohydrate. Individuals who do not pay attention to carbohydrate intake during prolonged competitions or exercise may be forced to stop because of hypoglycemia.

> ### ▶ K E Y   P O I N T ◀
>
> Most of the world measures concentrations of glucose, cholesterol, lactate, and other substances in metric units—that is, millimoles per liter (mmol/L or mM). Conversion of a glucose concentration in millimolar units to mg/dL can be worked out using a known relationship: A 1 mM glucose solution contains 18 mg of glucose per deciliter. For lactate, a 1 mM concentration equals 9 mg/dL. For cholesterol, a 1 mM concentration is equivalent to 38.5 mg/dL.

A gradient exists for glucose entry into cells because the glucose concentration in the blood and extracellular fluid is much higher than inside cells. Glucose is a polar molecule with five hydroxy (OH) groups. It is therefore a poor substance for crossing the hydrophobic cell membrane. To get glucose inside cells, a transporter is needed, that is, a protein molecule that will allow glucose entry across the cell membrane. The process of transporting a substance down its concentration gradient across a membrane is known as facilitated diffusion. As mentioned in chapter 2, membrane transport is characterized by simple Michaelis-Menten kinetics with a V_{max} and a K_m. As such, membrane transport will exhibit saturation kinetics similar to the effect of increasing substrate concentration on the activity of an enzyme. A family of membrane transporters exists for glucose. These **glucose transporters** are given the abbreviation GLUT and a number to identify the particular transporter isoform. These transporters exhibit tissue-specific locations and unique kinetic parameters as shown in table 5.1.

To be metabolized, glucose must diffuse into a cell through a glucose transporter. Entry into some cells is regulated; entry into others is unregulated. We would expect the unregulated entry of glucose, which depends only on the relative concentration gradient of glucose across the membrane, to occur in cells that rely primarily on glucose as an energy source. In fact, this is what happens for red blood cells, brain cells, and kidney cells. Liver cells, which store excess glucose as glycogen, also have unregulated glucose uptake; but, as we will see, there is another level of control. Large tissues such as skeletal muscle or fat, as well as the heart, have regulated glucose uptake. Glucose transport across cell membranes in regulated tissues occurs primarily by the GLUT-4 transporter, which, unlike the other transporter isoforms, is regulated by insulin. Skeletal muscle accounts for about 70% to 80% of the insulin-stimulated uptake of glucose.

Insulin, a polypeptide hormone secreted by the beta cells of the pancreas, is the main regulator of glucose transport. When blood glucose concentration is elevated (e.g., following a meal), blood insulin concentration increases to help glucose enter the regulated tissues. Insulin binds to an insulin receptor that spans the cell membrane. Through a complicated mechanism involving other intracellular subunits and protein kinases, GLUT-4 transporters are translocated from intracellular storage vesicles to the cell membrane to aid glucose entry. Thus, insulin increases the

Table 5.1   Mammalian Hexose Transporters

GLUT isoform	Approximate K_m (mM) for main substrate	Tissue specificity and characteristics
GLUT-1	5 for glucose	Erythrocytes, neurons, muscle, for resting glucose uptake in most cells
GLUT-2	10-15 for glucose	Liver, pancreatic beta cells, small intestine, and kidney
GLUT-3	1-2 for glucose	Brain mainly, but expressed to small extent in other cell types
GLUT-4	3-5 for glucose	Expressed in insulin-sensitive tissues (muscle, heart, adipose tissue)
GLUT-5	6 for fructose	Jejunum (part of small intestine)

V_{max} of glucose transport, but only in those cell types (i.e., muscle and fat) expressing the GLUT-4 transporter gene. The GLUT-4 transporters are inserted into the surface membrane but more prominently in the T-tubules, where the glucose that is taken up can be distributed more easily to the interior of the fiber. Muscle does contain GLUT-1 transporters that are consistently present in the sarcolemma. These do not respond to insulin, and because they are not present significantly in the sarcolemma, they have a limited effect on glucose uptake. A more detailed examination of the effects of insulin binding to cells is presented later in this chapter.

Exercising skeletal muscle also has an increased ability to take up glucose from the blood, independent of the effect of insulin. The exercise effect also involves a stimulation of GLUT-4 transporter translocation to the cell membrane from intracellular storage sites that are different from those affected by insulin. The mechanism for the exercise effect involves a different signaling pathway, employing the elevated calcium concentration caused by activation of the muscle fiber through its motor neuron. Interestingly, the effects of exercise and insulin are additive, supporting the fact that the signaling systems are different. This muscle contraction effect persists into the early postexercise period in order to rebuild depleted stores. During prolonged exercise tasks or games, one must ingest glucose to keep blood levels maintained because exercising muscle has an augmented capacity to take up glucose from the blood. Failure to supply enough glucose to the body during prolonged physical activity can lead to problems associated with hypoglycemia.

▶ KEY POINT ◀

People who engage in endurance exercise training programs increase the total content of GLUT-4 transporters in the trained muscle. This means that there is an increased maximal capacity for glucose transport in the trained muscle. Interestingly, when subjects are compared during the same exercise intensity after training versus before, they have more total muscle GLUT-4 transporters, but fewer in the sarcolemma, to aid in glucose transport. One of the adaptations that takes place after training is the use of less carbohydrate and more fat to fuel the same exercise level.

Type 1 diabetes mellitus or insulin-dependent diabetes mellitus (T1DM or IDDM) is a condition in which there is no blood insulin due to an autoimmune destruction of the insulin-secreting beta cells of the pancreas. Type 1 diabetes mellitus is also known as juvenile diabetes since it typically presents in childhood. Individuals with T1DM need exogenous insulin to survive, but because of the exercise effect on glucose uptake, they must carefully balance the type, intensity, and amount of exercise with blood glucose levels, carbohydrate intake, and insulin dose. Type 2 diabetes mellitus or non-insulin-dependent diabetes mellitus (T2DM or NIDDM) usually is diagnosed in adults, is often associated with obesity, and is one of the most pressing health problems in wealthy countries. It is typically characterized by hyperglycemia, hyperinsulinemia (elevated blood insulin, at least in the early stages), and insulin resistance. Insulin resistance means that for a given blood insulin concentration, less glucose is taken up into muscle and fat cells. It also means that the insulin cannot suppress output of glucose from the liver, as it does in healthy people, when the blood glucose concentration is increased.

In early stages, T2DM can be characterized by excess secretion of insulin from the pancreas in response to an increase in blood glucose. As the condition progresses, insulin secretion from the beta cells of the pancreas is reduced, suggesting some failure in the beta cells. At least one-third of T2DM patients need to take insulin. Figure 5.2 illustrates blood glucose and insulin concentrations in response to an oral glucose challenge in the fasted state. For comparison purposes, representative data are shown from a healthy adult, a person in a pre-T2DM condition, and a person with a more advanced stage of T2DM. A major source of the problem with T2DM is resistance to insulin in muscle and adipose tissue, although abnormalities in lipid metabolism, such as increased levels of fatty acids in the blood as well as elevated levels of fat in skeletal muscle, also coexist. New research shows a common link between obesity and diabetes in that each is characterized by markers of cellular inflammatory stress that appears physiologically as insulin resistance and T2DM (Wellen and Hotamisligil 2003). Health problems with T2DM include increased risks for heart attack, stroke, some cancers, blindness, kidney failure, and problems due to inadequate blood flow in the legs that can require amputation.

The term *metabolic syndrome* is prominently used by health professionals. A person is defined as having

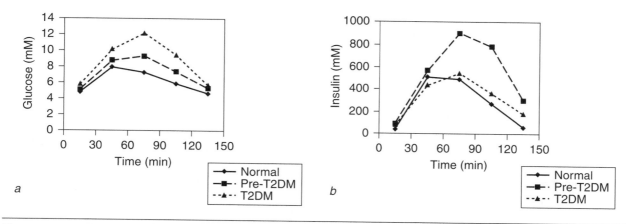

Figure 5.2 Representative responses in *(a)* blood glucose and *(b)* blood insulin concentrations in fasted subjects following an oral glucose challenge. Values are shown for a healthy normal individual, a person with insulin resistance but not yet classified as having type 2 diabetes mellitus (pre-T2DM), and a person with T2DM. Note that the blood glucose concentration increases more with T2DM, yet the insulin response is blunted.

metabolic syndrome if he or she has three or more of the following five disorders: fasting hyperglycemia; obesity, particularly in the abdominal area; high blood pressure; elevated blood triglyceride levels; and decreased high-density lipoprotein (HDL) cholesterol levels (Armitage et al. 2004). This syndrome may progress to T2DM and cardiovascular disease. Although we typically associate T2DM with adults, new evidence shows impaired glucose tolerance in obese children under 10 years of age (Rocchini 2002). Type 2 diabetes can be controlled to some extent with drugs. However, the strong association between T2DM and obesity points to a reduction in body fat as an important nondrug approach to treatment. With an understanding of the mechanism for exercise-induced blood glucose transport, health authorities are proposing daily exercise for those predisposed to or experiencing T2DM (Tucker et al. 2004).

PHOSPHORYLATION OF GLUCOSE

Once glucose enters a cell, it is covalently modified by transfer of the terminal phosphate from ATP to carbon atom 6 of glucose to make glucose 6-phosphate (glucose 6-P) as shown in the following equation:

$$\text{glucose} + \text{ATP} \rightarrow \text{glucose 6-phosphate} + \text{ADP}$$

A class of enzymes known as hexokinases catalyze this reaction. When glucose is phosphorylated, the product, glucose 6-P, is trapped inside the cell. This reaction is essentially irreversible because of the accompanying large free energy change since ATP is hydrolyzed. There are four hexokinase isoenzymes, identified as hexokinases I, II, III, and IV (HK I, HK II, HK III, and HK IV). Hexokinase IV is more commonly known as glucokinase and is found in the liver and beta cells of the pancreas.

The differences between glucokinase (HK IV, also referred to as GK) and the other hexokinase isozymes (HK I, HK II, and HK III) are as follows:

1. Glucokinase is found only in the liver and pancreas, whereas the other hexokinase isozymes are found in all cells.

2. The amount of hexokinases I, II, and III in cells remains fairly constant; they are thus constitutive enzymes. The amount of glucokinase in liver cells depends on carbohydrate content of the diet. Glucokinase is thus described as an inducible enzyme because a diet high in carbohydrate will induce liver cells to make more. This effect is attributed to insulin. Conversely, diabetes (with its attendant low insulin), starvation, or a low-carbohydrate diet will mean less glucokinase.

3. Hexokinase isozymes I, II, and III have a low K_m for glucose (0.02-0.13 mM), whereas glucokinase has a high K_m for glucose (~5-8 mM). Hexokinases I, II, and III are thus very sensitive to glucose, whereas glucokinase activity becomes important for phosphorylating glucose only when the intracellular concentration of glucose is elevated. (Figure 2.4 on

p. 20 illustrates the responses of the reaction velocities for glucokinase and a hexokinase to changes in glucose concentration.)

4. Hexokinase isozymes I, II, and III, but not glucokinase, can be inhibited by the product of the reaction, glucose 6-P. This is an example of feedback inhibition. Thus, if the concentration of glucose 6-P increases inside a cell, it inhibits the activity of hexokinase; glucose will not get phosphorylated, and its concentration will increase in the cell. This increase will reduce the gradient for transport, thus slowing down glucose entry into the cell.

5. In liver cells only, there is a glucokinase regulatory protein (GKRP) that binds GK and inhibits its activity, effectively stopping glucose phosphorylation in the liver. This inhibition can be overcome by an increase in liver cell glucose concentration as would occur following a meal containing carbohydrate.

For glucose to be metabolized in a cell, two processes must occur; it must be transported into the cell using a glucose transporter, and it must be phosphorylated to glucose 6-P. Which of these two processes limits subsequent metabolism of glucose can be hotly debated. However, under most normal physiological conditions, the concentration of free glucose inside skeletal muscle and most other cells except liver is very low. This points to glucose transport as **rate limiting** to glucose metabolism, since glucose is phosphorylated about as fast as it enters the cell. However, in some situations during exercise, blood flow in muscle is high and there are sufficient GLUT-4 transporters to

---

> ▶ **K E Y   P O I N T** ◀
>
> The fact that glucose transport is usually rate limiting to glucose entry into skeletal muscle helps to explain an important point regarding postexercise (competition) feeding. Feeding carbohydrate to athletes after training or competition should take place as soon as possible. One reason is the effect of the prior muscle activity on the content of GLUT-4 transporters in the muscle sarcolemma. This exercise effect, plus the independent effect of an increase in blood insulin with the carbohydrate feeding, will promote glucose transport and hasten the synthesis of muscle glycogen for the next exercise session or competition.

---

move blood glucose into the muscle cytosol. Under such conditions, glucose phosphorylation can limit glucose use (Fueger et al. 2003).

Diets high in sucrose or high-fructose syrups, common in many sweetened products, can provide considerable fructose. Dietary fructose is absorbed in the jejunum of the small intestine and transported, like glucose, via the portal vein to the liver. The liver and kidney have a fructokinase enzyme that specifically phosphorylates fructose to fructose 1-phosphate, which can enter glycolysis but at a step below the main regulatory enzyme we will encounter in the next section. The lack of regulatory control of fructose 1-phosphate is believed to promote elevated formation of a class of blood lipoproteins rich in triglycerides, the very low density lipoproteins (VLDLs). Unlike glucokinase activity, fructokinase activity is not regulated by insulin, so people with diabetes can metabolize it.

GLYCOLYSIS

Glycolysis is the principal route for carbohydrate breakdown in all cells. As we discussed in chapter 3, glycolysis is the breakdown of glucose or glycogen to pyruvate. Chapter 3 showed the two major starting materials for glycolysis, glucose and glycogen, and provided two reactions summarizing their conversion to pyruvate:

$$\text{glucose} + 2\ \text{ADP} + 2\ \text{Pi} + 2\ \text{NAD}^+ \rightarrow 2\ \text{pyruvate} + 2\ \text{ATP} + 2\ \text{NADH} + 2\ \text{H}^+$$

$$\text{glycogen}_n + 3\ \text{ADP} + 3\ \text{Pi} + 2\ \text{NAD}^+ \rightarrow \text{glycogen}_{n-1} + 2\ \text{pyruvate} + 3\ \text{ATP} + 2\ \text{NADH} + \text{H}^+$$

We will further classify glycolysis as **anaerobic** if the pyruvate is reduced to lactate or **aerobic** if the pyruvate is completely oxidized in mitochondria, first by being converted to acetyl coenzyme A (CoA), which enters the citric acid cycle (figure 5.3). Anaerobic glycolysis is the only source of ATP for erythrocytes (red blood cells) and a major source of ATP for skeletal muscle under severe exercise conditions or when muscle blood flow is compromised. Although cardiac muscle can utilize carbohydrate as a fuel, it is not well endowed to produce ATP by anaerobic glycolysis and thus has relatively poor ability to survive under ischemic (low blood supply) conditions. Cancer cells use anaerobic glycolysis to a significant extent for their ATP supply.

Figure 5.3 summarizes carbohydrate metabolism in a skeletal muscle cell. Blood glucose is taken up

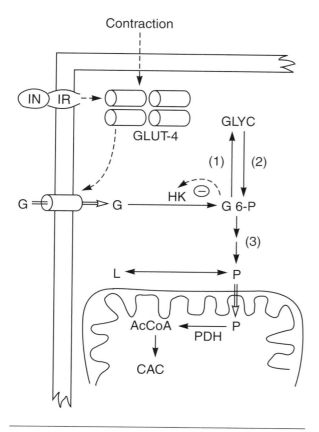

Figure 5.3 Overview of glucose metabolism in a muscle cell. Glucose uptake into muscle fibers through GLUT-4 transporters is greatly facilitated by insulin (IN) binding to the insulin receptor (IR) or by contractile activity. Through two mechanisms, GLUT-4 transporters are then translocated from internal storage sites to the muscle membrane. Once inside the cell, glucose is phosphorylated by hexokinase (HK) to glucose 6-phosphate (G 6-P), which can either be used to make glycogen through pathway 1 or immediately be broken down to pyruvate (P) in the glycolytic pathway (3). Breakdown of glycogen (2) during exercise is accelerated to produce G 6-P for glycolysis. Pyruvate has two major fates, reduction to lactate (L) or entry into the mitochondrion and conversion into acetyl CoA (AcCoA) using pyruvate dehydrogenase (PDH). Acetyl CoA enters the citric acid (TCA) cycle. The dotted arrow with the negative sign indicates that G 6-P, the product of the HK reaction, can inhibit HK if the concentration of G 6-P rises.

through GLUT-4 transporters whose content in the sarcolemma is increased by insulin binding to its receptor, or with active muscle contraction, or both. Once phosphorylated, the product glucose 6-P has two major fates in a muscle cell. Following a meal, and when the muscle fiber is inactive, glucose 6-P will be directed to the synthesis of glycogen. During exercise, stored glycogen is broken down to glucose 6-P and metabolized in the glycolytic pathway to pyruvate. Any glucose that enters an actively contracting muscle fiber is first converted to glucose 6-P, then broken down to pyruvate using the glycolytic

pathway. Figure 5.3 emphasizes the important central position occupied by glucose 6-P. Because our focus is on skeletal muscle, and since there are two sources of glucose 6-P, we will consider the glycolytic pathway to begin with glucose 6-P. As mentioned previously, glucose 6-P can inhibit its own formation by hexokinase if it is not removed in an adequate way through glycolysis or glycogen formation. As we will see, this is important because it prevents unnecessary entry and phosphorylation of blood glucose if stored glycogen can provide glucose 6-P at an adequate rate. One of the interesting questions we will look at later is how important blood glucose is as a source of glucose 6-P to fuel glycolysis during exercise.

Enzymes of Glycolysis

The conversion of glucose 6-P to pyruvate is catalyzed by glycolytic enzymes found in the cytoplasm of cells. In the past it was assumed that the enzymes of glycolysis are freely dissolved in the cytosol of the cell. This view is no longer held. Many, if not all, glycolytic enzymes may be bound to other structures in the cell, such as to structural filaments providing shape to the cell, to the endoplasmic reticulum (sarcoplasmic reticulum in muscle), to the outer membrane of the mitochondrion (as is the case for hexokinase), or to contractile proteins such as actin. Some of the enzymes catalyzing the reactions of glycolysis may also be physically linked to each other such that the product of one enzyme is immediately passed to the next enzyme as its substrate. In this case, the actual concentrations of the intermediates in glycolysis, except glucose 6-P, fructose 6-phosphate (fructose 6-P), pyruvate, and lactate, would not increase much, even if glycolysis were proceeding rapidly.

> **▷ KEY POINT ◁**
>
> The term *flux* refers to how fast substrates enter a biochemical pathway and become products at the other end. In terms of glycolysis, its flux may be rapid, but there may not be a large increase in the concentrations of intermediates within the pathway because, as fast as they are formed, they are acted on by the next enzyme in the pathway. We would expect the rate of formation of pyruvate and lactate to provide a useful index of the flux of glycolysis, since these are the major products.

Glycolysis has two major functions. One is to generate energy in the form of ATP. For red blood cells, glycolysis is the only energy-generating source; making ATP is its only role. The second function is to generate pyruvate for final oxidation in the mitochondrion, along with NADH. Figure 5.4 outlines the major reactions of the glycolytic pathway, showing all of the chemical structures. The individual reactions of glycolysis, starting at glucose 6-P and ending with pyruvate, are shown without chemical structures in table 5.2. If we start from glucose 6-P, there is one early reaction that requires ATP; this is the reaction catalyzed by phosphofructokinase-1. If we start from glucose and use the hexokinase reaction to make glucose 6-P, there is an additional step where ATP is used. It may seem absurd to use ATP to make ATP, but for help with this we can use the example of a roller coaster. The first part of the roller-coaster ride is a slow ascent to the highest point on the course, using a motor that catches the underside of the cars and moves them up. Thereafter, the potential energy of this high elevation is used to create the kinetic energy that makes a roller-coaster ride so thrilling and appealing. Overall, the $\Delta G°'$ for glycolysis for the sequence of reactions beginning with glucose 6-P and ending with pyruvate is −19 kJ/mol (−4.5 kcal/mol). Using the standard free energy change for each of these reactions, and putting in intermediate concentrations representative of normal physiological conditions, we find that the overall free energy

(ΔG) change for this sequence of reactions is −39 kJ/mol or −9.3 kcal/mol (Berg et al. 2002). From this sequence, three ATP are produced along with two NADH. The latter is a source of electrons for the electron transport chain, creating five more ATP.

Reactions of Glycolysis

As previously mentioned, the glycolytic pathway beginning from glucose 6-P and ending with pyruvate is shown with full structures in figure 5.4. For those who find chemical structures daunting, the individual reactions are shown in table 5.2. Starting from glucose, glycolysis can be considered to involve three stages. In the first or priming stage, two phosphorylation reactions produce a hexose with two phosphate groups attached. This is called a hexose bisphosphate. Stage 2 is the splitting stage, conversion of the hexose bisphosphate into two triose phosphates—three-carbon sugar molecules with an attached phosphate group. Stage 3 consists of oxidation–reduction reactions and the formation of ATP.

The *glucose phosphate isomerase* reaction converts glucose 6-P into another hexose phosphate, fructose 6-P. This is called an isomerase because glucose 6-P and fructose 6-P are isomers.

The next step is the phosphorylation of fructose 6-P, using a phosphate group from ATP and catalyzed by the enzyme *phosphofructokinase-1* (abbreviated PFK-1). We add the number 1 to the enzyme because

Table 5.2 The Individual Reactions of Glycolysis—the Pathway From Glucose 6-Phosphate to Pyruvate

Number	Detailed reaction	Enzyme
	First stage: priming	
1	Glucose 6-phosphate ↔ fructose 6-phospate	Glucose phosphate isomerase
2	Fructose 6-phosphate + ATP → fructose 1,6-bisphosphate + ADP + H⁺	Phosphofructokinase-1
	Second stage: splitting	
3	Fructose 1,6-bisphosphate ↔ glyceraldehyde 3-phosphate + dihydroxyacetone phosphate	Aldolase
4	Dihydroxyacetone phosphate ↔ glyceraldehyde 3-phosphate	Triose phosphate isomerase
	Third stage: oxidizing and reducing	
5	Glyceraldehyde 3-phosphate + Pi + NAD⁺ ↔ 1,3-bisphosphoglycerate + NADH + H⁺	Glyceraldehyde 3-phosphate dehydrogenase
6	1,3-bisphosphoglycerate + ADP ↔ 3-phosphoglycerate + ATP	Phosphoglycerate kinase
7	3-phosphoglycerate ↔ 2-phosphoglycerate	Phosphoglycerate mutase
8	2-phosphoglycerate ↔ phosphoenolpyruvate + H₂O	Enolase
9	Phosphoenolpyruvate + ADP + H⁺ ↔ pyruvate + ATP	Pyruvate kinase

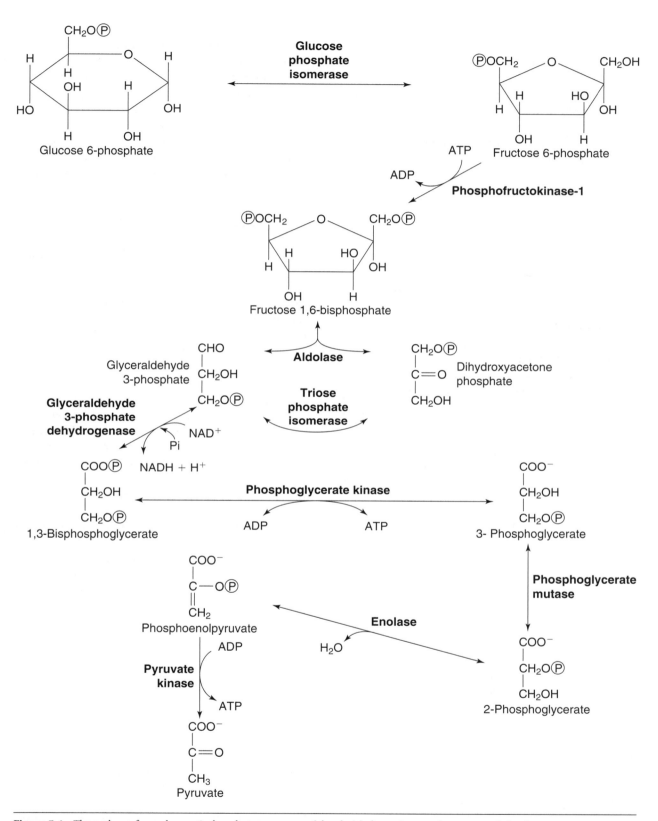

Figure 5.4 The pathway from glucose 6-phosphate to pyruvate (glycolysis) shown in complete structural detail.

there is another enzyme, phosphofructokinase-2, that will transfer a phosphate from ATP to the 2 position of fructose 6-P. As already mentioned, this is a priming reaction, increasing the chemical potential energy of the product fructose 1,6-bisphosphate. The reaction is strongly exergonic, and we show this with the arrow going in one direction only. Phosphofructokinase-1 catalyzes the committed step of glycolysis, committing the cell to glucose degradation. Phosphofructokinase is under tight regulation, and its activity controls the flux of glycolysis.

> ▶ **K E Y   P O I N T** ◀
>
> The prefix *bis* means that two phosphates are attached to separate locations on the same molecule. If there were three phosphates attached to separate places on the same molecule, we would use the prefix *tris*. Note the difference between these prefixes and *di-* and *tri-*, which are used when two phosphate and three phosphate groups are attached to each other as in adenosine diphosphate and triphosphate, respectively.

Stage 2 begins with the splitting of a hexose bisphosphate. *Aldolase* splits fructose 1,6-bisphosphate into two triose phosphates, that is, glyceraldehyde 3-phosphate and dihydroxyacetone phosphate. This reaction is freely reversible in the test tube, but is driven toward hexose bisphosphate splitting in glycolysis. As we will see, some tissues can make glucose from three carbon precursors. When they do, some of the reactions of glycolysis go the opposite way, including that catalyzed by aldolase. Only glyceraldehyde 3-phosphate has a further role in glycolysis. Therefore, the enzyme *triose phosphate isomerase* catalyzes the reversible interconversion of the isomers dihydroxyacetone phosphate and glyceraldehyde 3-phosphate. In the pathway toward pyruvate, this enzyme ensures that all of the carbon atoms in fructose 1,6-bisphosphate are funneled through the glycolytic pathway via glyceraldehyde 3-phosphate. At this point in the pathway, all of the original carbon atoms in each glucose molecule are in the form of two molecules of glyceraldehyde 3-phosphate.

The third stage begins with a complicated reaction, producing 1,3-bisphosphoglycerate from glyceraldehyde 3-phosphate. *Glyceraldehyde 3-phosphate dehydrogenase* carries out the following: (a) It oxidizes

the aldehyde group of glyceraldehydes (carbon 1) to a carboxylic acid group, and in the process reduces NAD^+ to $NADH + H^+$; and (b) it reacts the acid group with a phosphate (Pi) to make a mixed anhydride bond between a carboxylic acid and phosphoric acid. This bond is energy rich. The NADH generated in the glyceraldehyde 3-phosphate dehydrogenase reaction can be used to reduce pyruvate to lactate, or the electrons can be shuttled into the mitochondrion where they will be used to reduce oxygen and generate ATP.

The next reaction involves capturing the energy-rich bond of the mixed anhydride bond in 1,3-bisphosphoglycerate. *Phosphoglycerate kinase* catalyzes a substrate-level phosphorylation reaction in which an ATP is generated from ADP via removal of the phosphate group from the mixed anhydride, producing 3-phosphoglycerate. This represents the first energy-generating reaction of glycolysis; and based on a starting point of glucose, the balance for ATP used minus that created is zero. Remember, two 1,3-bisphosphoglycerates are obtained from each glucose.

Phosphoglycerate mutase catalyzes the movement of a phosphate group from carbon 3 to carbon 2 of the glycerate molecule. Mutases catalyze intramolecular phosphate transfer reactions. There is an intermediate in this phosphate transfer reaction known as 2,3-bisphosphate glycerate (i.e., 2,3-BPG). In red blood cells, 2,3-BPG is important because it can reduce the affinity of hemoglobin for oxygen, thus allowing more oxygen to leave its binding to hemoglobin at the tissue level. When a person moves from lower to higher altitude, one of the adaptations is an increase in the content of 2,3-BPG in red blood cells. This allows the tissues to obtain more oxygen than they would normally.

Enolase catalyzes the dehydration of 2-phosphoglycerate to form the energy-rich molecule phosphoenolpyruvate (PEP). Phosphoenolpyruvate is an energy-rich molecule because of the enol phosphate. As such, it is capable of phosphorylating ADP. Next, *pyruvate kinase* carries out the second substrate-level phosphorylation reaction, generating ATP from ADP and leaving the product pyruvate. Even though ATP is formed in this reaction, there is a large free energy change, to the extent that the reaction is considered to be essentially irreversible.

As shown in figure 5.3, pyruvate has two major fates: reduction to lactate or entry into the mitochondrion for complete oxidation. If the former occurs, the $NADH + H^+$ generated in the glyceraldehyde phosphate dehydrogenase reaction is oxidized to NAD$^+$ and the pyruvate is reduced to lactate using the enzyme *lactate dehydrogenase.* If the pyruvate enters the mitochondrion, then the NADH + H$^+$ generated in the glyceraldehyde phosphate dehydrogenase reaction must be converted back to NAD$^+$ by one of two shuttle mechanisms. This must occur, or glycolysis will come to a halt due to a lack of NAD$^+$. We discuss the shuttles later in this chapter.

In the glycolytic sequence of reactions in which glucose 6-P is converted to pyruvate, a net of one proton (H$^+$) is generated for each glucose 6-P that we start out with (see table 5.2). One proton is produced when fructose 1,6-bisphosphate is formed in the PFK-1 reaction. A single proton is generated when glyceraldehyde 3-phosphate is converted to 1,3-bisphosphoglycerate. Finally a proton is used when PEP is converted to pyruvate and ATP. Since there are two of each of the three-carbon intermediates, beginning with glyceraldehydes 3-phosphate, for each glucose 6-P we start out with, the net is one proton formed. Interestingly, if the pyruvate is reduced to lactate, a proton is consumed for each lactate formed. Thus the pathway from glucose 6-P to lactate actually removes a proton from the medium, in essence having a buffer effect. We will talk about this later.

If the final product of glycolysis is lactate (as opposed to pyruvate, which enters the mitochondrion), the lactate is not a waste product. It can have a number of functions. For example, lactate may leave the cell in which it is formed and enter the blood, where it may be taken up by another type of tissue and oxidized to pyruvate, and then converted into acetyl CoA for terminal oxidation in the citric acid cycle. Similarly, lactate from one muscle fiber may be transported into an adjacent fiber and oxidized. Blood lactate may be taken up by liver cells and be a precursor for making glucose (i.e., gluconeogenesis), or it may remain in the cell where it is formed and either be used as a source of energy or be changed into glycogen by reversal of glycolysis.

Regulation of Glycolysis

Glycolysis is a pathway that can generate ATP. It also produces cytosolic reducing equivalents in the form of NADH and plays a major role in providing substrate for the citric cycle as pyruvate. If pyruvate is the main fate of (aerobic) glycolysis, it produces a pair of electrons in the form of NADH for the electron transfer chain. Glycolysis is a major energy-yielding pathway during three general states of muscle. (1) It is important during rapid transitions in muscle activity from rest to exercise or from one exercise intensity to a higher level of intensity. (2) It is important during exercise when oxygen is limited as a substrate for oxidative phosphorylation. This can occur during isometric exercise when intramuscular pressure occludes blood flow. (3) Aerobic glycolysis, producing pyruvate, is important during steady state exercise when the $\dot{V}O_2$ is about 60% or more of $\dot{V}O_2$max.

Glycolysis does not occur haphazardly; its rate is governed by the energy needs of the cell. In part, regulation depends on the concentration of substrates for the various reactions in the pathway: glucose, glycogen, glucose 6-P, and ADP. The need for glycolysis depends on the rate at which ATP is hydrolyzed to drive endergonic reactions. With ATP hydrolysis, ADP increases in the cytosol and is available for the two substrate-level phosphorylation reactions where it will be converted to ATP, the phosphoglycerate kinase and pyruvate kinase reactions. However, the major form of control is in regulating the rate of key enzymes.

Normally, regulated enzymes are those that catalyze key irreversible reactions near the beginning of a pathway. There are three irreversible reactions in the pathway from glucose to pyruvate: the hexokinase, the PFK-1, and pyruvate kinase reactions. Of these, only the hexokinase and phosphofructokinase reactions are major points of control, at least for skeletal muscle. In liver, the pyruvate kinase reaction is subject to control, as will be discussed later.

Hexokinase can be regulated by the product of its reaction, glucose 6-P. If the concentration of glucose

---

> ▶ **K E Y   P O I N T** ◀

Fructose may be a significant part of the carbohydrates in our diets. It is phosphorylated on carbon 1 to fructose 1-phosphate by a fructokinase in the liver. A specific aldolase, aldolase B, splits fructose 1-phosphate into dihydroxyacetone phosphate and glyceraldehyde. The former can now continue on in glycolysis. The glyceraldehyde is phosphorylated to glyceraldehyde 3-phosphate by a triokinase and then proceeds along the glycolytic pathway. Although fructose may be phosphorylated by hexokinase in other tissues to make fructose 6-P, the fructokinase pathway in liver is the major way we metabolize dietary fructose.

6-P increases too much, either because the pathway is being slowed down at another step or because the breakdown of glycogen is producing the glucose 6-P at a sufficiently rapid rate, then glucose 6-P binds to a site on the hexokinase enzyme and slows down the reaction rate. We call this process feedback inhibition or product inhibition. As mentioned previously, inhibition of hexokinase results in a sharp decrease in glucose uptake into a cell.

Phosphofructokinase (PFK-1) is the major regulatory enzyme for the pathway flux from glucose 6-P to pyruvate and is under complex control. We will look at two tissues in which control of PFK-1 is critical. In liver, PFK-1 is a tetramer, identified as L4. In skeletal muscle it is identified as M4. In liver, the role of PFK-1 is primarily to control overall carbohydrate metabolism. We will discuss the control of PFK-1 in liver later in this chapter. In skeletal muscle, PFK-1 must respond very quickly from its low level of activity at rest to rapid activity at the onset of vigorous contractions in order to provide ATP. As an enzyme catalyzing a nonequilibrium reaction, PFK-1 has only two substrates, ATP and fructose 6-P. Since ATP is present in cells in millimolar quantities, with little variation, it poses no limitation to PFK-1 activity as a substrate. On the other hand, fructose 6-P concentration can play a significant role as a substrate. It can be generated rapidly by the breakdown of glycogen, as we will see.

Phosphofructokinase-1 control is primarily based on changes in the concentrations of a number of allosteric effectors. Table 5.3 lists the allosteric effectors for PFK-1 and summarizes their sources.

Adenosine triphosphate is a substrate, but it can also bind to a negative allosteric site on the enzyme and inhibit its activity. Protons (H^+) enhance the inhibition by ATP. Citrate, the product of the citrate synthase enzyme in the mitochondrion, is also present in the cytosol. The level of citrate provides an index of fuel and ATP sufficiency. For example, in rested muscle, the principal fuel is fatty acids. When fatty acids are actively providing the needed ATP in a skeletal muscle under rest or low-intensity activity, citrate concentration is elevated, and this inhibits PFK-1 at a negative allosteric site.

Adenosine monophosphate, ADP, and Pi are important positive allosteric effectors, and the changes in their concentrations reflect the intensity of muscle activity. We say that these metabolites provide a direct level of feedback control signaling the extent to which the glycolytic pathway must be adjusted. We have seen that the ATP concentration does not change markedly as a percentage of its rest value. However, even small changes in ATP can have much larger effects on ADP and AMP concentrations as seen in table 3.3. Further, the magnitude of their changes reflects the need for glycolytic ATP production. We also know that, at the onset of vigorous contractile activity, phosphocreatine (PCr) is critically important to regenerate ATP, and the decline in PCr is inversely related to the rise in free Pi. Further, when the phosphate group on PCr is transferred to ADP, a proton is consumed (see chapter 3). As a result, there is an alkalization effect early in intense contractions, signaling a decrease in $[H^+]$ (i.e., increase in pH). This has the effect of relieving the inhibition by ATP on PFK-1, which is exactly what is necessary. The formation of ammonia by the enzyme adenylate deaminase also has an acid-buffering effect and therefore a positive effect on PFK-1 activity.

Phosphofructokinase-1 is also positively affected by an increase in *fructose 2,6-bisphosphate* (F2,6-P_2). This is formed from fructose 6-P by an enzyme known as *phosphofructokinase-2* (PFK-2), which transfers a phosphate group to carbon 2 of fructose 6-P. Epinephrine and muscle contraction increase the activity of PFK-2, which increases the F2,6-P_2 concentration. F2,6-P_2 binds to an allosteric site to increase the activity of the PFK-1 for both the muscle PFK-1 isoform and the liver PFK-1 isoform (Rider et al. 2004). There

Table 5.3 Allosteric Modulators of Muscle Phosphofructokinase-1

Allosteric effector	Source	Effect on PFK
ATP	PCr, glycolysis, oxidative phosphorylation	Necessary substrate, but allosteric inhibitor
H^+	ATP hydrolysis	Enhances ATP inhibition
Citrate	Citrate synthase (TCA cycle)	Negative allosteric effector
ADP	ATP (ATPases)	Allosteric activator
AMP	ADP (adenylate kinase)	Allosteric activator
Pi	PCr (creatine kinase)	Allosteric activator
Fructose 2,6-P_2	PFK-2	Allosteric activator

may be other levels of control of PFK-1 that involve its binding to certain proteins in the cell, which can alter its sensitivity to inhibitors. It is difficult to determine how effective such binding is by doing a laboratory assay. The actual contraction process itself seems to be very important in both initiating glycolysis at the start of exercise and rapidly turning it off at the end (Crowther et al. 2002a, 2002b).

GLYCOGEN METABOLISM

Glycogen is a polysaccharide, composed of hundreds of glucose molecules (monosaccharides) joined end to end, with prevalent branches (figure 5.5). Glucose units are joined end to end through the alpha (α) configuration of carbon atom 1 of one glucose unit to carbon atom 4 of the next, producing an α 1-4 *glycosidic bond*. Branches occur when carbon atom 6 of one glucose is joined to carbon atom 1 of another glucose unit in the α configuration, producing the α 1-6 glycosidic bond. As shown in figure 5.5, a glycogen molecule consists of proteins and glycogen and is known as a glycogen granule or **glycosome** (Shearer and Graham 2004). The glucose units are referred to as glucosyl units. The glycogen part is built in tiers around a central protein molecule known as **glycogenin.** It is estimated that the maximum number of glucosyl units in the largest, 12-tiered glycogen particle is 5,500, and the glycogen particles exhibit a range of sizes in rested skeletal muscle (Shearer and Graham 2004). In addition to glycogenin, the glycosome contains enzymes that are involved in the formation and degradation of the glycogen particle, including enzymes that regulate these processes.

Glycogen is a convenient way to store glucose inside cells without making an impact on cell **osmotic pressure.** Osmotic pressure depends on the number, not the size, of dissolved substances. For example, one molecule of glycogen may contain 5,000 glucose units yet produce a miniscule influence on osmotic pressure compared to 5,000 individual glucose molecules. In this section, we look at the synthesis and breakdown of glycogen, integrating the latter with the glycolytic pathway in muscle. Glycogen contains a number of OH groups and therefore interacts with water in the cell. It is estimated that each gram of glycogen has between 2.5 and 3.0 g of associated water. This means that, on a weight basis, glycogen is a heavy fuel.

a   b

Figure 5.5 (*a*) A schematic of a glycogen particle showing the central glycogenin protein and three tiers of glycogen surrounding it. The glucose (glucosyl) units in the glycogen molecule are joined together by α 1-4 glycosidic bonds. Branches are formed by α 1-6 glycosidic bonds. (*b*) A section of the glycogen molecule with three individual glucosyl units. It illustrates the two kinds of glycosidic bonds that give rise to chains and branches.

Reprinted, by permission, from J. Shearer and T. Graham, 2002, "New perspectives on muscle glycogen storage and utilization during exercise," *Canadian Journal of Applied Physiology* 27(2): 179-203.

Mechanism of Glycogen Breakdown

Glycogenolysis, or the breakdown of glycogen, is the phosphorolytic cleavage of glucose units, one at a time, from glycogen molecules through the introduction of Pi. This way of breaking a bond is unusual, as the introduction of a water molecule or hydrolysis is what we typically see. The phosphate group breaks the α 1-4 bond between two glucosyl units, one of

which is at the end of the chain. This releases the terminal glucosyl unit as glucose 1-phosphate (figure 5.6). *Glycogen phosphorylase,* often simply called phosphorylase, acts on chains of glucose units, producing glucose 1-P. This is converted into glucose 6-P by *phosphoglucomutase.* A *debranching enzyme* is necessary to remove branches because phosphorylase can cleave off only glucosyl units that are joined in a linear fashion. The fact that there are numerous branches

Figure 5.6 *(a)* The phosphorolytic cleavage of the α 1-4 bond between the last glucosyl residue and the second-last glucosyl residue. This generates glucose 1-phosphate and a glycogen molecule shortened by one glucose unit. *(b)* Summary of the two reactions of glycogenolysis, beginning with glycogen and ending with glucose 6-phosphate. *(c)* Summary of the two major fates of glycogen in muscle and liver. Glucose 6-phosphate in muscle is directed to glycolysis, whereas the phosphate group on liver glucose 6-phosphate is removed and the free glucose can be released to the blood.

in glycogen enhances its exposure to phosphorylase, increasing the rate at which it can be broken down. Glucose 6-P produced by glycogenolysis in muscle feeds into the glycolytic pathway. In liver, glycogenolysis mainly provides glucose for the blood. Therefore, glucose 6-P from the phosphoglucomutase reaction is dephosphorylated to free glucose using the enzyme *glucose 6-phosphatase*. The glucose can be released from liver to help maintain blood glucose levels.

Glycogen Storage

The major stores of glycogen are in liver and skeletal muscle. Table 5.4 provides approximate values for the amounts of glycogen in liver and skeletal muscle for men and women under three different dietary conditions. As shown, the amount of carbohydrate in the diet influences the amount of stored glycogen. The normal, mixed diet commonly eaten by North American people contains about 45% carbohydrate, but many people exist on a much higher proportion of carbohydrate in their diet. Glycogen is stored in both liver and muscle following a meal. However, after exercise, when muscle glycogen levels are reduced, glycogen is stored preferentially in the exercised muscles. This is in part attributable to an increase in insulin sensitivity in skeletal muscles with low glycogen content because of the exercise.

The concentration of glycogen in muscle is typically determined from muscle biopsies. The usual procedure is to hydrolyze the glycogen molecule with acid or an enzyme, amyloglucosidase, to produce glucose units. Then the glucose concentration is determined and the glycogen is expressed as millimoles of glucose or glucosyl units. As we discussed in chapter 3, the concentration of glucose or glucosyl units can be expressed in three ways. We could make our measurement on the basis of the wet weight of tissue as extracted from the biopsy needle. We could also take the biopsy sample

and freeze-dry it to remove the water, or we could express the glucose concentration in millimolar (mM) units using the relationship that the fraction of intracellular water in muscle is 0.7. An earlier technique precipitated the glycogen and then determined its mass by a reaction that generated a colored product. In this case, the glycogen content was expressed in grams of glycogen per 100 g of muscle tissue. Muscle glycogen concentration can also be measured noninvasively using nuclear magnetic resonance spectroscopy, based on the signal from the ^{13}C isotope naturally present in carbon compounds (Price et al. 2000).

In glycogen storage or **glycogenesis,** glucose units are added one at a time to existing glycogen molecules, creating long unbranched chains with α 1-4 glycosidic bonds. A *branching enzyme* then creates α 1-6 bonds that lead to the highly branched final structure (see figure 5.5). Glucose enters the cell and is phosphorylated to glucose 6-P by hexokinase (muscle) or glucokinase (liver). Figure 5.7 illustrates the three-step process of glycogen formation, starting with an existing glycogen particle and glucose 6-P, the basic precursor for increasing the size of a glycogen molecule.

Glucose 1-phosphate is made as the phosphate group on glucose 6-P is moved to the 1 position in a reaction catalyzed by *phosphoglucomutase*. To make glycogen, the glucose unit must be activated by a reaction between glucose 1-phosphate and uridine triphosphate (UTP), making UDP-glucose. This reaction is catalyzed by *UDP-glucose pyrophosphorylase*, and the product is the precursor form of glucose to be added to the glycogen primer. Finally, *glycogen synthase* adds glucosyl (glucose) units from UDP-glucose to the nonreducing end of the primer, releasing UDP and producing a glycogen molecule enlarged by an added glucose. Overall, glycogen synthesis is irreversible, due in part to subsequent hydrolysis of inorganic pyrophosphate (PPi) by *inorganic pyrophosphatase*. The glycogen synthase reaction is also irreversible.

Table 5.4   Approximate Glycogen Stores in Liver and Muscle for Adults Following Normal, High-Carbohydrate (CHO), and Low-CHO Diets

Storage site	Tissue weight (kg)	Total glycogen content (g)		
		Normal diet	High-CHO diet	Low-CHO diet
Man, 70 kg				
Liver	1.2	40-50	70-90	0-20
Muscle	32.0	350	600	250
Woman, 55 kg				
Liver	1.0	35-45	60-70	0-15
Muscle	22.0	242	410	170

Figure 5.7 Glycogen synthesis. Glucosyl units in the form of uridine diphosphate (UDP)-glucose are added to a preexisting glycogen primer molecule with n glucosyl units, creating a glycogen molecule with n + 1 glucosyl units. Precursor to make glycogen is typically glucose that is phosphorylated to glucose 6-phosphate.

Glycogen synthase produces long chains of glucose molecules that the branching enzyme transforms into the treelike structure of glycogen found in vivo. The UTP needed to make glycogen is generated by transfer of the terminal phosphate from ATP on to UDP in a reaction catalyzed by *nucleoside diphosphate kinase.*

The actual primer needed to make a glycogen particle is glycogenin. This is a self-glycosylating enzyme that transfers glucosyl units to itself from UDP-glucose until there are 7 to 11 glucosyl units. Glycogen synthase now adds glucosyl units from UDP-glucose to the glycogenin primer, and branching enzyme creates the branched structure. Two physiologically distinct forms of glycogen have been identified; *proglycogen* is much smaller than *macroglycogen,* with the difference due to the much larger number of glucosyl units in macroglycogen. Whether these are actually distinctly different kinds of glycosomes or whether they simply differ on the basis of the size of the actual glycogen molecule is unclear. The proglycogen seems to be preferentially used during muscle activity, while the macroglycogen may be more of a reserve supply of carbohydrate for prolonged exercise (Shearer and Graham 2004).

Storage of glycogen following training or competition is a key consideration for athletes because glycogen is such an important fuel for exercising muscle. Moreover, there are limited stores of glycogen in muscle. Research has provided us with some valuable information. For example, the synthesis of glycogen following exercise is most rapid in the first hour. Thereafter, the rate is elevated until glycogen stores at least reach pre-exercise levels. Enhanced insulin-stimulated glucose uptake and glycogen synthesis seem to be sensitive to glycogen content. Glycogen storage beyond pre-exercise levels or supercompensation can take place for several days after glycogen stores are severely depleted. The slower storage of glycogen beyond normal levels seems to be insulin independent. The proglycogen fraction is replenished first, but after several hours the macroglycogen fraction begins to increase (Battram et al. 2004). It was once believed that ingestion of caf-feine would impair glycogen storage after exercise, but Battram and colleagues (2004) showed that this had no effect on either pro- or macroglycogen synthesis. A controversial issue is whether carbohydrate alone or carbohydrate supplemented with protein is better in terms of postexercise glycogen replenishment. Ivy and colleagues (2002) showed that the inclusion of protein with carbohydrate significantly increased the synthesis of glycogen compared to carbohydrate alone, by 4 h postexercise. As we will see in chapter 9, ingestion of protein enhances the synthesis of skeletal muscle protein following a resistance training workout.

> **KEY POINT**
>
> The literature is clear that feeding carbohydrate (or better, carbohydrate plus protein) as soon as possible following training enhances both glycogen resynthesis and protein synthesis in the exercised muscles. The sources of carbohydrate can be simple sugars (the monosaccharides glucose and fructose, the disaccharide sucrose, or both), which enter the general circulation quickly, or starches, which require more digestion and thus make a slower and more prolonged appearance in the blood. Starch-rich foods are generally rich in other nutrients and contain small amounts of protein.

REGULATION OF GLYCOGEN METABOLISM

As mentioned earlier, glycogen is found as a particle or glycosome including the large glycogen molecule along with the enzymes glycogen phosphorylase, synthase, branching and debranching enzymes, enzymes that regulate glycogen metabolism, glycogenin, and other proteins (Shearer and Graham 2004). In muscle, most glycogen particles are found in close association with the sarcoplasmic reticulum, between myofibrils near mitochondria, and under the sarcolemma.

Since the enzymes to synthesize and degrade glycogen are found in the same particle, it would seem that glycogenesis and glycogenolysis could occur simultaneously. If this were the case, a futile cycle would be set up that accomplished nothing except the hydrolysis of energy-rich phosphates such as UTP. To avoid the futile cycling of glycogen, glycogen phosphorylase should be active when glycogen synthase is inactive or vice versa. In the liver, phosphorylase should be inactive following a meal, but glycogen synthase should be active to store the glucose obtained from food. Between meals, liver phosphorylase should be active to provide glucose for the blood, whereas glycogen synthase should be inactive. In rested muscle, synthase should be active and phosphorylase inactive following a meal; but if the muscle starts to work, the phosphorylase should be active and the synthase inactive. In fact, we might expect the activity of phosphorylase to be graded during exercise; activity should be greatest during very hard exercise when carbohydrate is the most needed fuel and much less when exercise intensity is low enough for oxidation of fat to maintain ATP levels in the exercising muscle. Control of glycogen phosphorylase can thus be tied to the whole strategy of using carbohydrate for oxidative phosphorylation when plenty is available, or when the muscle demand for ATP can be adequately satisfied only by carbohydrate degradation.

One enzyme can be active and the other simultaneously inhibited if the two respond in opposite directions to the same stimulus. The main regulation of these enzymes is covalent attachment of a phosphate group, or phosphorylation, which requires a protein kinase. To remove the phosphate groups requires a phosphoprotein phosphatase. The regulation of glycogen metabolism in liver and muscle represents another form of signal transduction, introduced in chapter 2. For the control of glycogen metabolism, external signaling molecules (e.g., the hormones epinephrine, insulin, and glucagon) bind to specific cell membrane receptors and activate protein kinases that phosphorylate specific proteins—glycogen phosphorylase (GP) and glycogen synthase (GS) in this case. Phosphorylation activates phosphorylase (GPa form), but inhibits glycogen synthase (GSb form). Removal of the phosphate groups (dephosphorylation) by a phosphoprotein phosphatase inactivates glycogen phosphorylase (GPb), but activates glycogen synthase (GSa). An additional level of complexity in the control of glycogen metabolism lies in the fact that liver and muscle glycogen phosphorylase and synthase, as well as the enzymes that control them, are sensitive to the action of a number of molecules or ions that signal the nutritional and metabolic state of the cell.

Figure 5.8 summarizes the control of GP and GS by the phosphorylation and dephosphorylation mechanisms. Inactive glycogen phosphorylase (GPb) is phosphorylated on a single serine residue to make the active GPa form. Removal of the single phosphate group is accomplished by phosphoprotein phosphatase 1 (PP1). Control of glycogen synthase by phosphorylation and dephosphorylation is more complex. First, there are at least 10 sites on GS where a phosphate group can be attached. Second, there are at least seven protein kinases that can phosphorylate GS in total, including glycogen phosphorylase kinase (GPKa); glycogen synthase kinase-3 (GSK3); two casein kinases, I and II (casein K I and II); protein kinase A (PKA); protein kinase C (PKC); and a

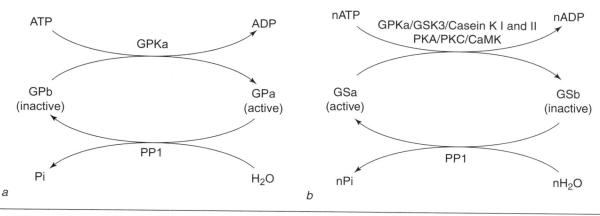

Figure 5.8 Regulation of glycogen phosphorylase and glycogen synthase by reversible phosphorylation. *(a)* The inactive form of glycogen phosphorylase (GPb) is phosphorylated by the active form of glycogen phosphorylase kinase *a* (GPKa) to make active GP (GPa). *(b)* Glycogen synthase is active when it is in the dephosphorylated state (GSa).

calmodulin kinase (CaMK). However, only a single phosphoprotein phosphatase (PP1) removes these phosphates. The ability to phosphorylate GS on multiple sites suggests that glycogen synthase activity can be graded more sensitively to cell needs.

Regulation of Glycogenolysis in Muscle

In our distant past, hundreds of thousands of years ago, we survived in a hostile world by being able to flee or fight. For both responses, rapid generation of ATP for muscles was essential. Glycolysis is indispensable for this purpose, and glycogen is the primary source of glucose 6-P to fuel the glycolytic pathway. Glycogenolysis is regulated at the level of glycogen phosphorylase. It can be controlled through alteration of the proportion of enzyme in the active (GPa) versus inactive (GPb) form. Two mechanisms accomplish this—one activated by the hormone epinephrine and the other based on changes in intracellular calcium concentration. Glycogenolysis can be regulated by substrate concentrations (glycogen and Pi), and it can be regulated by the concentrations of positive and negative allosteric effectors for both the GPa and GPb forms. Table 5.5 summarizes the regulation of glycogenolysis in muscle.

Role of Epinephrine

When the hormone epinephrine binds to its β-adrenergic receptor on a skeletal muscle fiber, it unleashes a cascade of activation events that results in the conversion of phosphorylase b to phosphorylase a. The overall scheme for regulation of glycogen phosphorylase by epinephrine in skeletal muscle is summarized in figure 5.9 and described in detail in the following.

- When one is aroused by a sudden or startling event or anxiousness before a competition, **epinephrine** is released to the blood from the adrenal medulla.
- Epinephrine binds to β-**adrenergic receptors** (βAR) on skeletal muscle. The higher the receptor concentration and the higher the epinephrine concentration in the blood, the more the binding.
- Epinephrine binding to the βAR results in a change in a specific membrane-attached **G protein** (G_s). The G refers to GTP (guanosine triphosphate) since the activated G protein binds GTP.
- The activated G protein interacts with a membrane protein, **adenylyl cyclase** (AC) (some call it adenylate cyclase), which increases its ability to change an ATP molecule into a **cyclic AMP (cAMP)** and inorganic pyrophosphate (PPi). Cyclic AMP has a single phosphate group that is joined to both the 5' and the 3' carbon of ribose. It is sometimes designated 3', 5'-AMP.
- Cyclic AMP activates *protein kinase A*, which then transfers a phosphate group from ATP to glycogen phosphorylase kinase b, making it glycogen phosphorylase kinase a (GPKa).
- Glycogen phosphorylase kinase a transfers a phosphate group to a serine residue in each of the two subunits of glycogen phosphorylase b, making it the active, glycogen phosphorylase a.
- Activated glycogen phosphorylase a now breaks down glycogen to make glucose 6-P for glycolysis.

The role of epinephrine to increase the activity of glycogen phosphorylase even before any activity has taken place is an example of a *feedforward* mecha-

Table 5.5 Mechanisms Regulating Glycogenolysis in Muscle

Mechanism of Regulation	Factor inducing	Source	Effect
Phosphorylation of GPb	Calcium	Sarcoplasmic reticulum	Increases glycogenolysis
	Epinephrine	Adrenal medulla	Increases glycogenolysis
Substrate concentration	Glycogen	Glycogenesis	Increases glycogenolysis
	Pi	Creatine phosphate	Increases glycogenolysis
Allosteric effectors	AMP	Adenylate kinase	Increases GPb activity, smaller effect on GPa activity
	ADP	ATPases	Increases GPa activity a little
	IMP	Adenylate deaminase	Increases GPb activity
	ATP	Energy systems	Inhibits GPb

GPa, glycogen phosphorylase *a*; GPb, glycogen phosphorylase *b*.

Figure 5.9 Conversion of inactive glycogen phosphorylase *b* (GPb) into the active (GPa) form by epinephrine (E). The beta-adrenergic receptor is βAR; G_s is an activating G protein intermediate; AC, the enzyme adenylyl cyclase, converts ATP into cyclic AMP (cAMP). PKA or protein kinase A converts inactive glycogen phosphorylase kinase b (GPKb) into the active GPKa form.

nism. In contrast to a feedback mechanism, which responds to a change in physiological condition, a feedforward mechanism prepares the person for what will follow.

Role of Calcium

When a muscle is activated by a nerve, the activation induces calcium ions to be released from sarcoplas-

mic reticulum of skeletal muscle fibers. The Ca^{2+} ions can separately activate glycogen phosphorylase kinase as shown in figure 5.10. Glycogen phosphorylase kinase has four different kinds of subunits: α, β, γ, and δ. The α and β subunits can each be phosphorylated by protein kinase A, so that the GPK is changed to the active GPKa form as discussed previously. The δ subunit is **calmodulin,** a small protein that can

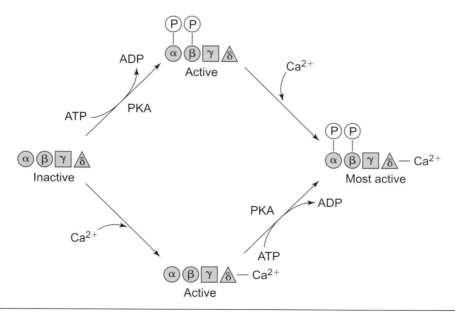

Figure 5.10 Glycogen phosphorylase kinase (GPK), which can be activated by the epinephrine cascade as well as by binding Ca^{2+}. GPK has four different kinds of subunits. The α and β subunits can be phosphorylated by protein kinase A (circled P). The δ subunit is calmodulin and can bind four Ca^{2+} ions. Activation of GPK is strongest when it is phosphorylated and also contains bound calcium ions.

bind up to four Ca^{2+} ions. When the δ subunit binds four Ca^{2+} it becomes active and can phosphorylate phosphorylase *b* to make it phosphorylase *a*. The highest level of activity for glycogen phosphorylase kinase occurs when it has both α and β subunits phosphorylated and the δ subunit with four calcium ions as shown in figure 5.10.

Reversing the Activation Steps

When the competition or fearful event is over and there is no need to maintain a high rate of glycogenolysis, the activation process is reversed quickly. We can summarize the various reversal steps as follows:

- Epinephrine concentration drops quickly after the event, so the activation of adenylyl cyclase ceases and cAMP no longer is formed.

- At the same time, there is an enzyme, cAMP phosphodiesterase, that changes cAMP to 5'-AMP. This means that protein kinase A is no longer activated.

- Phosphoprotein phosphatase 1, which is inhibited when cAMP concentration is elevated, becomes active and immediately removes the phosphate groups from both glycogen phosphorylase *a* and glycogen phosphorylase kinase *a*.

- The $[Ca^{2+}]$ in the muscle fiber quickly drops as Ca^{2+} ions are pumped back into the sarcoplasmic reticulum.

- Overall, glycogenolysis is rapidly reduced.

Regulation of Glycogenolysis by Substrate and Allosteric Effectors

Glycogen phosphorylase has two substrates, glycogen and Pi. We know that the K_m of glycogen phosphorylase for glycogen is approximately 1 mM or about 1% of the actual concentration of glycogen in rested muscle (Rush and Spriet 2001). This tells us that phosphorylase is saturated with glycogen even when the concentration of glycogen is low as it is at the end of prolonged exercise. However, the K_m of phosphorylase for its other substrate, Pi, is approximately two times the resting level of Pi in muscle (e.g., about 2 mM). This suggests that phosphorylase activity is very sensitive to changes in [Pi], responding progressively to an increase in Pi that parallels the intensity of exercise (see table 3.3 on p. 53).

Some interesting experimental observations point to even more complexity in the control of glycogen metabolism in humans. For example, epinephrine-deficient humans can exercise and break down glyco-

gen at rates similar to those seen in control subjects. If epinephrine-deficient patients are infused with epinephrine before exercise, no additional glycogen breakdown occurs for the same exercise task (Kjaer et al. 2000). Glycogen breakdown can occur in exercise even when the percentage of glycogen phosphorylase in the active form (GPa) is not higher than it is in rest (Kjaer et al. 2000). Such data suggest two possibilities. Phosphorylase *b* must play a role in glycogenolysis during exercise, or GPa must be capable of increasing its activity in response to changes in the concentrations of key metabolites in muscle (by a combination of lowered K_m for Pi and increase in reaction velocity), or both.

As evidenced by many exercise studies, the activation of glycogen phosphorylase to the *a* form (e.g., %GPa) occurs early in exercise, but then exercise can proceed with the proportion of phosphorylase in the active form at levels only slightly above those at rest. This has been a puzzling observation but is now easy to rationalize. Glycogen phosphorylase *b*, while basically inactive at rest, can be allosterically activated by AMP levels as would be seen in an active muscle (see table 3.3). This activation can be such that rapid glycogenolysis can take place even when the proportion of glycogen phosphorylase in the *a* form is low. In addition, phosphorylase *b* activity is increased allosterically by inosine monophosphate (IMP) and decreased by glucose 6-P and ATP. During moderate-intensity exercise, neither ATP nor IMP concentrations are likely to change sufficiently from rest values to allosterically alter phosphorylase *b* activity. However, for high-intensity exercise, an increase in IMP and decrease in ATP could act together to enhance rates of glycogenolysis solely by an increase in phosphorylase *b* activity. Glycogen phosphorylase *a* activity can also be increased up to threefold by concentrations of AMP and ADP found in active muscle (Rush and Spriet 2001).

The actual concentration of glycogen in muscle can influence the rate of glycogenolysis when all other circumstances are the same. For example, there can be differences of more than twofold in the use of glycogen for the identical exercise task in subjects if they start out with low versus high initial glycogen levels (Weltan et al. 1998; Arkinstall et al. 2004). As muscle glycogen content decreases during exercise, its use is proportionally reduced. This suggests that the rate of glycogenolysis is sensitive to the size of the glycogen molecule, perhaps because glycogen is more accessible to both the debranching enzyme and glycogen phosphorylase. Exercise training can reduce the reliance on glycogen as a fuel when the same exercise task is performed after, compared to before, training. After a seven-week training program, LeBlanc and colleagues (2004) found that glycogenolysis was decreased compared to pretraining levels, and this was not due to the extent of activation of phosphorylase or to a training-induced change in phosphorylase activity. They found lower concentrations of ADP, AMP, and Pi in muscle samples after training, pointing to changes in substrate and allosteric effectors that could account for the glycogen-sparing action with training. In summary, the regulation of glycogenolysis can be rapid and proportional based on activation mechanisms that overlap and reinforce each other to ensure that the provision of glucose 6-P meets the energy needs of the fiber. Exercise training results in reduced levels of the substrate Pi and allosteric effectors ADP and AMP so that the ATP needs are met less by carbohydrate and more by fat oxidation.

> ## ▶ K E Y   P O I N T ◀
>
> The body contains limited supplies of carbohydrate but plenty of fat. The use of glycogen as a fuel is therefore graded to the intensity of activity, responding to activating calcium levels and to metabolites whose concentrations increase during exercise (ADP, Pi, and AMP). On the other hand, if muscle glycogen content is high in muscle, mechanisms to regulate its breakdown appear to be less sensitive, with the result that more glycogen is used.

Since muscle glycogen is the principal source of substrate for glycolysis, regulation of glycogen phosphorylase is critical to appropriately meet the demands for ATP provision during exercise. This is especially important during exercise at altitude, where the oxygen content of the inspired air is reduced and glycolytic flux is therefore increased. Exercise training has a glycogen-sparing effect, with the result that less carbohydrate and more fat will be oxidized after training.

Regulation of Glycogenolysis in Liver

The liver stores glucose for the blood by converting it into liver glycogen. Between meals, when there is no source of absorbed glucose from the gut to maintain blood glucose concentrations, liver breaks down glycogen at a rate needed to maintain euglycemia. Regulation of glycogenolysis in liver is similar to that in muscle in that there is an active (phosphorylase a) and an inactive (phosphorylase b) form of liver phosphorylase. There are notable differences, however. Liver phosphorylase is controlled more by phosphorylation and dephosphorylation and much less by allosteric mechanisms. **Glucagon** is the primary hormone that stimulates the rise in liver cAMP concentration and conversion of liver phosphorylase b to liver phosphorylase a. Glucagon is produced in and released from the alpha cells of the pancreas in response to a decrease in blood glucose. Therefore, a rise in glucagon in the blood signals a need to break down liver glycogen to glucose and release this to the blood. Glucagon binds to specific receptors on liver cells, and this binding results in a cascade of events including a rise in cAMP concentration. Following a meal, when blood glucose and insulin concentrations are elevated, insulin activates protein phosphatase-1, increasing the conversion of liver phosphorylase a to liver phosphorylase b, while glucose binds to and inactivates any existing liver phosphorylase a.

Regulation of Glycogenesis

To store glycogen in liver and muscle, there must be an increase in glucose uptake into muscle and liver cells and glucose phosphorylation to glucose 6-P. In muscle, glucose transport plays a very important role. In liver, transport is less important than the activity of glucokinase. Like glycogen phosphorylase, glycogen synthase exists in two forms. Glycogen synthase a (GSa) is the unphosphorylated form that is normally active. Phosphorylation converts GSa to GSb, which is inactive. Glycogen synthase a has also been referred to as GSI, where the I stands for independent; that is, GSI is independent of the glucose 6-P concentration. Glycogen synthase b has also been referred to as glycogen synthase D (GSD); it is normally inactive, but can become activated by an allosteric mechanism if the glucose 6-P concentration increases.

We saw in figure 5.8 that phosphorylation and thus inactivation of GSa can be performed by a number of protein kinases. Dephosphorylation, which activates glycogen synthase, is catalyzed by a single enzyme, PP1. Figure 5.11 summarizes the regulation of glycogen synthase, focusing on just the key regulators and enzymes that are responsible for controlling the synthesis of glycogen. Key signaling substances (cAMP and Ca^{2+}) that activate the breakdown of glycogen are responsible for enhancing phosphorylation of GSa to make it inactive in glycogen synthesis. Cyclic AMP concentration is increased by epinephrine in muscle and by glucagon in liver. Removal of the phosphate groups using PP1 is stimulated by insulin and opposed by cAMP. This is just as we should expect, because we don't want the synthesis and breakdown to be simultaneously active. Following a meal, when blood glucose and insulin are increased, we would expect glycogen synthesis to be favored, and this is how it is. The role of glucose 6-P to allosterically activate the GSb (GSD) form, overriding the inhibition by phosphorylation, is entirely appropriate. Let's see how. Suppose you are doing intermittent or interval training. You want the phosphorylase to be active during the activity and the glycogen synthase to be inactive. However, when you stop your exercise bout and are recovering, glucose 6-P concentration should still be elevated. During this rest period you don't need glycogenolysis and glycolysis, but it is a good time to allow some of the glucose 6-P to be converted to glycogen, and this is what happens.

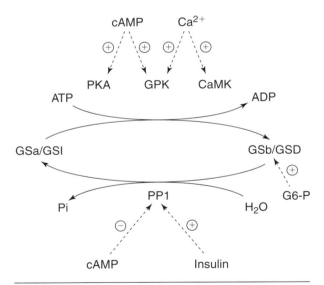

Figure 5.11 The major regulators of glycogen synthase. Conversion of the active, glucose 6-phosphate (G6-P)-independent form (GSa or GSI) to the inactive, glucose 6-phosphate-dependent (GSb or GSD) form takes place by phosphorylation of serine residues. Three key enzymes that phosphorylate GSa/GSI are protein kinase A (PKA), glycogen phosphorylase kinase (GPK), and calmodulin kinase (CaMK). These enzymes are activated by cAMP whose formation is stimulated by epinephrine (muscle) or glucagon (liver). Dephosphorylation of GSb/GSI is accomplished by phosphoprotein phosphatase 1 (PP1), whose activity is enhanced by insulin and inhibited by cAMP. An increase in G6-P can activate GSb/GSD. Dotted lines and a positive or negative sign indicate activation or inhibition, respectively.

The regulation of glycogenesis is very complex, and for simplicity's sake, the following summarizes the key control points.

- Provision of precursor in the form of glucose is critical to rapidly inducing glycogen resynthesis. An increase in blood glucose induces an increase in blood insulin concentration. Insulin enhances glucose transport and is known to increase PP1 activity, thereby enhancing the conversion of GSb to GSa.

- Muscle glycogen synthesis is most rapid immediately following exercise (Price et al. 2000). The proportion of glycogen synthase in the active form (GSa) is higher at low glycogen levels.

- Researchers have identified an early "insulin-independent" phase of glycogen synthesis, followed by an insulin-dependent phase. The former occurs in the immediate postexercise period, whereas the latter phase is associated with postactivity feeding.

- The more glycogen is depleted following exercise, the more rapid is the resynthesis rate, and the more this rate is sensitive to insulin and glucose. It is suggested that the smaller glycogen particle in the exercised muscle enhances GS and PP1 activity.

- Glycogen synthesis is acutely sensitive to glycogen concentration, but in a reciprocal way. This

KEY POINT

Following glycogen-depleting exercise, when resynthesis of glycogen is critical for subsequent activity, early and frequent carbohydrate feeding is essential. While this can ensure that glycogen replenishment starts early, when the rate of resynthesis is highest, there is a limit to the amount of glycogen that can be stored in muscle, despite the continued presence of high blood glucose and insulin. When some threshold amount of glycogen is reached, glycogen synthase activity is inhibited. Although elevated glucose transport into muscle still occurs, the glucose 6-P is channeled into the glycolytic pathway such that lactate is produced at an accelerated rate even in a rested muscle. Similarly, pyruvate oxidation is increased at the expense of fat oxidation.

suggests that glycogen exerts a feedback inhibition on glycogen synthase such that there is an upper limit for glycogen storage in muscle.

- After glycogen-depleting exercise, trained individuals can resynthesize glycogen faster than the untrained (Hickner et al. 1997).

LACTATE METABOLISM

There is probably more misunderstanding of lactate and lactic acid than any other area in exercise metabolism. For example, statements such as the following are routinely made by those who really do not understand the metabolism: Lactic acid is produced by anaerobic glycolysis; the lactic acid, being an acid, dissociates and generates a proton, thus acidifying muscle and blood and causing fatigue. It is also believed by many that lactate/lactic acid as a product of anaerobic glycolysis must be indicative of an oxygen insufficiency at the cellular level. Using a treadmill or cycle ergometer, equipment to measure oxygen and carbon dioxide levels and to provide $\dot{V}O_2$ and $\dot{V}CO_2$ data, and a blood lactate analyzer, it is easy to generate data as shown in figure 5.12. The data in the figure show $\dot{V}O_2$ and blood lactate concentration measures for a single subject during cycle ergometer exercise at progressively increasing intensity. Although $\dot{V}O_2$ increases linearly with exercise intensity, blood lactate concentration breaks from an early linear increase to become an exponential

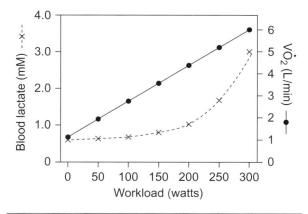

Figure 5.12 Blood lactate concentration and $\dot{V}O_2$. Values are for a single subject cycling against progressively greater resistance starting from loadless cycling (0 W). The resistance was increased every 2 min in 50 W increments, and measurements were made every 50 W. Physiologists define a "lactate threshold" or break point where lactate concentration changes from a linear increase with workload to an exponential increase. In this figure, blood lactate data are insufficient to define a specific break point.

increase. The break point is often described as the "lactate threshold" (Robergs and Roberts 1997). It is important to make frequent measures of lactate and to keep the step increases in intensity small to clearly identify the break point. One of the misleading assertions is that the lactate threshold represents the point where the muscle ATP yield from oxidative phosphorylation is compromised by insufficient oxygen. To maintain work output requires the addition of anaerobic glycolysis as a source of ATP. Thus the lactate threshold has been defined as an "anaerobic threshold" (Wasserman et al. 1973). This does not agree with what we now know.

Lactate Formation, Accumulation, and Disappearance

The summary that follows outlines what research has told us about lactate metabolism. Lactic acid is a modestly strong acid with a pK_a of 3.8. Thus at physiological pH of ~7, it exists almost exclusively as the anion (La^-) and not as the undissociated acid (HLa). Moreover, as we discussed previously, the formation of lactate from either stored glycogen or glucose taken up from the blood is not associated with net proton formation (Robergs et al. 2004). It has been assumed for years that lactate formation causes muscle fatigue. Part of this assumption is that lactate is elevated in fatigued muscles, but this does not prove a cause. Recently, it has been reported that lactate and acidification may actually defend muscle from the fatiguing effects of a loss of intracellular potassium ion from very active muscle fibers (Nielsen et al. 2001). A review by Westerblad and colleagues (2002) concludes that it is the increase in Pi, caused by a precipitous decline in PCr, that is the "major cause" of intracellular fatigue. Muscle fatigue is a very complex phenomenon, attributable to a variety of sites and factors, often depending on the nature of the muscle activity that caused it. An accumulation of intracellular lactate may possibly defend the muscle fiber against a force reduction. Nonetheless, lactate is transported from the fiber accompanied by a proton (see "Lactate Transport and Lactate Shuttles" on page 123). Perhaps it is extracellular acidification that is responsible for the pain associated with intense muscle activity (Allen and Westerblad 2004). If so, lactate does not cause the proton formation, but does provide the mechanism through its cotransport with H^+ to decrease extracellular pH.

Lactate is generated by the reduction of pyruvate in the cytosol of cells. Addition of electrons on NADH to pyruvate produces lactate. We describe this as anaerobic glycolysis. Pyruvate can be produced in glycolysis at a rate far in excess of the ability of pyruvate to be transported down its concentration gradient into mitochondria and there to be oxidized by pyruvate dehydrogenase. The activity of the enzyme lactate dehydrogenase is generally high in skeletal muscle. In addition, on purely energetic grounds, reduction of pyruvate by NADH to make lactate is the favored direction. Therefore, whenever glycolysis produces pyruvate, lactate should also be produced. Pyruvate can also accept an amino group from the amino acid glutamic acid in a **transamination** reaction. The pyruvate is then changed to the amino acid alanine. Exercising muscle releases alanine to the blood, a topic we consider in chapter 7.

Glycogenolysis produces glucose 6-P for the glycolytic pathway. Agents representing the principal mechanisms activating glycogenolysis (epinephrine, calcium, Pi, AMP) are graded to the intensity of contractile activity. Moreover, AMP, ADP, Pi, and fructose 2,6-bisphosphate, the key regulators of phosphofructokinase, are also increased in muscle in proportion to the relative intensity of contractile activity. Therefore we would expect an exercise-induced increase in glucose 6-P formation from glycogenolysis and a matching increase in the conversion of this glucose 6-P to pyruvate in the glycolytic pathway. Indeed, the activation of both glycogenolysis and glycolysis increases exponentially with a linear increase in exercise intensity. We also discussed the fact that glycogenolysis and therefore glycolysis are more active if glycogen stores are elevated as opposed to low. Finally, as exercise becomes more intense, more muscle fibers with a higher glycogen phosphorylase activity and glycolytic potential, and lower oxidative capacity become involved. Therefore, we should expect glycolysis to become increasingly important early in exercise and as exercise increases in intensity. It is not necessary for the relative intensity of exercise to be even moderate for lactate to be produced. Exercise at only 40% of maximum aerobic capacity can increase the rate of glycolysis by more than 20-fold compared to rest (Katz and Sahlin 1990).

Lactate generated by anaerobic glycolysis can accumulate within the fiber, but it can also leave the fiber, passing through a monocarboxylate transporter down a concentration gradient to the extracellular fluid. From here it could enter an adjacent fiber with a lower intracellular lactate concentration or enter a nearby capillary and be transported, dissolved in the blood. Lactate is also produced by red blood cells,

so blood lactate reflects erythrocyte metabolism and lactate generated through glycolysis, principally by muscle. As we have discussed, lactate is a fuel with the capability of generating six pairs of electrons for the electron transport chain and one GTP if the lactate is oxidized to pyruvate, the pyruvate oxidized to acetyl CoA, and the acetyl group is metabolized by the citric acid cycle. Complete oxidation of one lactate can generate approximately 15 ATP. As the lactate circulates in the blood throughout the body, lactate can diffuse down a concentration gradient, enter cells, and be used by the heart, liver, kidney, and nonworking skeletal muscle. As we will discuss in the later section on lactate shuttles, blood lactate levels during exercise reflect the balance between lactate-producing cells (predominantly active skeletal muscle fibers) and those tissues and cells that consume lactate as a fuel (heart and nonworking muscle) or convert it to glucose (principally liver and kidney). Any imbalance in the production or removal of lactate from the blood is reflected by an increase or decrease in its concentration.

Within a single muscle fiber, the fate of pyruvate produced by glycogenolysis and glycolysis is primarily entry into the mitochondrion for oxidation or reduction to lactate. Which of these two processes predominates depends for the most part on the following:

1. The rate of pyruvate formation or the glycolytic flux rate. This may depend both on the intensity of the activity and on the availability of glycogen, as we have discussed.

2. The cytosolic redox state, which is the concentration ratio of [NADH] to [NAD^+]. NADH and NAD^+ participate in oxidation–reduction reactions as substrates; thus major relative changes in their concentrations influence the direction of reversible redox reactions. An increase in the concentration ratio of NADH to NAD^+ would favor lactate formation. In the cytosol, the concentration ratio of NADH to NAD^+ is less than 0.01, as opposed to the situation in mitochondria where this ratio is closer to 0.5 (Sahlin et al. 2002). Thus the cytosol is highly oxidized, which helps to drive the reaction catalyzed by glyceraldehyde phosphate dehydrogenase and so to generate pyruvate. The more highly reduced mitochondrion favors electron transfer from NADH to oxygen.

3. The number and size of mitochondria, which reflect the potential for oxidative phosphorylation. More mitochondria would favor pyruvate entry and oxidation.

4. The availability of oxygen, which is the final acceptor of electrons in oxidative phosphorylation. Even though the cell can adjust to relatively low oxygen concentrations by increasing the phosphorylation and redox potential, if oxygen concentration decreases below a threshold level it can limit oxidative phosphorylation, forcing the cell to rely on anaerobic glycolysis. At present, it is unclear what the actual concentration of oxygen at the cytochrome c oxidase complex must be to prevent oxygen-limited ATP formation.

5. The total activity of lactate dehydrogenase (LDH), since this will be competing for pyruvate formed in glycolysis.

In chapter 3, we discussed the three common fiber types in human muscle based on their myosin composition (table 3.1 on p. 39). If we compare slow-twitch (ST) and two types of fast-twitch (FTA and FTX) skeletal muscle fibers with heart (cardiac) muscle fibers, we learn some important metabolic lessons. First, their typical activities differ greatly. Heart muscle is continuously active because the heart is always working. Slow-twitch (ST or type I) muscle fibers are involved in low-intensity activity and are also active during strong contractions. For the same cross-sectional area, these fibers generate nearly the same tension as fast-twitch muscle fibers, but they cannot shorten as rapidly. Fast-twitch (FT or type II) fibers are usually active during intense activity but are not normally active during low-intensity activity. They can shorten faster than slow-twitch muscle fibers primarily because they have fast myosin heavy chains.

Table 5.6 summarizes important differences between the heart muscle, ST, FTA, and FTX skeletal muscle fibers. Both FT fibers can hydrolyze ATP at a maximum rate that exceeds that of ST fibers and heart muscle. Thus, the regeneration of ATP must also be faster in FT fibers. However, ATP regeneration in FT fibers involves more glycolysis and less fuel oxidation than in the ST and heart muscle fibers because the FT fibers have a greater capacity for glycolysis and a poorer blood supply—thus less oxygen and a lower capacity for oxidative phosphorylation. As table 5.6 summarizes, there are also subtle differences in the metabolic capacities of the FT fibers, with FTA fibers having a higher oxidative capacity but a lower glycolytic capacity. In summary, either type of FT fiber would produce more lactate and consume less oxygen when regenerating ATP compared to an ST fiber or heart muscle.

Table 5.6 Relative Comparisons for HM, ST, and Two FT Skeletal Muscle Fibers

Metabolic factor	Fiber type comparison
Maximum rate of ATP hydrolysis	FTX > FTA ST > HM
Maximum glycolytic flux rate	FTX > FTA ST > HM
Blood supply or availability of oxygen	HM > ST > FTA > FTX
Fiber size	FTX > FTA > ST > HM
Maximum oxidative capacity	HM > ST > FTA > FTX

HM, heart muscle; ST, slow-twitch; FT, fast-twitch.

Lactate Transport and Lactate Shuttles

During exercise with the blood flow occluded, as in modestly strong isometric contractions, glycolysis is the major source of ATP. As we have seen, glycolysis is also significant during exercise at intensities beyond $\dot{V}O_2$max. For these two exercise conditions, lactate concentration will increase in the muscle fiber, and lactate will appear in the blood, well beyond the normal 1 mM level. Lactate can diffuse across cell membranes as the undissociated acid, lactic acid, but this is of very minor importance since there is little undissociated lactic acid at a pH of 7. Therefore, it is the lactate ion that must cross the cell membrane down its concentration gradient. Because it is a charged ion, lactate needs a lactate transporter. However, this lactate transporter has been found to transport other negatively charged ions such as pyruvate and therefore has been named **monocarboxylate transporter** (MCT). Recently, isoforms of the MCT have been observed and identified with numbers, much as in the numbering system for the glucose transporters. MCT-1 and MCT-4 are in skeletal muscle, with the former more prominent in slow oxidative fibers and the latter more common in fast-twitch muscle fibers (Coles et al. 2004). The MCT acts as a symport, transferring lactate down its gradient, accompanied by a proton (H^+).

Lactate transport across the muscle sarcolemma can occur in both directions; the favored direction depends only on the lactate and proton gradient. Thus we might expect that during exercise at an intensity near $\dot{V}O_2$max, an active FT fiber may be producing significant lactate and H^+, which are transported together through MCT-4 and MCT-1 into the extracellular fluid surrounding the fiber. As shown in figure 5.13, a neighboring ST fiber with a low lactate concentration may take up the lactate and a proton using an MCT isoform because the lactate concentration outside this fiber is higher than inside. In the ST fiber, lactate can be oxidized to pyruvate, which can enter mitochondria where pyruvate is converted to acetyl CoA for entry into the citric acid cycle. This is an example of intramuscular shuttling of lactate from a net-producing to a net-consuming fiber.

There is little doubt that lactate can be shuttled from one fiber to another in the same muscle, as in an *intercellular lactate* shuttle. Lactate can also be shuttled between an active muscle that produces it and other tissues such as heart or inactive muscle that can consume the lactate as a fuel. As we will see, lactate is an important source for glucose formation in the liver, a process known as gluconeogenesis. Brooks and colleagues (1999) made a compelling case for an *intracellular lactate shuttle*. In their hypothesis, lactate produced in the cytosol of a muscle fiber could be transported into a mitochondrion of the same fiber, oxidized to pyruvate, and then converted to acetyl CoA for the citric acid cycle. Such a process would need to have a mitochondrial monocarboxylate transporter (e.g., MCT-1) that would take it into the mitochondrion; but more importantly, there would need to be a mitochondrial lactate dehydrogenase that would oxidize the lactate to pyruvate. Two studies cast serious doubt

on the intracellular lactate shuttle. Rasmussen and colleagues (2002) were unable to find evidence of lactate dehydrogenase in mitochondria, nor were they able to demonstrate that lactate could be used as fuel for isolated mitochondria as pyruvate can. Sahlin and colleagues (2002) too were unable to show that lactate could support mitochondrial respiration, and they also argued against mitochondrial lactate oxidation on theoretical grounds. They argued that pyruvate reduction to lactate is energetically favored in cytosol even when the NADH concentration is low (high NAD^+ concentration). Why then would mitochondria oxidize lactate with a far lower substrate concentration (NAD^+) than exists in the cytosol?

Interaction of Exercise Intensity, Diet, and Training on Lactate Metabolism

In rested muscle, the activity of the glycolytic pathway is quite low, as evidenced by the observation that the RQ (respiratory quotient) of rested muscle is close to 0.75, indicating the preponderance of fat as the fuel for oxidative phosphorylation. During moderate-intensity dynamic exercise, the rate of glycolysis may increase by 30- to 40-fold or more above the resting level in some fibers, but less in others. Differences will reflect the metabolic profile

Figure 5.13 A lactate shuttle between two neighboring fibers in the same active muscle. The relative thickness of each arrow reflects the importance of the major pathways or membrane transport process to ATP formation. The low oxidative FTX fiber has active glycogenolysis and glycolysis, producing pyruvate at a high rate. Most of the pyruvate is reduced to lactate, which can leave the fiber and be taken up by an adjacent high oxidative ST fiber. In the ST fiber, lactate is oxidized to pyruvate and used as a source of acetyl CoA (AcCoA). Oxidation of fatty acids (FAs) occurs mainly in the ST fiber.

of the fibers and the extent to which the fiber is active during the exercise. During moderate exercise intensity, lactate is generated in the active muscle; but the concentration of lactate within the muscle and blood leaving it may not increase significantly, as much of the pyruvate is oxidized in mitochondria, or the lactate formed is consumed in oxidative fibers within the same muscle, or both of these processes occur. As exercise intensity increases linearly beyond 50% to 60% of $\dot{V}O_2$max, the rate of glycolysis, expressed as the rate of pyruvate formation, increases at an even faster rate. Although exercise may still be described as submaximal, that is, at an intensity less than $\dot{V}O_2$max, pyruvate can be formed at such a rate that a significant fraction will be reduced to lactate. As we have discussed, this does not mean that there is a lack of oxygen to act as the electron acceptor, because measures of cytosolic PO_2 reveal that sufficient O_2 is present to meet the needs of the electron transport chain at exercise intensities approaching $\dot{V}O_2$max. During supramaximal exercise, glycogen utilization and lactate production can increase exponentially. In well-motivated subjects, muscle lactate and blood lactate concentrations can exceed 30 and 20 mM, respectively.

We noted in a previous section that the rate of glycogenolysis during submaximal exercise depends on the concentration of glycogen. Since the glucose 6-P produced during glycogen breakdown feeds into the glycolytic pathway, we might expect the amount of lactate formed during submaximal exercise also to be directly related to muscle glycogen concentration. Indeed, a number of experiments have demonstrated that the same person doing exactly the same submaximal exercise task can have a muscle lactate concentration significantly higher if starting muscle glycogen stores are high, compared to low. Figure 5.14 illustrates the relationship between starting glycogen concentration, the rate of glycogen utilization, and muscle lactate concentration during submaximal (70% of $\dot{V}O_2$peak) exercise on a cycle ergometer. The relationships shown in figure 5.14 are configured from studies that have demonstrated that muscle glycogen utilization and lactate production can vary significantly if subjects carry out the same exercise task with high and low starting muscle glycogen concentrations (Spencer et al. 1992; Baldwin et al. 2003; Arkinstall et al. 2004). Compared to values from the same exercise performed with elevated glycogen, overall carbohydrate oxidation is reduced and fat utilization is increased when pre-exercise glycogen stores are low (Spencer et al. 1992). We will discuss

Figure 5.14 The effect of muscle glycogen concentration in vastus lateralis muscle on glycogen utilization (circles) and lactate production (triangles) during cycle ergometer exercise at 70% of peak $\dot{V}O_2$. Filled symbols refer to the exercise condition with elevated pre-exercise glycogen concentration, and open symbols refer to the same exercise condition for the same subjects when pre-exercise glycogen concentration was reduced.

more of the interaction between fat and carbohydrate as fuels in the next chapter.

Endurance training can have a profound effect on fuel utilization when the same absolute exercise task is performed after, compared to before, training. Considering that the same absolute exercise intensity means that the rate of energy expended would be the same, there are still profound changes in the fuels utilized. At the same absolute intensity, there would be smaller reductions in PCr and glycogen, along with smaller increases in ADP, AMP, ammonia, Pi, Cr, and lactate. Given that ADP, AMP, Pi, and ammonia all promote glycogenolysis and glycolysis, the use of carbohydrate would be reduced. Training increases the content of citric acid cycle and electron transport chain proteins, as well as the enzymes involved in fat metabolism. This means that relative changes in ratios of NADH to NAD^+, ATP to ADP, and acetyl CoA to CoA would be less after training and that the stimulus for pyruvate dehydrogenase activation would be reduced (LeBlanc et al. 2004). In all, the adaptations with training would act to reduce carbohydrate utilization while enhancing fat oxidation. As few as five days of training can attenuate the decreases in PCr and glycogen and the increase in lactate during submaximal exercise (Phillips et al. 1996). These data demonstrate that the intracellular milieu can be altered in such a way as to spare carbohydrate utilization even when there is no evidence of a change in muscle oxidative capacity or whole-body $\dot{V}O_2$max.

OXIDATION OF CYTOPLASMIC NADH

During the formation of pyruvate in glycolysis, NAD^+ is reduced by the enzyme glyceraldehyde phosphate dehydrogenase, forming NADH. If the fate of pyruvate is anaerobic glycolysis, NADH is oxidized by lactate dehydrogenase, regenerating NAD^+. On the other hand, when pyruvate enters mitochondria and is oxidized there, or if pyruvate is converted to the amino acid alanine (a minor but still significant fate in muscle), no immediate regeneration of NAD^+ occurs. We have already mentioned that the cytosol is much more highly oxidized (i.e., $[NAD^+]$ is much greater than [NADH]) compared to mitochondria and that this helps to drive glycolysis. This gives rise to the question: How is cytoplasmic NAD^+ regenerated if not by lactate dehydrogenase?

The simplest solution to this metabolic problem would be for cytoplasmic NADH to enter the mitochondrial matrix and be oxidized in the electron transport chain. However, the inner mitochondrial membrane is impermeable to NADH; this is critical to maintaining a highly oxidized cytosol and more reduced mitochondrial matrix in order to facilitate the specific metabolism for each compartment. To get around this roadblock, there are two shuttle systems that transfer electrons (reducing equivalents) on cytoplasmic NADH into the mitochondrion without actual inner membrane crossing by NADH.

Glycerol Phosphate Shuttle

The glycerol phosphate shuttle (see figure 5.15) transfers electrons on cytosolic NADH to FAD, then to ubiquinone (coenzyme Q) in the mitochondrial inner membrane. The cytosolic NADH forms during the glyceraldehyde phosphate dehydrogenase reaction of glycolysis (reaction 1 in figure 5.15). This and subsequent reactions in glycolysis are not affected by this shuttle system. In reaction 2, the cytosolic NADH, not reoxidized back to NAD^+ by the lactate dehydrogenase reaction, transfers a hydride ion and proton to dihydroxyacetone phosphate, changing it to glycerol 3-phosphate (glycerol 3-P) and oxidizing NADH to NAD^+. The enzyme catalyzing this reaction (reaction 2 in figure 5.15) is *cytosolic glycerol phosphate dehydrogenase*. Glycerol 3-P diffuses to the outer side of the inner mitochondrial membrane and is oxidized by *mitochondrial glycerol phosphate dehydrogenase* (reaction 3). This enzyme is located in the inner membrane but faces the intermembrane space, so glycerol 3-P need not penetrate the inner membrane. Instead of using the NAD^+ coenzyme, the mitochondrial form of glycerol phosphate dehydrogenase uses FAD. The FAD binds tightly to the mitochondrial form of the glycerol phosphate dehydrogenase and is another flavoprotein dehydrogenase similar to succinate dehydrogenase (complex II) in the electron transport chain. In reaction 3, the products are $FADH_2$ and dihydroxyacetone phosphate. The electrons on $FADH_2$ transfer to ubiquinone (coenzyme Q) in the

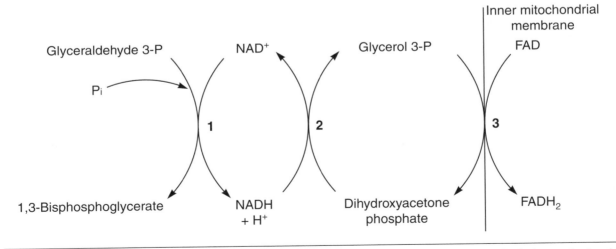

Figure 5.15 The glycerol phosphate shuttle. This shuttle transfers electrons from cytosolic NADH to FAD in the inner mitochondrial membrane, generating $FADH_2$. The NADH generated in reaction 1, catalyzed by the glycolytic enzyme glyceraldehyde 3-phosphate dehydrogenase, is used to reduce dihydroxyacetone phosphate in reaction 2, catalyzed by a cytoplasmic glycerol phosphate dehydrogenase. The resulting glycerol 3-phosphate is oxidized by a FAD-dependent mitochondrial glycerol phosphate dehydrogenase located in the inner mitochondrial membrane (reaction 3). Subsequently, electrons on $FADH_2$ are transferred to coenzyme Q (ubiquinone). This shuttle is irreversible because of reaction 3.

electron transport chain and then to oxygen. The dihydroxyacetone phosphate is then able to accept electrons from cytoplasmic NADH.

The glycerol phosphate shuttle exhibits no net consumption or production of dihydroxyacetone phosphate or glycerol 3-P. It simply transfers electrons from cytosolic NADH to the electron transport chain. This shuttle generates approximately 1.5 ATP per pair of electrons transferred from cytosolic NADH to oxygen and is irreversible because reaction 3 goes in only one direction. Although not the more prevalent of the two major shuttle systems, this system is easier to explain. The glycerol phosphate shuttle operates to a minor extent in a variety of tissues such as brain but is very important in fast-twitch (type II) skeletal muscle fibers.

Malate-Aspartate Shuttle

The malate-aspartate shuttle is more complicated and is reversible (see figure 5.16). It is the dominant shuttle for the liver, the heart, and slow-twitch (type I) muscle fibers. This shuttle system transfers electrons (reducing equivalents) on NADH in the cytosol to NAD$^+$ in the mitochondria (as opposed to electron transfer to FAD in the glycerol phosphate shuttle). As a result, for each two cytosolic electrons on NADH

transferred to oxygen in the mitochondria, we get approximately 2.5 ATP.

Cytosolic NADH is converted to NAD$^+$ through reduction of oxaloacetate to malate. This allows the NAD$^+$ to be used in the glyceraldehyde phosphate dehydrogenase reaction in glycolysis. The malate is transported into the matrix using an antiport mechanism with α-ketoglutarate. In the matrix, the malate now transfers its electrons to NAD$^+$, generating NADH and oxaloacetate. In this way, cytosolic reducing equivalents become matrix electrons on NADH. The complication of the malate-aspartate shuttle is that there is no mechanism to directly transport oxaloacetate from the matrix to the cytosol across the inner membrane. Instead, oxaloacetate is converted into the amino acid aspartic acid (aspartate) as the amino group on glutamic acid (glutamate) is moved to oxaloacetate. Amino group transfers are catalyzed by a class of enzymes known as **aminotransferases** (also known as transaminases). The resulting aspartate is transported across the inner membrane by an *aspartate-glutamate transporter* (antiport) that simultaneously transports glutamate from the cytosol to the matrix. As in the glycerol phosphate shuttle, there is no net production or consumption of any of the intermediates in this shuttle. Direction is provided to this potentially

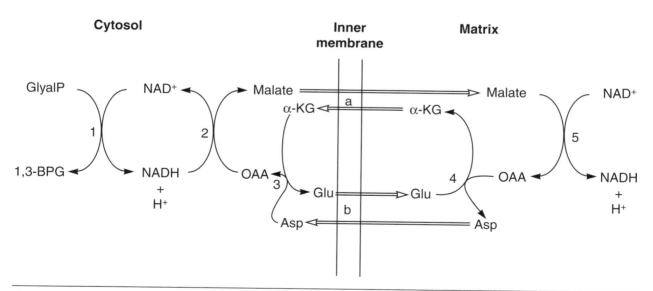

Figure 5.16 The malate-aspartate shuttle. This shuttle transfers electrons from cytoplasmic NADH to mitochondrial NADH. Reduction of NAD$^+$ in the cytosol by electrons on glyceraldehyde phosphate (GlyalP) is catalyzed by glyceraldehyde phosphate dehydrogenase (1). The resulting NADH is used to reduce oxaloacetate (OAA), producing malate and NAD$^+$ catalyzed by cytosolic malate dehydrogenase (2). Malate is transported into the matrix using an antiport (a) that simultaneously transports α-ketoglutarate (α-KG) from the matrix to the cytosol. The α-KG is converted into glutamate (Glu) by accepting an amino group from aspartate (Asp); the latter is converted into oxaloacetate using cytosolic aspartate aminotransferase (3). Glutamate is transferred from the cytosol to the matrix using the antiport (b) that simultaneously transfers Asp to the cytosol. In the matrix, NADH is produced by the oxidation of malate to OAA using mitochondrial malate dehydrogenase (5). The oxaloacetate is converted to aspartate using mitochondrial aspartate aminotransferase.

reversible shuttle by the formation of NADH in the cytosol and its oxidation in the electron transport chain of the mitochondria.

GLUCONEOGENESIS

As mentioned previously, glucose is the principal fuel for the brain and also erythrocytes (red blood cells) and some other tissues. In the brain, the glucose is almost exclusively oxidized. In red blood cells, anaerobic glycolysis is the fate of the glucose, since these cells lack mitochondria. The amount of glucose in the blood is very small, and it is used constantly by a variety of tissues. To maintain blood glucose levels, liver glycogen is broken down through glycogenolysis, and the glucose is released to the blood. At rest, in the postabsorptive state, glucose release from the liver is about 10 to 14 μmol per kilogram per minute (about 125-175 mg per minute for a 70 kg [154 lb] person). For a person with 70 g of liver glycogen, the liver would be exhausted of glycogen in about 8 h if no food was eaten or if there were no other way to produce glucose for the blood. This R_a (rate of appearance) of glucose can increase up to 10-fold during hard exercise (Angus et al. 2002). However, there are limited supplies of glycogen in liver to maintain blood glucose. For example, an overnight fast can markedly reduce liver glycogen, while a 24 h fast can deplete liver glycogen completely. Consider what can happen with exercise and no food intake. Without exogenous glucose to help maintain blood glucose, and in the face of an increased uptake by exercising muscle, blood concentration could fall to hypoglycemic levels during prolonged exercise, seriously impairing performance. Obviously, there must be another mechanism to make glucose for the body when glycogenolysis of liver glycogen becomes limited.

As the name suggests, gluconeogenesis is the formation of new glucose from noncarbohydrate precursors such as lactate, pyruvate, glycerol, propionic acid, and particularly the carbon skeletons of amino acids. The carbon skeletons of amino acids are what is left of the amino acids when the amino group is removed. Humans cannot oxidize amino groups on amino acids. These are mainly converted to the molecule **urea** using the **urea cycle,** and the urea is excreted in the urine. We will discuss the urea cycle in chapter 7. Gluconeogenesis occurs mainly in the liver and to a much smaller extent in the kidney.

Gluconeogenesis becomes important when liver glycogenolysis cannot produce glucose for the blood at a necessary rate. The central nervous system (principally the brain) needs about 125 g of glucose each day, and other tissues that rely exclusively on glucose need an additional 30 to 40 g of glucose a day. This means we need about 160 g or so of glucose just for the glucose-dependent tissues. If we exercise or do work, we use blood glucose in the contracting muscles. Fat cells use glucose to act as a source of glycerol (as glycerol 3-phosphate) to make triglyceride molecules.

Whenever dietary carbohydrate (glucose, fructose, and galactose) is inadequate to supply the body with the glucose it needs, gluconeogenesis makes up the difference. This becomes particularly significant during fasting or starvation, when one is eating a low-carbohydrate diet, or during prolonged exercise when the working muscles use a lot of blood glucose. Gluconeogenesis also occurs whenever the blood lactate concentration rises, as in moderate to severe exercise. Although lactate is used as a fuel by other skeletal muscles and the heart, a considerable amount is extracted from the blood by the liver and used to make glucose. The **Cori cycle** describes the process by which carbon cycles between liver and muscle. As shown in figure 5.17, lactate released from muscle circulates to the liver where it is converted to glucose. The glucose is released to the blood and taken up by active muscle. Through glycolysis, glucose is converted to lactate, released from the muscle, and recycled again by the liver. Heart muscle also oxidizes lactate (figure 5.17).

> **KEY POINT**
>
> During prolonged exercise, gluconeogenesis gradually increases in activity to compensate for the diminishing ability of liver glycogenolysis to produce glucose as the liver glycogen particles decrease in size.

Reactions of Gluconeogenesis

For the most part, new glucose is made from simple precursors by the reversal of the glycolytic pathway, as figure 5.18 reveals. In this figure, you can tell where the glycolytic pathway fits in, although only the reactions that contribute to glucose formation are shown. Most of the reactions of glycolysis are reversible and so from this perspective can go in either direction, driven by mass action. However, there are three irre-

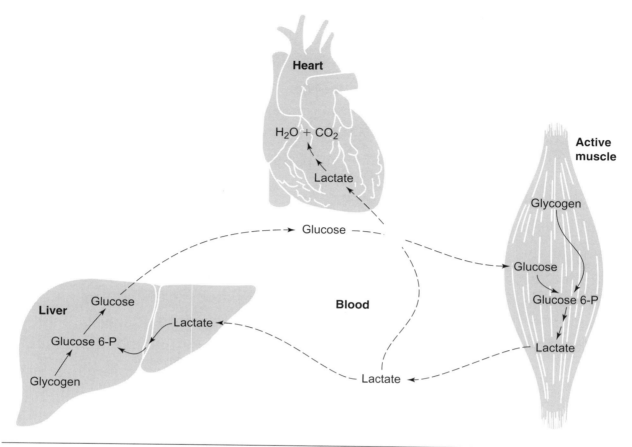

Figure 5.17 Lactate shuttling between tissues. Here, an active muscle releases lactate to the blood where it can be used as a fuel by the heart. The lactate can also be taken up by liver and converted to glucose in a process known as gluconeogenesis. The dotted arrows show lactate and glucose release to and uptake from the blood.

versible (nonequilibrium) glycolytic reactions: the pyruvate kinase reaction, the phosphofructokinase reaction, and the hexokinase or glucokinase reaction. These irreversible reactions act as one-way valves, making the glycolytic pathway flow only to pyruvate and lactate. In principle, glycolysis can go backward if there are alternate ways of getting around the three irreversible reactions.

Lactate is an important gluconeogenic precursor. It is taken up by liver cells and oxidized to pyruvate, as we have seen. However, the pyruvate must be converted to PEP, and this cannot be done through a driving backward of the pyruvate kinase reaction because pyruvate kinase catalyzes a nonequilibrium reaction. Therefore, to convert pyruvate to PEP, an alternate route is needed. This requires two new reactions with two new enzymes, *pyruvate carboxylase* and *phosphoenolpyruvate carboxykinase* (most people simply use the acronym PEPCK). Pyruvate is first converted to oxaloacetate using pyruvate carboxylase; then the oxaloacetate is converted to

PEP, catalyzed by PEPCK. The irreversible pyruvate kinase reaction is bypassed, but at a cost of the equivalent of two ATP. The process involves multiple steps and membrane crossings in the conversion of lactate or pyruvate (or both) to glucose as shown in figure 5.19. This figure emphasizes the role of the mitochondrion in the conversion of lactate/pyruvate to glucose.

Pyruvate carboxylase catalyzes a carboxylation reaction, using the tightly bound coenzyme **biotin.** This is a B vitamin that acts to bind and transfer CO_2 in the form of bicarbonate ion (HCO_3^-), which is needed in this reaction to change a molecule with three carbon atoms into one with four carbon atoms.

$$ATP + pyruvate + HCO_3^- \rightarrow oxaloacetate + ADP + Pi$$

This reaction takes place in the mitochondrial matrix.

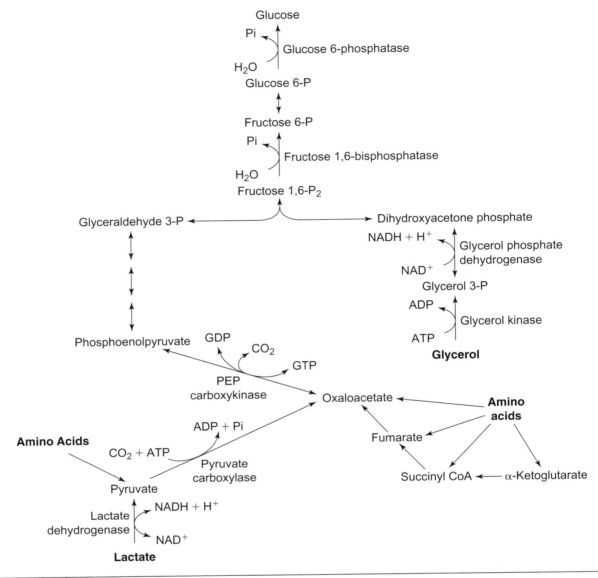

Figure 5.18 Formation of glucose from noncarbohydrate precursors (lactate, glycerol, and amino acids). A number of equilibrium (reversible) reactions of glycolysis are used; to bypass the remainder, new reactions are employed. Pyruvate carboxylase and PEP carboxykinase allow pyruvate to be converted into PEP. Fructose 1,6-bisphosphatase and glucose 6-phosphatase allow fructose 1,6-bisphosphate to be converted to glucose. Glycerol can be made into dihydroxyacetone phosphate using glycerol kinase and glycerol phosphate dehydrogenase. Many amino acids form citric acid cycle intermediates or pyruvate once they lose their amino groups. These can be converted to oxaloacetate using reversible citric acid cycle reactions.

The oxaloacetate is converted into PEP by phosphoenolpyruvate carboxykinase (PEPCK).

$$\text{oxaloacetate} + GTP \leftrightarrow \text{phosphoenolpyruvate} + GDP + CO_2$$

PEPCK is located in both the cytosol and mitochondrial matrix. If the oxaloacetate is converted to PEP in the matrix, the latter is then transported across the mitochondrial inner membrane where it will continue toward glucose. If oxaloacetate is not converted to PEP by the matrix PEP carboxykinase, it must be a substrate for the cytosolic PEP carboxykinase enzyme. However, as noted earlier, oxaloacetate cannot be transported across the inner mitochondrial membrane. Therefore, oxaloacetate must be converted into something that can cross the inner membrane via its own transporter. It can be transaminated to aspartate, which can use the aspartate-glutamate transporter (antiport). Alternatively, it can be reduced to malate, and the malate can be transported by an antiport with α-ketoglutarate.

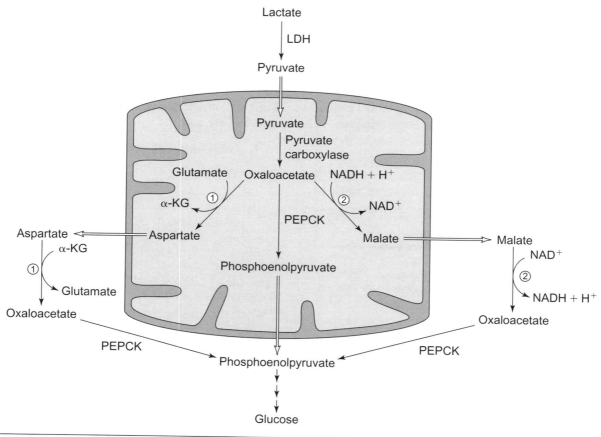

Figure 5.19 Specific steps and their cellular location as lactate and pyruvate are converted to glucose. The conversion of pyruvate to oxaloacetate takes place in the mitochondrial matrix. The next step is the formation of phosphoenolpyruvate, catalyzed by a mitochondrial PEPCK (phosphoenolpyruvate carboxykinase), which can be transported out of the mitochondrion. The oxaloacetate can also be converted to phosphoenolpyruvate by a cytosolic PEPCK. However, oxaloacetate cannot leave the mitochondria. It can undergo transamination using glutamate to form aspartate (reaction 1) or be reduced to malate (reaction 2). Both malate and aspartate can be transported out of mitochondria. In the cytosol, aspartate can undergo reverse transamination, forming oxaloacetate (reverse of reaction 1), and malate can be oxidized to oxaloacetate (reverse of reaction 2). Cytoplasmic oxaloacetate can be converted to PEP using cytosolic PEPCK.

Once the pyruvate kinase reaction is circumvented, glycolysis can run backward up to the fructose 1,6-bisphosphate step. At this point, an enzyme known as *fructose 1,6-bisphosphatase* removes one phosphate group, creating fructose 6-P. After this latter compound is converted to glucose 6-P, *glucose 6-phosphatase* removes the phosphate group; and the product, free glucose, is ready to be released by the liver (and, to a very small extent, the kidney) to the bloodstream.

Glycerol is also a key precursor for gluconeogenesis as shown in figure 5.18. Glycerol is released from fat cells when triglyceride molecules are broken down in a process known as lipolysis. Two glycerol molecules are capable of making one glucose molecule. In liver, the glycerol is first phosphorylated by *glycerol kinase* to make glycerol 3-P, which is oxidized to form dihydroxyacetone phosphate. The enzyme that

catalyzes this latter reaction is the cytosolic form of *glycerol 3-P dehydrogenase* that we saw in the glycerol phosphate shuttle. During the oxidation of glycerol 3-P, NAD^+ is reduced to NADH. During starvation, glycerol produced during lipolysis of stored fat can act as a significant gluconeogenic precursor, providing a third or more of the glucose produced by the liver.

Humans do not have the ability to form glucose from acetyl CoA. Therefore, the carbon atoms in fatty acids that are converted to acetyl CoA cannot be used as a source of glucose. Most fatty acids have an even number of carbon atoms. As we will see in the next chapter, all of the carbon atoms in these fatty acids appear as two-carbon acetyl groups, each attached to a CoA. Thus an 18-carbon fatty acid (stearic acid) gives rise to nine acetyl groups. There are some odd-chain fatty acids with an odd number of carbon atoms. During the breakdown of these to

> ### ▶ KEY POINT ◀
>
> If you start from pyruvate and make a molecule of glucose, you can see that this is an energy-requiring process. Adenosine triphosphate is consumed to make oxaloacetate; GTP (equivalent to ATP) is consumed in the PEPCK reaction; and an ATP is consumed when the phosphoglycerate kinase reaction goes backward. Add to this the loss of NADH when the glyceraldehyde 3-phosphate dehydrogenase reaction is driven backward. Now, it takes two pyruvate to make one glucose molecule, so the overall process is energetically expensive.

acetyl units, there will be one leftover three-carbon propionyl group, or *propionate*. This can be converted to glucose, so it is considered a gluconeogenic precursor. Though we will skip a detailed discussion of the metabolism of propionate (propionyl group), it is first converted to propionyl CoA at the cost of two ATP, then in three additional steps is converted to succinyl CoA. As a citric acid cycle intermediate, succinyl CoA can be converted to oxaloacetate using the last four steps in the citric acid cycle.

The Role of Amino Acids

Chapter 7 contains a detailed description of amino acid metabolism. For now, we will take a simple overview. Figure 5.18 reveals that amino acids can be used as a source of carbon atoms to make glucose. In fact, of the 20 common amino acids, 18 may have all or part of their carbon atoms directed to the formation of glucose; we call these **glucogenic amino acids.** The exceptions, in mammalian tissues, are leucine and lysine; these are said to be **ketogenic amino acids.**

Most adults are in a state that we call **protein balance** or, as it is also known, **nitrogen balance.** The former term means that over a long period of time, adults maintain a constant level of protein in their bodies. The latter term means that if the entire nitrogen intake into the body is measured (mainly in the form of amino acid amino groups) and all of the nitrogen lost by the body is measured (in urine, feces, sweat, etc.), these measures will balance. Whether we use the term protein balance or nitrogen balance, we are essentially talking about the same thing. The average adult takes in about 1 g or more of amino acids in dietary protein each day per kilogram of

body weight. If the person is in protein balance (or nitrogen equilibrium), he or she must get rid of the same amount of amino acids. We do not excrete these amino acids. Rather, the nitrogen in the form of the amino groups is removed and then excreted, mainly in the form of urea, while the remaining parts of the amino acids, which can be called the carbon skeletons, are used for

1. immediate oxidation to generate energy;
2. conversion to glucose and later oxidation; or
3. conversion into fat, which is stored and then oxidized later.

For all of these options, the final disposal of the carbon skeletons is oxidation, with the carbon atoms appearing as carbon dioxide and the hydrogen and electrons as water.

When the amino groups on most of the amino acids are removed, we get *citric acid cycle intermediates* such as oxaloacetate, α-ketoglutarate, fumarate, and succinyl CoA, as well as pyruvate. Of course, these four tricarboxylic acid cycle intermediates and pyruvate can be used to make glucose, as figure 5.18 reveals. If you review the citric acid cycle, you can see that α-ketoglutarate, fumarate, and succinyl CoA can all produce oxaloacetate, which is the precursor to forming PEP. The net formation of citric acid cycle intermediates through removal of the amino groups from amino acids is known as **anaplerosis.** Conversion of the amino acid glutamate to aspartate does not produce a net citric acid cycle intermediate because while it does generate α-ketoglutarate, it also consumes an oxaloacetate. While our focus here is on the formation of glucose in liver, it is appropriate to mention that these citric acid cycle intermediates can be formed in exercising muscle where they may increase the flux capacity for the cycle.

The carbon skeletons of amino acids, including the hydrogen and oxygen, are not wasted but are treated as fuel. The amino groups can be removed from most of the amino acids through transferal to other molecules. An example of this is shown in figure 5.20 for the **branched chain amino acids** (leucine, isoleucine, and valine). The amino groups on these amino acids are removed, primarily in muscle, through transferal to α-ketoglutarate to make branched chain keto acids (BCKAs) from the branched chain amino acids plus the amino acid glutamic acid (glutamate). Then, the amino group on glutamic acid is transferred to pyruvate, regenerating the α-ketoglutarate and forming alanine that is released from muscle to the blood. Liver extracts the alanine and removes the amino group by transferring it to oxaloacetate, making aspartate (aspartic acid). The carbon skeleton remaining, which is pyruvate, can then be used to make glucose. Figure 5.21 shows the cycling of amino groups from muscle to liver by way of alanine. The amino group ends up as urea, and the carbon skeleton (pyruvate) is converted into glucose in the liver. This has been described as the *glucose-alanine cycle,* and during exercise the rate of this cycle increases. The glucose-alanine cycle shows the importance of amino acids as a source of glucose. It also reinforces the fact that muscle protein can be broken down more rapidly under certain conditions, generating amino acids that can be a source of glucose. Muscle, the primary protein-containing tissue, is wasted to make glucose if the diet does not contain enough food energy.

Figure 5.20   Removal of amino groups from amino acids. In muscle, amino groups on the branched chain amino acids (BCAAs) such as leucine, isoleucine, and valine can be transferred to α-ketoglutarate, making glutamate and branched chain keto acids (BCKAs). The amino group on glutamate can be transferred to pyruvate, making alanine and regenerating α-ketoglutarate. Alanine can exit the muscle cell and travel through the blood to the liver.

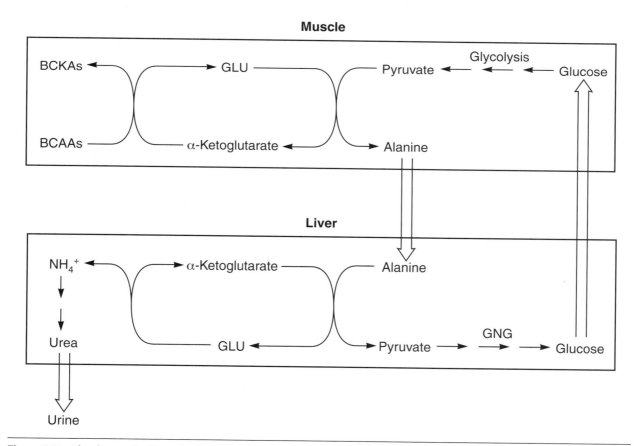

Figure 5.21   The glucose-alanine cycle. This cycle transfers amino groups (NH_2) and carbon in the form of alanine from muscle to liver. In the liver the amino group is used to make urea, whereas the carbon skeleton is used to make glucose by gluconeogenesis (GNG). Urea is removed from the body in the urine. The glucose leaves the liver and is a carbon source for pyruvate in muscle glycolysis. BCAAs are the branched chain amino acids; BCKAs are the corresponding branched chain keto acids that result from the loss of amino groups. GLU represents glutamic acid.

> **KEY POINT**
>
> Diets high in protein are favored by a number of athletes. Because there is a limit to how much protein can be synthesized in skeletal muscle and other tissues, and because there is no way to store excess amino acids, they are oxidized or used to make glucose. When the need for gluconeogenesis is minimal, the excess amino acids are oxidized. Thus, like carbohydrate, which promotes its own oxidation when present in excess, protein from an excessively high-protein diet promotes protein oxidation.

Regulation of Gluconeogenesis

Glucose is brain food, so we should not be surprised that gluconeogenesis is very carefully and comprehensively controlled. The conditions that favor the onset of gluconeogenesis are those in which blood glucose levels are threatened because liver glycogenolysis is declining as a result of low liver glycogen stores. To help explain the regulation of gluconeogenesis and show how the manner of its control is the opposite of that for the glycolytic pathway, the two are integrated in figure 5.22. Gluconeogenesis and glycolysis share the same enzymes in the middle portion of the glycolytic pathway, with the flux of gluconeogenesis going from pyruvate toward glucose and glycolysis going the opposite way (figure 5.22). As on a rail route going east to west and west to east and sharing the same section of track, traffic cannot move in both directions simultaneously. Thus, the control of gluconeogenesis and glycolysis is integrated, ensuring that one direction is favored while the other is inhibited. Regulation is complex and involves expression of genes for key enzymes in liver, control by protein phosphorylation, and allosteric regulation of enzymes. The net effect is to depress glycolysis and stimulate gluconeogenesis simultaneously under circumstances in which blood glucose levels are threatened and to stimulate glycolysis and inhibit gluconeogenesis under carbohydrate feeding conditions.

At rest, in the postabsorptive period, the brain accounts for 60% or more of the glucose output from the liver. Of the remainder, most can be accounted for by skeletal and heart muscle. During prolonged exercise, utilization of glucose by skeletal muscle can increase almost 20-fold. This helps to explain the need for careful control of liver glucose output between meals, particularly under exercise conditions. Fortunately, exercise results in an increase in two key precursors for gluconeogenesis, lactate and glycerol. The rate of appearance of lactate in the blood increases exponentially with exercise intensity. As we will discuss in the next chapter, glycerol released from fat depots during exercise increases as stored triglyceride is broken down to provide fatty acids as a fuel. Thus, the accelerated provision of key substrates for gluconeogenesis helps to maintain blood glucose concentration.

Role of Hormones

Under conditions in which gluconeogenesis is needed, specific hormones play a direct or participatory role in turning on and increasing the rate of gluconeogenesis. Glucagon is a polypeptide hormone secreted by the alpha cells of the pancreas. Its concentration rises as the blood glucose concentration decreases below normal (<5 mM). It acts to stimulate the breakdown of liver glycogen in the postabsorptive period. When liver glycogen concentration is low, glucagon stimulates gluconeogenesis. Glucagon has a variety of actions in liver to promote gluconeogenesis. It stimulates the formation of gluconeogenic enzymes. It promotes gluconeogenesis while inhibiting glycolysis by protein phosphorylation and allosteric mechanisms. Glucagon binds to its specific cell membrane receptor, and through a G protein mechanism activates the enzyme adenylyl cyclase, which produces cAMP. As we have seen already, cAMP activates protein kinase A, which increases glycogenolysis in liver. Cyclic AMP, acting through protein kinase A, also stimulates gluconeogenesis.

As mentioned earlier in this chapter, insulin is released by the beta cells of the pancreas. Its concentration rises when blood glucose concentration increases beyond normal levels (>5 mM). Insulin promotes glucose uptake into insulin-sensitive tissues and thereby helps lower blood glucose concentration. Insulin has a negative effect on gluconeogenesis, opposing not only the secretion of glucagon, but also its biochemical effects on liver. The mechanism of action of insulin is very complex. We will mention some of this at the end of the chapter. Briefly, insulin opposes the effects of glucagon by activating cAMP phosphodiesterase, which breaks down cAMP. It activates specific phosphoprotein phosphatases that remove phosphate groups added by protein kinase A.

Researchers have concluded that it is not so much the absolute levels of insulin and glucagon that play

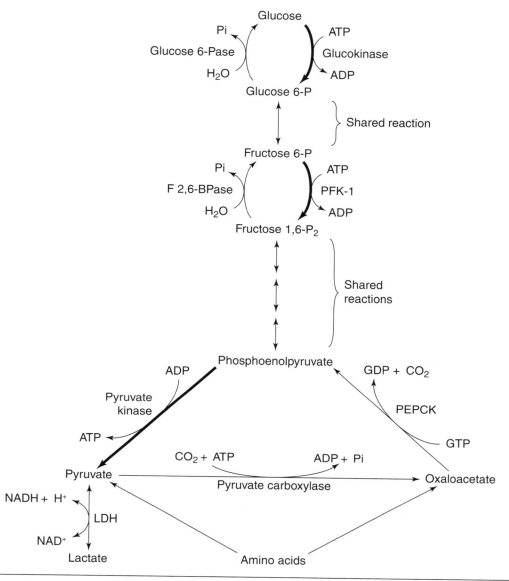

Figure 5.22 Glycolysis and gluconeogenesis in the liver. In the opposing pathways of glycolysis and gluconeogenesis in liver, the potential exists for the two processes to be active simultaneously. As the means of regulation, the three substrate cycles between glucose and glucose 6-phosphate (glucose 6-P), between fructose 6-P and fructose 1,6-P_2 (P_2 means bisphosphate), and between phosphoenolpyruvate and pyruvate operate in one direction only at any time. The glycolytic enzymes (glucokinase, phosphofructokinase-1 [PFK-1], and pyruvate kinase) and the gluconeogenic enzymes (glucose 6-Pase [glucose 6-phosphatase]), fructose 1,6-BPase [bisphosphatase], PEPCK [phosphoenolpyruvate carboxykinase], and pyruvate carboxylase) can be regulated by allosteric effectors or protein phosphorylation or through regulation of the expression of their genes to promote either glycolysis or gluconeogenesis, but not both at the same time.

a role in determining glucose production by the liver, but the insulin to glucagon ratio (I/G). During exercise, we know that there is a decline in blood insulin concentration and an increase in blood glucagon, thus magnifying the decline in the I/G ratio (Trimmer et al. 2002).

Under more extreme conditions of physical stress, the glucocorticoid cortisol may play an important role in gluconeogenesis. Cortisol is a steroid hormone secreted by the adrenal cortex under catabolic

conditions. We might suspect that its concentration would increase under severe or prolonged exercise conditions, particularly in the fasted state. Cortisol has a catabolic effect in that it promotes net protein breakdown in skeletal muscle. In this way, it increases the availability of amino acids to act as gluconeogenic precursors. In the liver, cortisol stimulates the expression of genes coding for gluconeogenic enzymes. If cortisol levels do not increase, the presence of even a low level of cortisol in the blood acts to assist the

---

> ▷ **K E Y   P O I N T** ◁
>
> Trimmer and colleagues (2002) reported that the maximal capacity of gluconeogenesis to support glucose release from the liver peaks at about 1.5 mg per kilogram per minute. This level of glucose production is insufficient to maintain blood glucose during exercise if there is not a substantial contribution from liver glycogenolysis. Accordingly, blood glucose concentration declines, leading to a decrement in performance. This emphasizes the need for people to eat some hours before prolonged exercise and to support blood glucose through regular carbohydrate intake during the activity.

process of gluconeogenesis; persons with cortisol deficiency can develop hypoglycemia more easily than others.

Control by Altering Gene Expression

Glycolysis and gluconeogenesis share the same equilibrium (reversible) enzymes in the central portion of the glycolytic pathway. Control of gluconeogenesis and glycolysis takes place outside of this central portion through three substrate cycles that regulate the overall rates of glycolysis and gluconeogenesis in the liver (see figure 5.22). These substrate cycles involve the intermediates glucose and glucose 6-P, fructose 6-P, and fructose 1,6-bisphosphate, as well as PEP and pyruvate. If you focus on these substrate cycles, you should be able to see that if the enzymes involved are simultaneously active, the net effect is no flux in either direction, but there would be a loss of energy-rich phosphates. The enzymes involved in these substrate cycles catalyze reactions that are irreversible (nonequilibrium) with large energy changes.

One strategy to ensure that one direction is favored is to alter the expression of the genes for the key reactions in the substrate cycles shown in figure 5.22. Glycolysis is favored by increased activity of glucokinase, phosphofructokinase-1, and pyruvate kinase via stimulation of the transcription of their genes and the simultaneous depression of transcription of genes for the gluconeogenic enzymes. On the other hand, gluconeogenesis can be favored through increased transcription of glucose 6-phosphatase, fructose 1,6-bisphosphatase, pyruvate carboxylase, and PEPCK and the simultaneous depression of transcription of the three nonequilibrium glycolytic enzymes. The

hormones insulin and glucagon play a major role in controlling gene expression, with glucagon promoting gluconeogenic enzymes and insulin promoting glycolytic enzyme formation (glucokinase, PFK-1, and pyruvate kinase) and suppressing expression of genes for PEPCK and glucose 6-phosphatase.

The initial stages of type 2 diabetes are accompanied by insulin resistance. This insulin resistance is not just with muscle; insulin's suppressive effect on gluconeogenic enzymes is also reduced, so that activities of glucose 6-phosphatase, fructose 1,6-bisphosphatase, PEPCK, and pyruvate carboxylase are increased. The net effect is increased glucose formation by gluconeogenesis, release of glucose from the liver, and hyperglycemia.

A number of signaling pathways are involved in the specific regulatory roles of glucagon and insulin. For example, we discussed earlier the activation of protein kinase A by glucagon. Protein kinase A can phosphorylate a nuclear protein known as *cyclic AMP response element binding protein* (CREB) (figure 5.23). Phosphorylated CREB now binds to a control region of a gene whose protein product is a coactivator in the formation of key liver enzymes. This coactivator is known as *peroxisome proliferator-activated receptor γ coactivator 1* or PGC-1γ. PGC-1γ now works with a special liver gene transcription regulator known as *hepatocyte nuclear factor-4* (HNF-4). Together, PGC-1γ and HNF-4 upregulate genes coding for gluconeogenic enzymes. Although this suggests enormous complexity in the regulation of gene expression, it helps to fine-tune the levels of enzymes that need to respond to the metabolic changes taking place in our bodies over a day.

Control of Enzyme Activity

Control of gluconeogenesis and glycolysis through regulation of the transcription of genes for key enzymes plays a powerful role. However, there is a considerable lag in time before changing the expression of specific genes can lead to significant changes in the amounts of functional enzymes. Therefore, rapid-acting mechanisms must be available to control glycolysis and gluconeogenesis.

Pyruvate kinase, the nonequilibrium glycolytic enzyme, can be inhibited by phosphorylation. The increase in glucagon under conditions favoring hepatic glucose output increases the activity of protein kinase A through an increase in cAMP. Among its many substrates (e.g., CREB), protein kinase A phosphorylates *pyruvate kinase,* leading to its inhibition (figure 5.23). This effectively blocks

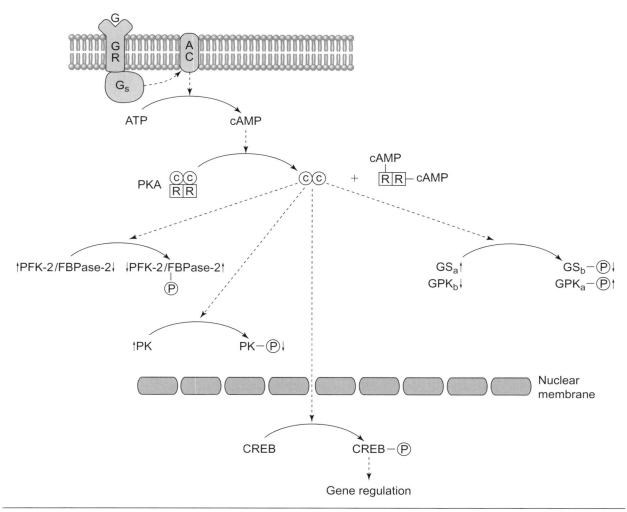

Figure 5.23 A summary of the many roles played by glucagon in the liver. Glucagon (G) binds to its receptor (GR), activating a stimulatory G protein (G_s), which results in activation of the membrane enzyme adenylyl cyclase (AC) that converts ATP into cyclic AMP (cAMP). Protein kinase A exists as an inactive tetramer with two catalytic (C) and two regulatory (R) subunits. The binding of cAMP to the R subunits allows the catalytic subunits to phosphorylate a number of protein substrates, changing their activity, shown as active (up arrow) or inactive (down arrow). A key substrate is the bifunctional enzyme phosphofructo-2-kinase/fructose 2,6-bisphosphatase (PFK-2/FBPase-2), where phosphorylation decreases the kinase activity while increasing the phosphatase activity, thus decreasing the overall concentration of the allosteric regulator fructose 2,6-bisphosphate. Other substrates are pyruvate kinase (PK), glycogen synthase *a* (GSa), phosphorylase kinase *b*, and cyclic AMP response element binding protein (CREB). A circled P indicates the attached phosphate. Not shown are the phosphoprotein phosphatases that are responsible for removing phosphate groups added by active protein kinase A.

glycolysis at the end of the pathway yet has no influence on gluconeogenesis. A glucagon-induced rise in hepatic cell cAMP concentration results in the phosphorylation of another remarkable bifunctional enzyme known as *6-phosphofructo-2-kinase/fructose 2,6-bisphosphatase* (PFK-2/FBPase-2). On this single polypeptide protein, there are two distinct enzyme activities. The PFK-2 part is responsible for converting fructose 6-P to fructose 2,6-bisphosphate (F2,6-P_2). We discussed F2,6-P_2 earlier in the context of the regulation of glycolysis. F2,6-P_2 is a potent activator of PFK-1, which is the glycolytic enzyme.

As well as activating glycolysis, F2,6-P_2 allosterically inhibits fructose 1,6-bisphosphatase. The other half of this bifunctional enzyme, FBPase-2, breaks down fructose 2,6-bisphosphate to fructose 6-P. Here is where glucagon comes in. Phosphorylation of PFK-2/FBPase-2 at a single serine residue results in the instantaneous inhibition of the kinase part of the bifunctional enzyme and activation of the phosphatase part with the result that the concentration of fructose 2,6-bisphosphate falls. The net effect of the phosphorylation of PFK-2/FBPase-2 is to inhibit glycolysis and allow gluconeogenesis. In

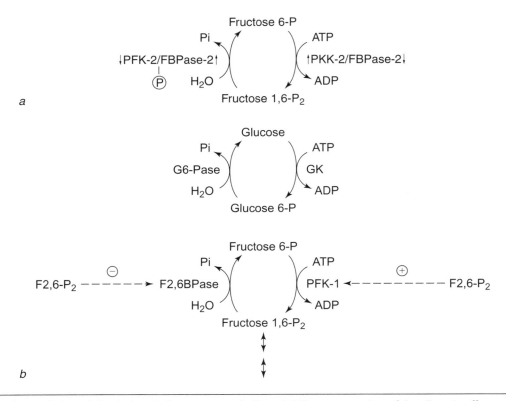

Figure 5.24 The regulation of glycolysis and gluconeogenesis in liver. *(a)* The concentration of the allosteric effector, fructose 2,6-bisphosphate (F2,6-P$_2$), is determined by the relative activity of the kinase and phosphatase activities of the bifunctional enzyme phosphofructo-2-kinase/fructose 2,6-bisphosphatase (PFK-2/FBPase-2). Phosphorylation of PFK-2/FBPase-2 by protein kinase A inhibits the kinase part while increasing the phosphatase part of the activity, leading to a decrease in F2,6-P$_2$ concentration. *(b)* The role of F2,6-P$_2$ in increasing PFK-1 activity while decreasing the activity of F2,6BPase. The net effect of an increase in F2,6-P$_2$ in liver is to increase glycolysis while simultaneously decreasing gluconeogenesis. PFK-1 is the glycolytic enzyme phosphofructokinase-1. F2,6BPase is the gluconeogenic enzyme fructose 2,6-bisphosphatase.

summary, phosphorylation of two key enzymes by a cAMP-dependent protein kinase A acts in switch-like fashion to inhibit glycolysis and to activate gluconeogenesis. This is summarized in figure 5.24.

Under conditions in which gluconeogenesis needs to be stimulated, there is an increase in the acetyl CoA level in liver. This arises from an increase in beta-oxidation of fatty acids, a topic we cover in the next chapter. Acetyl CoA from fatty acids is a substrate for the citric acid cycle. As well, acetyl CoA can act as an allosteric effector in two ways. First, an increase in cytosolic acetyl CoA inhibits pyruvate kinase at a negative allosteric site. Secondly, acetyl CoA is an obligatory activator for pyruvate carboxylase, which carboxylates pyruvate to make oxaloacetate. The rise in acetyl CoA signals that there is plenty of fuel for the citric acid cycle, so carbon should be shunted to make glucose.

THE PENTOSE PHOSPHATE PATHWAY

As we will see in the next chapter, most cells require a constant source of reducing equivalents for biosynthesis. These reducing equivalents are found in NADPH, a molecule differing from NADH in the fact that the former contains an additional phosphate group. As we have seen, electrons on NADH are used in the electron transport chain to reduce oxygen and generate ATP. Electrons on NADPH are used to synthesize new molecules such as fatty acids, amino acids, and steroids. The pentose phosphate pathway

---

▶ **K E Y   P O I N T** ◀

From a biochemical perspective, the brain could be characterized as the body's most important tissue. It is hard to argue with this statement given the complex, yet complementary, mechanisms that regulate the level of glucose in the blood to act as a fuel for the brain.

is the most common name, but this pathway is also known as the hexose monophosphate shunt or the 6-phosphogluconate pathway. Besides NADPH, the pentose phosphate pathway generates another important substance used in biosynthetic reactions. Ribose 5-phosphate is used in nucleotide biosynthesis and thus is needed to make ATP, DNA, RNA, FAD, coenzyme A, and NAD+ and NADP+.

Figure 5.25 summarizes in full structural detail the major reactions of the pentose phosphate pathway as they take place in tissues such as the liver, adipose tissue, adrenals, testes, and ovaries, but not skeletal muscle. The reaction begins with glucose 6-P, which is also a beginning point for glycolysis and glycogen synthesis. However, there are huge differences. In the pentose phosphate pathway, the glucose 6-P first undergoes an oxidation, catalyzed by *glucose 6-phosphate dehydrogenase*, generating NADPH plus H+. This first reaction is the key reaction in the pentose phosphate pathway, and it is under tight regulation by one of its products, NADPH. NADPH is also formed

in the reaction catalyzed by *6-phosphogluconate dehydrogenase*. The pentose phosphate pathway is regulated entirely at the first step. NADPH is an allosteric inhibitor for glucose 6-phosphate dehydrogenase. When the ratio of the reduced to oxidized form of this coenzyme is elevated (NADPH/NADP+), glucose 6-phosphate dehydrogenase is inhibited.

In many tissues, the pentose phosphate pathway stops at ribose 5-phosphate, having generated sufficient NADPH for biosynthetic reactions and sufficient ribose 5-phosphate for nucleoside and coenzyme synthesis. This is known as the *oxidative stage* of the pentose phosphate pathway. In some cells, the need for NADPH exceeds that for ribose 5-phosphate. In these instances, the *nonoxidative* part of the pathway can continue in a different direction from ribulose 5-phosphate. In this portion of the pathway, three ribulose 5-phosphates can be converted into two fructose 6-phosphates and one glyceraldehyde 3-phosphate. In this way, the carbons in ribulose 5-phosphate are conserved as glucose precursors, or the fructose 6-P

Figure 5.25 The pentose phosphate pathway (hexose monophosphate shunt), which can be divided into oxidative and nonoxidative stages. Detailed reactions reflecting the oxidative portion of the pathway are shown. Beginning with glucose 6-phosphate, this portion of the pathway generates two NADPH for biosynthesis and ribose 5-phosphate for nucleoside and coenzyme synthesis.

and glyceraldehyde 3-phosphate can enter the glycolytic pathway for pyruvate formation. If we include both the oxidative and nonoxidative portions of the pathway and remember that we lose one CO_2 for each glucose 6-P we start out with, we could summarize the pentose phosphate pathway as follows, starting with six molecules of glucose 6-P:

$$6 \text{ glucose 6-P} + 12 \text{ NADP}^+ + 7 \text{ H}_2\text{O} \rightarrow 5 \text{ glucose 6-P} + 12 \text{ NADPH} + 12 \text{ H}^+ + \text{Pi}$$

SIGNALING PATHWAYS

Chapter 2 introduced the concept of signal transduction, the mechanism by which external signals are communicated to and within cells in order to alter their metabolism. The preponderance of signaling pathways involve phosphorylation of proteins by protein kinases and removal of the attached phosphates by phosphoprotein phosphatases. To date we know of at least 500 genes in humans that code for protein kinases (Hardie 2004). In our discussion of carbohydrates so far, we have noted specific signaling mechanisms related to control of glycogenolysis and gluconeogenesis. In this section, we investigate in more detail how two signal transduction pathways affect carbohydrate metabolism.

AMP-Activated Protein Kinase

There is a protein kinase, activated by 5'-AMP, that plays a major role in metabolic regulation in a host of cell types. Its activity can be affected by nutrient supply and stress and can influence the metabolism of carbohydrates, fats, and protein, as well as ion channels and cell survival. It occupies a special niche in metabolism. We will look at this protein kinase here and also refer to it in subsequent chapters.

Adenosine monophosphate–activated protein kinase or AMPK consists of three different kinds of protein subunits. The α subunit is the catalytic subunit and acts to transfer a phosphate group from ATP to a protein substrate. The β and γ subunits play a regulatory role. In all tissues surveyed to date, there are two α subunits (α_1 and α_2), two β subunits (β_1 and β_2), and three γ subunits (γ_1, γ_2, and γ_3). As the name implies, regardless of the subunit composition (e.g., $\alpha_2 \beta_2 \gamma_2$, common in skeletal muscle), the protein kinase activity is increased when AMP binds to the γ subunit. The enzyme can also be activated when it is phosphorylated on its α subunit by the enzyme *AMP kinase kinase* or AMPKK. AMPKK is also activated by AMP. Adenosine monophosphate binding to AMPK makes it a better substrate for phosphorylation by AMPKK, thus reinforcing the activity of AMPK. While AMPK activity can be stimulated by about 10-fold with an increase in AMP concentration, phosphorylation by AMPKK can increase AMPK activity nearly 100-fold.

In skeletal muscle, the increase in [AMP] is brought about by *adenylate kinase*, which converts two ADP into an AMP and an ATP. AMPK is inhibited by PCr and ATP. Adenosine triphosphate is a substrate for AMPK and also appears to competitively inhibit AMP from binding to its allosteric site unless the latter increases significantly. Thus, the activation state of AMPK reflects the energy status of the cell. At rest, with [AMP] low and [PCr] and [ATP] elevated, we would expect the activity of AMPK to be low. With the onset of muscle activity, and in a manner directly reflecting the intensity of contractile activity, the concentration of AMP rises; ATP and especially PCr fall, making AMPK activity proportional to the metabolic stress induced by the contractile activity. AMPK activation is less when muscle glycogen stores are elevated, suggesting there is a site, possibly on subunit β, where glycogen can bind to inhibit the AMPK. Researchers have discovered that AMPK and AMPKK can be activated artificially by a molecule referred to as 5-amino-imidazole-4-carboxamide-riboside (AICAR). When given to animals, AICAR is taken up by cells and phosphorylated to make an analog of AMP, known as ZMP. ZMP is a powerful activator of AMPK and AMPKK, and studies using it have led to a great understanding of the array of effects that AMPK has in cells and whole organisms. Regulation of AMPK is summarized in figure 5.26.

There are other important regulatory roles for AMPK in skeletal muscle. When active, it enhances glucose transport by increasing the translocation of GLUT-4 transporter to the cell membrane. The mechanism for enhancing glucose transport is not the same as that caused by insulin, as we will discuss next. Active AMPK also increases the oxidation of fatty acids in muscle, as covered in the next chapter. AMPK increases the content of fructose 2,6-bisphosphate, thus increasing glycolysis. AMPK activity is also reported to be sensitive to endurance training, with increases in total AMPK activity and alterations in the subunit composition (Frøsig et al. 2004). In summary, AMPK activation parallels the cells' need to make ATP, thereby reinforcing some of the mechanisms we have already discussed. The drug

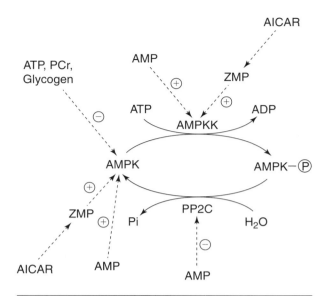

Figure 5.26 Adenosine monophosphate-activated protein kinase (AMPK) activation. AMPK can be activated allosterically by AMP or artificially by AICAR, which produces ZMP, an AMP analog. ATP, PCr, and glycogen all have a negative effect on AMPK activity. Phosphorylation of AMPK by AMPK kinase (AMPKK) results in the most active form of the enzyme. AMPKK is also activated by AMP. Removal of the phosphate group on AMPK is catalyzed by phosphoprotein phosphatases, such as PP2C.

metformin, one of the most widely prescribed drugs for type 2 diabetes mellitus, increases AMPK activity.

Two Mechanisms for Regulating Glucose Uptake

In the beginning of this chapter, we discussed the necessity to regulate glucose uptake in skeletal muscle, adipose tissue, and cardiac muscle. This regulation is based on the availability of GLUT-4 transporters in the cell membrane to transport glucose into the cell down its concentration gradient. In skeletal muscle, the number of GLUT-4 transporters active in transporting glucose into fibers is regulated both by insulin and by the contractile process itself. These operate by two different mechanisms as briefly reviewed next.

Role of Insulin

Insulin has potent metabolic effects in a host of tissues, including skeletal muscle. It has a predominantly anabolic role, stimulating glucose uptake and glycogen storage, fat synthesis and storage, and protein synthesis. Here we briefly review the role of insulin in stimulating glucose uptake in skeletal muscle.

The insulin receptor in the cell membrane consists of four subunits. Two α subunits protrude into the extracellular space and are responsible for insulin binding. Upon insulin binding to the α subunits, the two intracellular β subunits become active as a **protein tyrosine kinase,** such that they phosphorylate tyrosine residues on their own polypeptides (autophosphorylation). Once phosphorylated, the β subunits become active in phosphorylating tyrosine residues on other intracellular proteins, known as *insulin receptor substrates* (IRS). There are at least four of these, but IRS-1 is principally involved in insulin signaling in skeletal muscle. Phosphorylated IRS-1 becomes a docking site for an enzyme known as phosphatidylinositol 3-kinase or PI3K, which becomes active, creating a molecule known as phosphatidylinositol 3,4,5-trisphosphate (PI3,4,5-P_3). The PI3,4,5-P_3 activates two other protein kinases, protein kinase B, more commonly known as Akt, and an atypical member of the *protein kinase C* family, PKCξ. Activation of these latter protein kinases leads to translocation of GLUT-4 transporters from intracellular storage sites to the sarcolemma. The overall process of insulin-regulated glucose uptake is summarized in figure 5.27.

Role of Exercise

Exercising skeletal muscles can take up glucose in an insulin-independent way. As we have mentioned, exercise leads to a decrease in blood insulin concentration. Furthermore, the exercise effect does not operate through IRS-1 or PI3K, suggesting that the mechanism for the exercise effect must operate in a different way from the insulin-dependent pathway. Indeed, the effect of exercise is not just during the actual contractile activity, but extends to at least several hours after exercise. Although much is unknown, and the detailed mechanism(s) awaits additional research, we do know some factors. First, exercise leads to an increase in GLUT-4 transporters in the sarcolemma, but these are mobilized from a different storage site than those of the insulin-dependent pathway. Second, AMP kinase is involved. This is activated by exercise in an intensity-dependent fashion because the level of AMP reflects the increase in ADP, which depends on the rate of ATP hydrolysis. The increase in calcium concentration that accompanies muscle activity is also a likely signal that could increase GLUT-4 mobilization. Calcium can activate several kinase systems. The mitogen-activated protein kinase (MAPK) system also provides a mechanism that could influence GLUT-4 translocation from intracellular storage sites, during or after exercise.

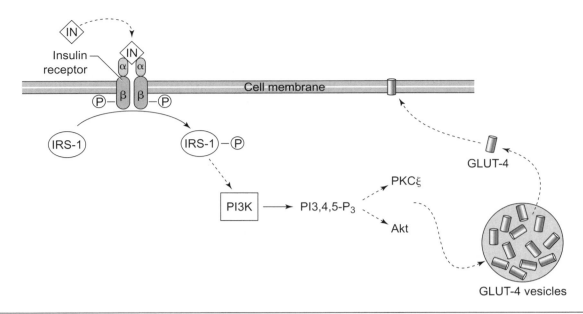

Figure 5.27 The role of insulin signaling in glucose transport in a muscle cell. Insulin (IN) binds to its receptor, consisting of two extracellular α subunits, inducing tyrosine autophosphorylation in the two intracellular β subunits. This activates a protein kinase activity in the β subunits that phosphorylate insulin receptor substrate-1 (IRS-1), which now provides a docking site for phosphatidylinositol 3-kinase (PI3K). PI3K produces phosphatidylinositol 3,4,5-trisphosphate (PI3,4,5-P_3), which leads to activation of two additional protein kinases, Akt and protein kinase Cξ, and release of GLUT-4 transporters from intracellular vesicles. The GLUT-4 transporters embed in the cell membrane, allowing glucose to enter the cell. Circled P represents phosphate groups, and dotted arrows show the flow of the signaling steps.

SUMMARY

Most dietary carbohydrate is broken down in the intestinal tract to glucose, fructose, and galactose, but glucose is by far the most important carbohydrate. In the form of blood glucose and liver and muscle glycogen, body carbohydrate is a major fuel, particularly for the brain. Blood glucose is carefully regulated so that its concentration is approximately 5 to 6 mM (90-110 mg per 100 ml of blood). Glucose from the blood enters cells down a concentration gradient through a specific glucose transporter. The GLUT-4 transporter is found in skeletal and cardiac muscle and fat cell membranes, and its content in these membranes is greatly increased by insulin. Exercise also increases the uptake of glucose into muscle. The principal metabolic pathway for breaking down carbohydrate is glycolysis. Pyruvate, a product of glycolysis, has two major fates. It can be reduced to lactate by the enzyme lactate dehydrogenase, or it can enter the mitochondrion, to be oxidized to acetyl CoA to enter the citric acid cycle.

Glycolysis is a carefully regulated process, with most of the control exerted by the enzyme phosphofructokinase (PFK). This enzyme is sensitive to the binding of a number of positive and negative modulators to allosteric sites. Rest or low-level muscle activity is associated with a low activity of glycolysis, whereas vigorous exercise greatly increases its rate in order to produce ATP. When pyruvate is oxidized in the mitochondrion and not reduced to lactate during glycolysis, NADH can potentially build up to the extent that there is too little NAD+ to permit glycolysis to continue. Oxidation of cytoplasmic NADH, other than by lactate dehydrogenase, occurs through the action of two shuttle systems. The glycerol phosphate shuttle and the malate-aspartate shuttle transfer electrons from the cytosol to the mitochondria for use by the electron transport chain.

Glycogen, the storage form of glucose in liver and muscle tissue, is synthesized from glucose taken up from the blood in a process that is greatly accelerated following a carbohydrate-rich meal. Liver glycogen is a reservoir of glucose for the blood. In muscle, the glycogen is ready to be fed into the glycolytic pathway when the muscle fiber becomes active. Since glycogen synthesis and glycogenolysis are opposing processes taking place in the cytosol of cells, one process must be inactive when the other is active in order to avoid a futile cycle. Two regulatory enzymes, glycogen phosphorylase and glycogen synthase, are thus affected in opposite ways by the same signal, which leads to their simultaneous phosphorylation. In addition to covalent modification, the activities of both synthase and phosphorylase are influenced by the binding of regulatory molecules. Following a meal, when blood insulin concentration is increased, the synthase is active and the phosphorylase inactive in both liver and muscle. Between meals, liver phosphorylase is activated by a rise in glucagon and a fall in insulin so that the glycogen is slowly broken down to glucose. During exercise, when epinephrine concentration is increased, muscle phosphorylase is active and synthase inhibited. The tight regulation of glycogen metabolism emphasizes the importance of controlling body carbohydrate stores.

When body carbohydrate content is decreased and glucose output from the liver is threatened, glucose is made from a variety of noncarbohydrate sources such as lactate, pyruvate, and the carbon skeletons of amino acids. This process, gluconeogenesis, takes place in liver and kidney and involves the reversible reactions of glycolysis. New reactions that bypass irreversible glycolytic reactions are catalyzed by pyruvate carboxylase, phosphoenolpyruvate carboxykinase, fructose 1,6-bisphosphatase, and glucose 6-phosphatase. These reactions allow the liver to produce glucose during times of need while glycolysis is virtually shut down. The pancreatic hormone glucagon promotes gluconeogenesis. Glycerol, produced when triglycer-ides are broken down, is an important gluconeogenic precursor in liver. In addition, when the amino group of any of the 20 common amino acids is removed, the carbon skeletons of 18 of these can be converted to glucose. The fact that skeletal muscle can be severely wasted during diets providing inadequate food energy demonstrates the importance of the brain as a tissue, since it can be a significant user of glucose derived from amino acids obtained from muscle proteins. The pentose phosphate pathway is not important in muscle; but in breaking down glucose 6-P in a variety of tissues, this pathway produces reducing equivalents in the form of NADPH and ribose 5-phosphate, a precursor for making nucleotides. Carbohydrate and lipid metabolism are regulated by the action of signaling systems that operate through protein kinases and phosphatases to alter enzyme activity within the cell. The AMP-activated protein kinase system is stimulated under conditions in which ATP provision is threatened. Thus exercise is a potent stimulus for AMP kinase. Insulin operates by a signal transduction pathway, involving insulin receptor substrates and the phosphatidylinositol 3-kinase activation. The net effect of insulin on muscle is to increase GLUT-4 transporters in muscle cell membranes. Muscle activity per se also increases GLUT-4 membrane transporters, likely through an AMP kinase pathway as well as other signaling pathways.

REVIEW QUESTIONS

1. Stored fat is considered an ideal fuel from an efficiency perspective because, being stored in an anhydrous form, it provides about 9 kcal (38 kJ) per gram, as opposed to 4 kcal (17 kJ) per gram for glycogen. From the same perspective, we could say that glycogen is not an efficient fuel because it is stored with water and has a lower fuel value. Calculate the approximate fuel value of glycogen in kilocalories and kilojoules per gram weight as stored, assuming that each gram of stored glycogen contains 3 g of associated water.

2. We have identified a variety of exercise and competition activities in which elevations in muscle glycogen concentration can enhance performance. Can you think of activities in which glycogen is an essential fuel but in which supercompensated levels may be detrimental to performance?

3. The concentrations of glucose and lactate in the blood may be described in units such as millimoles or in units of milligrams per deciliter of blood. If the molecular weight of glucose is 180 and of lactate is 90, convert the following concentrations of glucose and lactate to millimolar (mM) units: lactate concentration, 990 mg/dL; blood glucose concentration, 108 mg/dL.

4. From RER and other measures, it is determined that a distance runner is oxidizing carbohydrate at the rate of 4 g per minute and fat at the rate of 0.5 g per minute. If he is running at 4 min per kilometer, what percentage of his fuel is derived from fat if his total energy expenditure is 3.43 megajoules (MJ)? How far does he run?

5. A muscle glycogen concentration of 500 mmol of glucosyl units per kilogram of dry muscle weight is considered high. Convert this into

millimoles of glucosyl units per kilogram wet muscle weight and into millimolar units using the relationship that a muscle sample as removed from a biopsy needle contains 77% water. Also the proportion of intracellular water is 0.7 of the total muscle weight.

6. MCT-1 is found more in slow, high oxidative fibers and MCT-4 is found more in type II fibers. If you knew that the K_m values of these transporters for lactate differed by a factor of 5, which fiber type would have the lower K_m transporter? Give a rationale for your answer.

CHAPTER

6

LIPID METABOLISM

Most people are well aware that fats (or lipids) are very important fuels in the body. Most have probably heard that too much fat in the diet is harmful and that storage of excess lipid is a major health concern. Many may also know that lipids are not soluble in water. Few are aware, however, that lipids are also important structural elements and are involved in cell signaling. Fat is stored as **triacylglycerol** (triglyceride) in fat cells. Scientists describe the tissue containing fat cells as **adipose tissue,** usually white adipose tissue (WAT), to distinguish it from the brown adipose tissue (BAT) used as a heat source in small animals. Stored fat is considered a long-term energy store, in contrast to glycogen, which is considered a short-term energy store. To understand the importance of lipids one must learn about the various kinds of lipids and their metabolism. We will examine the storage of fat and how it and ketone bodies are oxidized to make ATP. We will review how humans can synthesize new fat. Fat stored in adipose tissue and muscle plays a key role in exercise metabolism. Moreover, there is competition between the oxidation of fat and that of carbohydrate, and we will address how this is regulated. We will also look at the newly discovered ability of fat tissue to secrete key regulatory peptides that act in a hormone-like fashion. The chapter concludes with a discussion of cholesterol.

TYPES OF LIPIDS

Lipids are categorized as fatty acids, triacylglycerols, phospholipids, and sterols, whose roles in the body are largely determined by their chemical structure.

We will address the role of cholesterol, the dominant sterol, at the end of this chapter. Lipids are hydrophobic, containing mostly nonpolar components. Some are amphipathic, meaning that they have some polar groups although they are largely nonpolar. We have already considered fatty acids in chapter 4, where we noted that the beta-oxidation of fatty acids provides acetyl CoA for the citric acid cycle. Fatty acids are therefore important fuel molecules.

Fatty Acids

Fatty acids are carboxylic acids, containing a long alkyl chain with a carboxylic acid group at one end. For the fatty acids in our body and food, the overwhelming majority contain at least 16 carbon atoms in total. This means they have a long hydrocarbon tail and a carboxyl group for the head. Short chain fatty acids are generally four to six carbon atoms in length; medium chain fatty acids are 8 to 12 carbon atoms in length; and long chain fatty acids have 14 or more carbon atoms. Fatty acids are usually known by their common or trivial names. Normally they have an even number of carbon atoms and can be saturated, with no carbon-to-carbon double bonds, or unsaturated, with one or more carbon-to-carbon double bonds. Figure 6.1 gives examples of a saturated and an unsaturated fatty acid. Because the fatty acids have pK_a values in the range 4.5 to 5.0, they exist predominantly as anions at neutral pH. Thus the fatty acids in figure 6.1 are named palmitate (from palmitic acid) and oleate (from oleic acid).

A shorthand notation for describing fatty acid structure has been devised. The first number represents the number of carbon atoms in the fatty acid.

Figure 6.1   Detailed structures for the saturated fatty acid palmitic and the unsaturated fatty acid oleic acid. Both are named as anions because this is how they would exist at pH 7. The carboxylate is carbon atom 1, and other key carbon atoms are also shown with small numbers. The location of the double bond in oleate begins at carbon 9. Beneath each structure is the shorthand way it can be described.

This is followed by a colon and a number that identifies the number of double bonds. After this, the actual location of the double bonds is shown either in parentheses or using the Greek letter Δ with one or more superscript numbers to show the location of the double bonds. For example, linoleic acid is 18:2 (9, 12), or 18:2 $\Delta^{9,12}$ if we are using the system with the Greek letters. The configuration of the double bonds in naturally occurring fatty acids is almost always cis, as shown in figure 6.1. If there is a trans double bond, it must be identified. In the manufacturing process in which unsaturated fatty acids are hydrogenated, to make them more solid and less liquidlike, trans double bonds can be introduced. We know that these trans fatty acids are not good for us.

Most of the fatty acids in the body come from our diet, as we have a limited capacity to synthesize fatty acids. From a human nutrition perspective, the three most common fatty acids are palmitic acid (palmitate at pH 7), oleic acid (oleate), and stearic acid (stearate). Fatty acids also are classified into families depending on the location of the last double bond and the number of carbon atoms. As shown in

table 6.1, α-linolenic acid is [18:3 (9, 12, 15)] and is called an omega-3 (ω-3) or n-3 fatty acid; that is, the last double bond begins three carbons from the end carbon. Note that γ-linolenic acid is an ω-6 or n-6 fatty acid. The n-3 (or omega-three) fatty acids are reputed to offer many health benefits, helping to lower bad blood lipid concentrations. They may also

> ### ▶ K E Y   P O I N T ◀
>
> Linoleic acid (18:2) is generally found as the all cis 9,12 isomer. Two other isomers can be obtained through cow's milk or in beef and lamb; these are identified collectively as conjugated linoleic acid (CLA). Produced in small amounts in ruminant animals, CLA represents two isomers of linoleic acid, identified as cis 9, trans 11 or trans 10, cis 12 forms. Considerable research attention is paid to CLA, as it is reported to decrease body fat and insulin sensitivity (Brown and McIntosh 2003). It can be obtained as a supplement.

Table 6.1 Common Fatty Acids of Biological Significance

Common name	Systematic name	Shorthand notation	Family
Stearic acid	Octadecenoic acid	18:0	NA
Linoleic acid	9, 12-Octadecadienoic acid	18:2 (9, 12)	ω-9 or n-9
α-Linolenic acid	9, 12, 15-Octadecatrienoic acid	18:3 (9, 12, 15)	ω-3 or n-3
γ-Linolenic acid	6, 9, 12-Octadecatrienoic acid	18:3 (6, 9, 12)	ω-6 or n-6
Arachidonic acid	5, 8, 11, 14-Eicosatetraenoic acid	20:4 (5, 8, 11, 14)	ω-6 or n-6
Timnodonic acid	5, 8, 11, 14, 17- Eicosapentaenoic acid	20:5 (5, 8, 11, 14, 17)	ω-3 or n-3
Cervonic acid	4,7,10,13,16,19-Docosahexaenoic acid	22:6 (4,7,10, 13, 16, 19)	ω-3 or n-3

help our nervous systems. Linoleic acid and α-linolenic acid are **essential fatty acids** that we must get in the diet. They are also known as polyunsaturated fatty acids, or PUFAs, with two (or more) double bonds. Note that linoleic and α-linolenic acid represent the ω-6 and ω-3 families. The brain is particularly rich in PUFAs. Oleic acid, a monounsaturated fatty acid, is not considered an essential fatty acid; but a diet rich in this fatty acid also confers important health benefits. Flaxseed and flaxseed oil, as well as the oils from fish found in cold waters, are rich sources of PUFAs. The essential fatty acids are important components of phospholipids, found in cell membranes. They are also precursors for the synthesis of a family of hormone-like compounds known as the eicosanoids. Eicosanoids are synthesized in cells from the fatty acid arachidonic acid [20:4 (5, 8, 11, 14)]. The eicosanoids are known as local hormones because they are released from cells where they may have an effect on the secreting cell itself (autocrine effect) or on the cells in the immediate environment of the secreting cell (paracrine effect). Examples of eicosanoids are prostaglandins, thromboxanes, and leukotrienes.

Triacylglycerols

Although they should be called triacylglycerols, abbreviated TAG, the term *triglyceride* is more commonly used; we will use *triacylglycerol* or its abbreviation TAG. An older term, *neutral fat,* is also used. Figure 6.2 shows these molecules as triesters, made from the combination of a three-carbon molecule with three alcohol groups, known as glycerol, and three fatty acids. Remember that the combination of an acid with an alcohol is known as an ester. In the body, stored triacylglycerols rarely have the same fatty acid attached at all three positions on the glycerol. Stored fats have both saturated and unsaturated fatty acids. **Diacylglycerols** (also known as diglycerides) have only two fatty acids attached to the glycerol. Monoacylglycerols (also known as monoglycerides) have only one. It is important to note that the major fatty acids we have just discussed are found in foods as triacylglycerols.

The physical state of a TAG depends on the length of the carbon chain in the fatty acids and the number of double bonds. Shorter-length fatty acids as well as

$$CH_2OH \quad + \quad CH_3(CH_2)_{14}COOH$$
$$CHOH \quad + \quad CH_3(CH_2)_7CH = CH(CH_2)_7COOH \longrightarrow$$
$$CH_2OH \quad + \quad CH_3(CH_2)_{16}COOH$$

Glycerol   Three fatty acids

$$CH_3(CH_2)_7CH = CH(CH_2)_7COOC\underset{\underset{H_2COOC(CH_2)_{16}CH_3}{|}}{\overset{\overset{H_2COOC(CH_2)_{14}CH_3}{|}}{H}} \quad + \quad 3 \ H_2O$$

Triacylglycerol

Figure 6.2 A triacylglycerol molecule, consisting of three fatty acids each joined by ester bonds (–COOC–) to each of the three carbons in glycerol.

more double bonds lower the melting point. Thus, vegetable fats with polyunsaturated fatty acids are liquids, whereas the fat on the side of a steak, which contains more saturated fatty acids, is a solid. Palm oil is a liquid, despite having mainly saturated fatty acids, because the fatty acids have 10 to 12 carbon atoms as opposed to 16 to 18 in most other triacylglycerols. Triacylglycerols and the fatty acids found in triacylglycerols are insoluble in water due to the large degree of hydrophobic hydrocarbon components. The hydrophobic nature of triacylglycerols makes them ideal for storing energy because, by weight, they have more chemical potential energy than other fuel molecules such as carbohydrates or protein. In fact, because the fatty acids in triacylglycerols are so reduced, the oxidation of 1 g of triacylglycerol has a standard free energy change of more than 9 kcal per gram (38 kJ/g).

Phospholipids

Many **phospholipids** are derivatives of **phosphatidic acid** (or phosphatidate, since it is in anion form), whose structure is illustrated in figure 6.3. You should be able to see that there is a glycerol part, with two fatty acids joined to it by ester bonds. Also attached is a phosphate group. Different groups can be attached to the phosphate in phosphatidic acid. If choline is attached, the molecule is called phosphatidylcholine or, commonly, lecithin (see figure 6.3). Phospholipids have a hydrophilic (water liking) part and a hydrophobic part. The hydrophilic part is due to polar chemical bonds and charged groups on the phospholipid. The hydrophobic part is the long hydrocarbon tail of the fatty acids, which can contain more than 16 carbon atoms. Phospholipids are major components of cell membranes and are shown in figures using a circle at the top, representing the hydrophilic part, with two squiggly lines running downward to represent the long, hydrophobic alkyl parts of the fatty acids.

Phosphatidylinositols are found in membranes, where they play an important role in cellular regulation as part of signaling systems. Figure 6.4 shows the structure of inositol. When phosphorylated at carbons 4 and 5 and attached to phosphatidic acid, it is known as phosphatidylinositol 4,5-bisphosphate, abbreviated PI 4,5-P_2. Figure 6.5 shows the hydrolysis of PI 4,5-P_2 by phospholipase C, producing inositol 1,4,5-trisphosphate (1,4,5-IP_3) and a diacylglycerol (DAG). The latter remains within the membrane, but 1,4,5-IP_3 is mobile to some extent in the cytosol. Hydrolysis of PI 4,5-P_2 by phospholipase C plays an

Figure 6.4 The chemical structure for inositol. The numbers identify its six carbon atoms.

Figure 6.3 Phosphatidic acid and phosphatidylcholine. *(a)* Phosphatidic acid (or phosphatidate, since it is charged) consists of one glycerol with two fatty acids (R_1 and R_2 represent the long alkyl chains of each fatty acid) and a phosphate attached to the third carbon of glycerol. *(b)* Phosphatidylcholine is created when choline is combined with phosphatidic acid. Phospholipids such as phosphatidylcholine have a hydrophilic part, based on the positive and negative charges and the polar ester bonds.

Figure 6.5 Hydrolysis of PI 4,5-P_2 by phospholipase C. Phosphatidylinositol 4,5-bisphosphate (PI 4,5-P_2) has its two fatty acid chains buried in the hydrophobic core with the hydrophilic part, including the phosphorylated inositol, pointing into the interior of the cell. When activated, phospholipase C hydrolyzes PI 4,5-P_2, producing a diacylglycerol (DAG), which remains in the membrane, and inositol 1,4,5-trisphosphate (1,4,5-IP_3), which has finite mobility in the cytosol. The circled P represents phosphate groups.

important role in signal transduction. The 1,4,5-IP_3 can act on the endoplasmic reticulum of the cell to release calcium ions, which can then activate other enzymes. Calcium ions and DAG can activate typical protein kinase C enzymes that phosphorylate other proteins to alter their function. In the previous chapter, we noted the involvement of an atypical protein kinase C, which when activated in an insulin-dependent mode can lead to translocation of GLUT-4 transporters to the cell membrane of skeletal muscle fibers.

FAT STORES

Fat is stored as triacylglycerol in fat cells. It is considered a long-term energy store in contrast to glycogen, which is considered a short-term energy store. A normal adult can store about 8,000 to 12,000 kJ (2,000 to 3,000 kcal) of glycogen in liver and muscle, but even a lean male or female has more than 300,000 kJ (75,000 kcal) stored as triacylglycerol in fat cells. As shown in figure 6.6, the major component of a fat cell (adipocyte) is a liquid droplet of triacylglycerol occupying much of the volume of the cell.

Fatty acids are obtained mainly from food fat. After the fatty meal has been digested and the fatty acids have been absorbed, triacylglycerols are formed inside intestinal cells, combined with some protein and phospholipid, and released into the lymphatic system as lipoprotein particles known as **chylomicrons.** Another source of triacylglycerol is the liver. The liver makes and secretes triacylglycerol-rich **lipoproteins** known as **very low density lipoproteins (VLDLs).** The fatty acids in chylomicrons and VLDLs

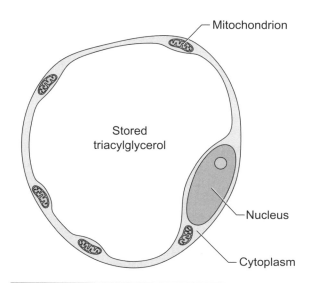

Figure 6.6 Cross section of a roughly spherical fat cell or adipocyte. Most of the volume of the cell is composed of a droplet of triacylglycerol. Cytoplasm is pushed to the periphery by the stored lipid. Fat cells have mitochondria and a single nucleus.

are released from their triacylglycerols through hydrolysis by an enzyme known as *lipoprotein lipase* (LPL). This enzyme is synthesized in adjacent fat cells, secreted from the cell, and attached to the endothelial lining of a nearby capillary. Fatty acids freed in the capillaries by LPL can flow down a concentration gradient into adjacent adipocytes with the aid of a specific carrier. Lipoprotein lipase is also present in capillaries in skeletal muscle, where it can generate fatty acids for muscle fibers.

Formation of Triacylglycerols

Fat synthesis is favored following a meal when chylomicrons and blood glucose levels are elevated. An elevated insulin concentration in the blood facilitates entry of glucose into fat cells using the GLUT-4 transporter. Triacylglycerol is also stored inside muscle cells, although the quantity is much less than in fat cells. Formation of TAG is a simple process as shown in figure 6.7. We will consider this as three main steps.

1. The fatty acid must be activated by becoming attached to coenzyme A (CoA). This process involves the hydrolysis of ATP to AMP and PPi and is catalyzed by the enzyme *acyl CoA synthetase*. Even though the product fatty acyl CoA is an energy-rich molecule, the reaction that produces it is considered nonequilibrium because of the irreversible hydrolysis of inorganic pyrophosphate by the enzyme *inorganic pyrophosphatase*.

2. The synthesis of triacylglycerols requires glycerol 3-phosphate. In adipose tissue the major pathway is to form dihydroxyacetone phosphate starting from glucose and using glycolytic reactions, followed by reduction of dihydroxyacetone phosphate using the enzyme *glycerol phosphate dehydrogenase*. This is an enzyme we encountered in the previous chapter in the glycerol phosphate shuttle.

There are other sources of glycerol 3-phosphate that we will describe later.

3. The formation of a triacylglycerol requires three fatty acyl CoA molecules and a glycerol 3-phosphate. The actual formation of triacylglycerol is a simple four-step process involving the sequential addition of two fatty acyl groups to glycerol 3-phosphate using a *glycerol phosphate acyltransferase* enzyme (or simply acyltransferase) to form phosphatidate (anionic form of phosphatidic acid), hydrolysis of the phosphate group catalyzed by a *phosphatidate phosphatase*, and then addition of another fatty acyl group.

Lipolysis

The hydrolysis of triacylglycerols is best known as **lipolysis.** In this process, a triacylglycerol molecule yields glycerol and three fatty acid molecules. Mobilization of fat (lipolysis) is favored under conditions of increased energy need, such as exercise, low-calorie dieting, or fasting (starvation), or when the body is cold. We usually view the process as one in which triacylglycerol in adipose tissue is broken down and the fatty acids and glycerol are released to the blood where they are available to the energy-requiring cells. Skeletal muscle fibers also contain stored triacylglycerol, known as intramuscular triacylglycerol. When hydrolyzed, the fatty acids generated are oxidized within the contracting fiber. An enzyme inside fat cells and other fat-storing cells (e.g., muscle), known as *hormone-sensitive lipase*, is responsible for triacylglycerol hydrolysis. Lipolysis involves three hydrolysis reactions, each catalyzed by hormone-sensitive lipase (see figure 6.8). The last hydrolysis reaction is also assisted by a *monoacylglycerol lipase*. The numbers refer to the carbon atoms of glycerol. For example, a 2-monoglyceride (2-monoacylglycerol) has a single fatty acid attached to the glycerol molecule at the middle carbon atom.

Lipolysis takes place in locations other than in fat cells. For example, hydrolysis (digestion) of dietary triacylglycerol occurs in the small intestine, catalyzed by **pancreatic lipase.** As already mentioned, hydrolysis of triacylglycerol in blood lipoproteins is catalyzed by lipoprotein lipase. Finally, intramuscular triacylglycerol is hydrolyzed by a muscle-specific hormone-sensitive lipase that generates fatty acids that are immediately available as a fuel for that fiber.

Step 1: Formation of fatty acyl CoA

$$R_1COOH + ATP \xrightarrow{\text{Acyl CoA synthetase}} R_1\overset{\overset{\displaystyle O}{\|}}{C}\text{—}CoA + AMP + PPi$$

Step 2: Formation of glycerol 3-phosphate

$$\text{Glucose} \xrightarrow{\quad} \underset{\text{Glycolysis}}{\rightleftarrows} \xleftarrow{\quad} \text{Dihydroxyacetone phosphate}$$

$$\text{Dihydroxyacetone phosphate} + NADH + H^+ \xrightarrow[\text{dehydrogenase}]{\text{Glycerol phosphate}} \text{Glycerol 3-phosphate} + NAD^+$$

Step 3: Triacylglycerol synthesis

CH$_2$OH
|
CHOH   +   R$_1$CCoA (with =O) $\xrightarrow[\text{acyltransferase}]{\text{Glycerol phosphate}}$ CH$_2$OCR$_1$ (with =O)
| |
CH$_2$OPO$_3^{2-}$ CHOH   +   CoA
                                                                                    |
Glycerol 3-P   Fatty acyl CoA   CH$_2$OPO$_3^{2-}$

Monoacylglycerol 3-P

R$_2$CCoA (with =O) $\xrightarrow[\text{acyltransferase}]{\text{Glycerol phosphate}}$

CH$_2$OCR$_1$ (with =O)
|
CHOCR$_2$ (with =O)   +   CoA
|
CH$_2$OPO$_3^{2-}$

Phosphatidate

H$_2$O $\xrightarrow[\text{phosphatase}]{\text{Phosphatidate}}$

CH$_2$OCR$_1$ (with =O)
|
CHOCR$_2$ (with =O)   +   Pi
|
CH$_2$OH

1,2-Diacylglycerol

R$_3$CCoA (with =O) $\xrightarrow[\text{acyltransferase}]{\text{Glycerol phosphate}}$

CH$_2$OCR$_1$ (with =O)
|
R$_2$COCH (with =O)
|
CH$_2$OCR$_3$ (with =O)

Triacylglycerol

Figure 6.7 Storage of triacylglycerol as a three-step process. First, fatty acids are activated through attachment of a CoA. In the second step, glucose is partially broken down in glycolysis to dihydroxyacetone phosphate, which is subsequently reduced to make glycerol 3-phosphate. In the third step, three fatty acyl groups are added using the enzyme glycerol phosphate acyltransferase. Note that the phosphate group must be removed by a specific phosphatase.

$$\text{Triacylglycerol} + H_2O \xrightarrow{\text{HSL}} \text{1,2-Diacylglycerol} + \text{Fatty acid}$$

$$\text{1,2-Diacylglycerol} + H_2O \xrightarrow{\text{HSL}} \text{2-Monoacylglycerol} + \text{Fatty acid}$$

$$\text{2-Monoacylglycerol} + H_2O \xrightarrow{\text{HSL/MAGL}} \text{Glycerol} + \text{Fatty acid}$$

Figure 6.8 The breakdown of stored triacylglycerol in fat cells, involving three steps. Hormone-sensitive lipase (HSL) can catalyze all three hydrolyses, but a monoacylglycerol lipase (MAGL) plays a major role in the last hydrolysis.

> ### ►► K E Y   P O I N T ◄◄
>
> The processes of triacylglycerol synthesis (esterification) and breakdown (lipolysis) take place in the cytosol of fat cells and other fat-containing cells, including skeletal muscle. This might seem to suggest that since these are opposing processes occurring in the same location, they should not be simultaneously active—that if they were, the result would be wasteful ATP hydrolysis in the formation of fatty acyl CoA and waste of glucose to produce glycerol 3-phosphate. This suggestion is not correct, as the two processes can be active simultaneously. Indeed, there are advantages in their both being active at the same time, although not to the same extent. As when a drag racer guns the engine with a foot on the brake in order get away faster, if both processes take place simultaneously (although not maximally), a faster response can occur when there is a need for fatty acids, such as during exercise.

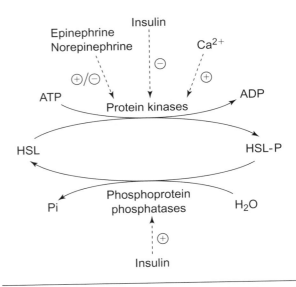

Figure 6.9 The major regulation of lipolysis through control of hormone-sensitive lipase (HSL). Phosphorylation of sites on HSL by a variety of protein kinases makes the enzyme active (HSL-P). Dephosphorylation by phosphoprotein phosphatases deactivates it. The influence of some important hormones is shown with dotted arrows and a plus or negative sign to indicate the effect. Epinephrine and norepinephrine can both activate or block HSL activity.

Regulation of Triacylglycerol Turnover

Formation of triacylglycerol molecules and their degradation take place in the cytosol of adipose tissue and other types of cells (skeletal muscle). Although different enzymes are used for synthesis and degradation of triacylglycerol, theoretically both processes could be fully active at the same time, which would result in ATP hydrolysis and glucose consumption but little else. Although there are some advantages to continuous lipolysis and esterification, as we will see, typically one process is favored over the other, depending on circumstances in the body. For example, during exercise, when fatty acids are needed as a fuel, esterification is reduced whereas lipolysis is accelerated. Following a meal, when there is a supply of precursors to make triacylglycerol, lipolysis is depressed whereas esterification is accelerated. As shown in figure 6.9, regulation takes place primarily at the level of hormone-sensitive lipase (HSL), controlled by phosphorylation and dephosphorylation, although other mechanisms operate.

Regulation of Lipolysis in Adipose Tissue

Mechanisms behind the regulation of lipolysis in adipose tissue are shown in figure 6.10.

At rest, HSL exists mostly in an inactive unphosphorylated form, and lipid droplets in a fat cell are surrounded by a protein family—the **perilipins** that render the droplet inaccessible to HSL. When both HSL and the perilipins are phosphorylated, the HSL becomes active and the perilipins cannot block lipolysis. Phosphoprotein phosphatases remove the phosphate groups from both HSL and the perilipins, and lipolysis is attenuated. The overall process of lipolysis is controlled primarily by the action of two hormones and a neurotransmitter. Epinephrine, released from the adrenal medulla, activates lipolysis when it binds to a β-adrenergic receptor on the fat cell. Norepinephrine, a neurotransmitter in the sympathetic nervous system, generates a positive effect on lipolysis by binding to β-adrenergic receptors (β_1 or β_2 subtypes). Both epinephrine and norepinephrine can bind to α_2-adrenergic receptors on fat cells, where the effect is to reduce lipolysis. Finally, insulin can bind to its own receptor on fat cells. Besides increasing glucose transport into fat cells by mobilizing GLUT-4 receptors, insulin can activate an enzyme that functions to decrease lipolysis. The details are as follows.

The catecholamines, epinephrine and norepinephrine, bind to a β_1- or β_2-adrenergic receptor and activate adenylyl cyclase through an activating

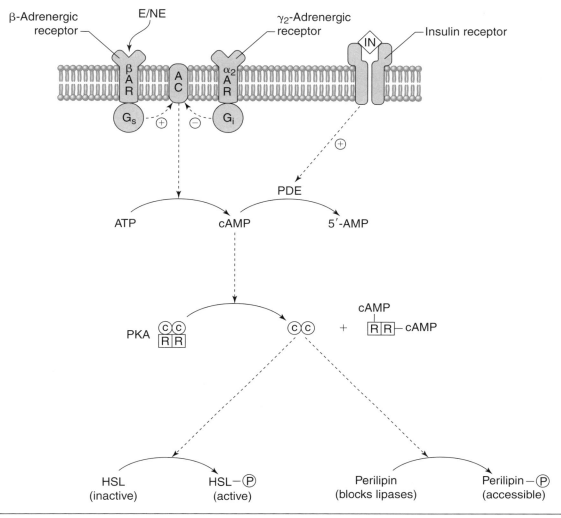

Figure 6.10 Mechanisms behind the regulation of lipolysis in fat by the actions of the catecholamines epinephrine (E) and norepinephrine (NE) and insulin (IN). Binding of E and NE to a β-adrenergic receptor (βAR) activates adenylyl cyclase (AC), mediated by a stimulatory G protein (G_s), leading to an increase in cyclic AMP (cAMP). When cAMP binds to the regulatory subunits (R) of protein kinase A (PKA), catalytic subunits (C) are released to phosphorylate hormone-sensitive lipase (HSL) and perilipin. E and NE can also bind to an α_2-adrenergic receptor (α_2AR), leading to a decrease in cAMP mediated by an inhibitory G protein (G_i). Insulin opposes lipolysis by activating cAMP phosphodiesterase (PDE), which breaks down cAMP.

G protein (G_s). As we saw in the previous chapter, *adenylyl cyclase* catalyzes the conversion of ATP into cyclic AMP (cAMP) and inorganic pyrophosphate. The cAMP binds to the regulatory subunits of protein kinase A, releasing the catalytic subunits. These catalytic subunits are now active as protein kinases, phosphorylating substrates such as HSL and perilipins, which coat lipid droplets in fat cells. Once phosphorylated, perilipins can no longer inhibit the action of HSL on the stored triacylglycerols; perhaps phosphorylation causes the perilipin to dissociate from the lipid droplet. Hormone-sensitive lipase, phosphorylated on three serine residues, is translocated to the lipid droplet where it catalyzes the

hydrolysis of triacylglycerols. Phosphorylation of adipose tissue HSL alone increases its activity about twofold. However, the combination of HSL and perilipin phosphorylation increases lipolysis more than 90-fold, pointing to the concerted interaction between these two regulatory proteins. Interestingly, in older animals, HSL translocation from cytosolic sites to the lipid droplet under conditions favoring lipolysis is depressed. This may explain the reduced ability to mobilize lipids in older people (Holm 2003; Yeaman 2004).

Insulin is inhibitory to this process by activating an enzyme, *cAMP phosphodiesterase*, which changes cAMP to 5'-AMP. As we saw in the previous chapter, occupancy of the insulin receptor by insulin leads to the activation of other protein kinases. In the case of adipose tissue, insulin activates a protein kinase B (also known as Akt), which phosphorylates cAMP phosphodiesterase, making it active in breaking down cAMP. Thus insulin acts to reduce the activation of protein kinase A. Insulin also activates phosphoprotein phosphatases, which remove phosphate from glycogen synthase *a* and glycogen phosphorylase *a* in both liver and skeletal muscle. It is quite likely that phosphoprotein phosphatases in fat cells are also activated by insulin, removing phosphates from HSL and perilipins and thus inhibiting lipolysis by another mechanism (Holm 2003).

Inhibition of lipolysis through catecholamine binding to the α_2-adrenergic receptors is mediated by an inhibitory G protein (G_i) that decreases the activity of adenylyl cyclase, thus reducing the level of cAMP. The balance between the prolipolysis β-adrenergic receptors and the antilipolysis α_2 receptors determines how easily fat can be mobilized from each adipocyte. For example, we know that adipose tissue just beneath the skin (subcutaneous adipose tissue, SCAT) is less responsive to exercise-induced adrenergic stimulation in obese compared to normal-weight men (Stich et al. 2000). Such a condition is not permanent, since overweight men who exercise can markedly improve exercise-induced lipolysis by decreasing the antilipolytic α_2-adrenergic receptor effect (De Glisezinski et al. 2003).

Protein kinase A is not the only enzyme that phosphorylates and activates HSL. The **mitogen-activated protein kinase** or **MAP kinase** family is involved in a host of regulatory processes in cells. Although a discussion of the roles and mechanisms of this pathway is beyond the scope of this book, we cannot neglect to mention its role in HSL phosphorylation. One of the MAP kinase signaling pathways involves an *extracellular signal-related kinase* or ERK. Specifically, ERK1/2 is activated when it is phosphorylated by cAMP-dependent kinase (protein kinase A); and once activated, ERK1/2 can phosphorylate sites on HSL. The exact role that phosphorylation by protein kinase A and ERK1/2 plays in terms of activation of HSL is difficult to establish in fat cells because phosphorylation leads to both an increase in HSL activity and HSL translocation to the lipid droplet. Moreover, phosphorylation of perilipin also plays a huge role in activating the overall process of lipolysis.

Other hormones can also stimulate lipolysis. **Growth hormone** from the anterior pituitary and *cortisol* from the adrenal cortex are secreted in response to stressful conditions, such as fasting and exercise. Both increase the rate of whole-body lipolysis in an additive fashion, likely by increasing the cAMP concentrations beyond the levels induced by catecholamines. It is possible that these hormones decrease the activity of the inhibitory G protein pathway or increase insulin resistance so that the effect of insulin to downregulate lipolysis is reduced. One of the reasons athletes inject themselves with growth hormone is to reduce the size of adipose tissue stores while promoting skeletal muscle growth. *Testosterone* and related anabolic steroids have strong effects on skeletal muscle growth, but they also have much milder effects to increase lipolysis. Castration is associated with a blunted adrenergic-dependent response and often an increase in adipose tissue mass. This can be reversed by testosterone administration. Androgens influence fat cell metabolism by binding to intracellular receptors.

In addition to insulin, there are other negative effectors for lipolysis. *Adenosine*, produced locally from ATP, can bind to its own receptor on the adipocyte membrane. This receptor is coupled to the G_i, the inhibitory G protein that depresses the activity of adenylate cyclase, resulting in inhibition of lipolysis. **Estrogens** also function by binding to intracellular receptors. The number of these receptors varies in fat depots from different sites, pointing to the role of estrogen in promoting fat deposition at specific locations. For example, the gluteofemoral fat deposits are augmented with the rise in estrogen in pubescent females. Estrogens lead to a relatively small reduction in the lipolytic effect of catecholamines on fat cells.

Regulation of Lipolysis in Skeletal Muscle

Skeletal muscle contains variable amounts of intracellular lipid droplets that tend to be located between myofibrils, close to mitochondria (Watt et al. 2002). There is good evidence that this TAG provides a source of fatty acids when the muscle fiber is active. Intramuscular triacylglycerol (IMTG, also known as intramyocellular lipid or IMCL) is hydrolyzed by an HSL in much the same way as in adipose tissue. How skeletal muscle HSL is regulated is incompletely known. We do know that skeletal muscle HSL is activated when it is phosphorylated at sites similar to those found in the adipose tissue. Like adipose tissue HSL activity, skeletal muscle HSL activity is increased by elevated epinephrine through a cAMP-dependent mechanism involving protein kinase A. However, there is no perilipin in skeletal muscle, so the extra level of regulation of lipolysis afforded by this protein is absent in muscle. Compared to one at rest, an active muscle fiber has a calcium concentration increased by approximately 100-fold as well as an increased AMP concentration. As we have seen, elevated Ca^{2+} levels can activate several kinases, including protein kinase C. The increase in AMP concentration in active muscle affects AMP-activated protein kinase (AMPK) activity, which is in part responsible for the contraction-induced increase in muscle glucose uptake. In addition, exercise, as a stressor, can lead to phosphorylation and activation of the MAP kinase, ERK. On the basis of current understanding, phosphorylation of HSL by protein kinase A and phosphorylation by ERK are the two most likely mechanisms for increasing lipolysis in skeletal muscle cells (Watt et al. 2003; Roepstorff et al. 2004). The roles, if any, played by other protein kinases remain uncertain.

Fate of Fatty Acids and Glycerol

Let us review the role of fat cells in the human body. These represent the principal storage site for triacylglycerol in the body. Triacylglycerol is a long-term energy source and turns over very slowly. If a person has 10 kg (22 lb) of TAG in his or her body, only about 50 to 70 g per day is turned over. This means that the lifetime of a TAG molecule stored in a fat cell is more than six months (Strawford et al. 2004). Following feeding, dietary fatty acids are taken up and esterified to make triacylglycerol. Esterification is reduced and lipolysis accelerated when the body is under stress, as during exercise. This leads to increased release of fatty acids to the blood. Lipolysis and esterification are active simultaneously, although at any time, one of the two is usually dominant, through, as we have seen, the action of hormones. Lipolysis produces fatty acids and glycerol, but a portion of the fatty acids can be re-esterified back to TAG after they are first activated by attaching a CoA at the expense of ATP (figure 6.11). This continuous circle of lipolysis and re-esterification or recycling within a cell constitutes an intracellular triacylglycerol-fatty acid cycle (Reshef et al. 2003). Adipose tissue is the principal source of blood fatty acids that can be used for energy by other tissues, including the heart and skeletal muscles. According to new research, fat cells produce and secrete some important substances that can influence our overall metabolism and food intake behavior. Abnormalities in fat metabolism can underlie conditions such as obesity, type 2 diabetes, and cardiovascular disease.

Figure 6.11 summarizes the opposing processes of lipolysis and esterification in an adipocyte in the postabsorptive state. In this state, fat cells play a major role to provide fatty acids for oxidation by other tissues. Virtually all of the glycerol generated by TAG lipolysis is released to the blood because the gene for *glycerol kinase* is scarcely expressed in fat cells. Glycerol kinase is needed to attach a phosphate from ATP to glycerol to make glycerol 3-phosphate, essential for re-esterification. Accordingly, glycerol release from adipose tissue can be used as a marker for the rate of lipolysis. If only lipolysis, and not re-esterification, were active, we would expect to see three fatty molecules released for each glycerol because each TAG molecule has three fatty acids attached to a glycerol. However, a ratio of three fatty acids released to the blood for every glycerol is not present under any circumstances because there are always some fatty acids that stay in the fat cell and that are reutilized to make TAG. Indeed, in the fasted state it is estimated that about 30% of the fatty acids released during lipolysis undergo re-esterification.

Figure 6.11 shows that there are two major sources of glycerol 3-phosphate (glycerol 3-P) for re-esterification. The classical view of TAG formation in the fasting state is that glucose is the source of glycerol 3-phosphate needed for re-esterification. This is not likely. During fasting, both blood glucose and insulin concentrations are low, and insulin is needed to stimulate glucose transport into fat cells in just the same way it does in muscle cells. The question, then, is what is the source of glycerol 3-phosphate to re-esterify fatty acids? As shown in figure 6.11, there is a recently identified process that converts gluconeogenic precursors (lactate, pyruvate, and some amino acids) into glycerol 3-P. This process, known as **glyceroneogenesis,** operates in much the same way gluconeogenesis does in the fasted state. The importance of this process is that it provides a source of glycerol 3-phosphate under conditions when blood glucose should be directed to those tissues that need it as a critical energy source—that is, the brain. Detailed biochemical studies have shown that if the glyceroneogenesis pathway operates during fasting conditions, then a limiting enzyme, *phosphoenolpyruvate carboxykinase* (PEPCK), must be expressed to a greater extent, allowing glycerol 3-phosphate to be made. Indeed, this is the case, as PEPCK gene expression can be turned on rapidly in the postabsorptive period and turned off when glucose is available. This response to feeding and fasting in adipose tissue is much like the response in liver

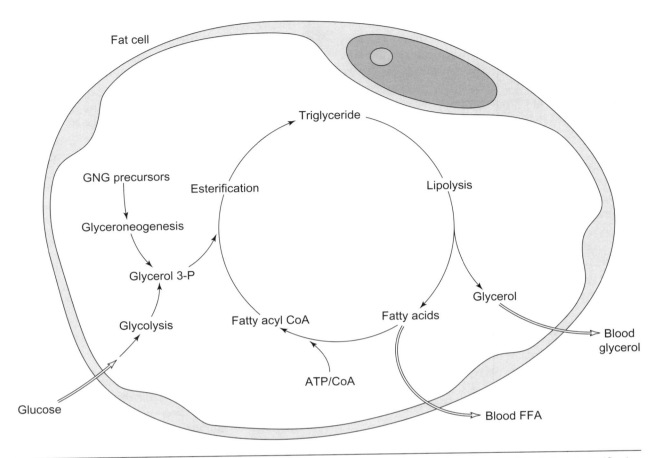

Figure 6.11   Recycling of triacylglycerol in a fat cell (adipocyte). This occurs when the opposing processes of lipolysis and esterification take place at the same time. Triacylglycerol molecules are broken down to fatty acids and glycerol during lipolysis. The glycerol is released from the fat cell to the blood, while the fatty acids have two fates. They may leave the fat cell and enter the blood, or they may be converted to fatty acyl CoA molecules within the fat cell where they can be used to make new triacylglycerol. Glycerol 3-phosphate (3-P) is derived from glycolysis using glucose taken up from the blood. Open arrows show membrane transport. In the blood, fatty acids travel bound to the protein albumin and are known as FFA (free fatty acids).

for the genes coding for enzymes of gluconeogenesis (GNG). There are differences between GNG in liver and fat cells, however. Cortisol, a stress hormone from the adrenal cortex, upregulates the PEPCK gene in liver, stimulating GNG, but inhibits the PEPCK gene in adipose tissue. This differential response is exactly as we might predict, because the net effect of cortisol is to stimulate glucose production in liver during stress while allowing more fatty acids to be released from the adipose cells by downregulating their re-esterification.

The dominant fate for the glycerol released when a triacylglycerol is hydrolyzed is release to the blood. Since glycerol has three OH groups, it is quite soluble in the blood. It is metabolized by tissues that contain a *glycerol kinase,* an enzyme that phosphorylates the glycerol to form glycerol 3-phosphate. The major tissue in which this occurs is the liver. As we saw in the previous chapter, glycerol is an important source

of glucose during periods of fasting or starvation. When fatty acids leave a fat cell and enter the blood, they become attached to the blood protein albumin since they are not soluble in the aqueous plasma. We call fatty acids attached to albumin in the blood **FFA (free fatty acids).** In many sources, FFA are also referred to as nonesterified fatty acids or NEFA. The FFA circulate in the blood and may enter other cells where they can be used as an energy substrate, as we shall soon see. Transfer of FFA from adipose tissue to skeletal muscle during exercise requires sufficient perfusion of adipose tissue by blood. Therefore, adipose tissue blood flow (ATBF) could pose a limitation to the delivery of fatty acids from fat tissue to exercising skeletal muscle.

Not all FFA circulating in the blood are taken up by cells and immediately oxidized. The liver has a high capacity to take up fatty acids from the blood. Some FFA are used to make other lipid compounds, as we will see, but others can be re-esterified in the liver and released as part of the VLDL particles secreted by the liver. Circulating in the blood, the triacylglycerol-rich VLDLs are hydrolyzed by lipoprotein lipase in capillaries, and the fatty acids can enter an adjacent cell. If this is an adipocyte, the fatty acids are likely to be incorporated into TAG molecules and stored as part of the lipid droplet in the fat cell. Even a skeletal muscle fiber stores fatty acids as TAG, although when the fiber is active, the more likely fate is oxidation to support the contractile activity. Thus there is also a triacylglycerol-fatty acid cycle that operates between cells in different organs.

Fatty acids are an ideal fuel for oxidation as they have a high energy density. Thus, the levels of FFA in the blood are elevated during stressful circumstances, such as exercise. Obese individuals with a larger adipose tissue mass have higher [FFA] under most circumstances. However, a consistently high [FFA] has drawbacks, not only because in high concentrations fatty acids are toxic, but also because there is a convincing relationship between elevated blood FFA and insulin resistance. As discussed previously, insulin resistance is a hallmark of type 2 diabetes. Elevated blood FFA levels lead to increased fatty acid uptake and storage as TAG in skeletal muscle, which renders muscle less sensitive to insulin. One new class of drugs developed to treat type 2 diabetes is

the thiazolidinediones (TZDs). Among their effects is a lowering of plasma FFA concentrations. Thiazolidinediones act as agonists for the peroxisome proliferator-activated receptor γ (PPARγ), which is a member of a gene regulatory family. In fat cells, the PPARγ receptor is part of the control region for two key genes; one codes for glycerol kinase and the other for PEPCK. While the gene for glycerol kinase is weakly expressed in fat cells, the TZDs increase this expression so that glycerol kinase plays a significant role in producing glycerol 3-phosphate for TAG formation. The PEPCK gene is normally activated by fasting, but the TZDs ensure that there is consistently more PEPCK enzyme. The net effect of TZDs in adipose tissue is to increase the re-esterification of fatty acids by providing more glycerol 3-phosphate. In this way, the TZDs lower blood [FFA] and help to relieve the insulin resistance (Tordjman et al. 2003). Figure 6.12 illustrates triacylglycerol-fatty acid cycling in a fat cell, including the sources of glycerol 3-phosphate and the role played by the antidiabetic drugs TZDs.

OXIDATION OF FATTY ACIDS

Skeletal muscle is a major site for oxidation of long chain fatty acids (LCFAs), those fatty acids containing 14 or more carbon atoms and the dominant

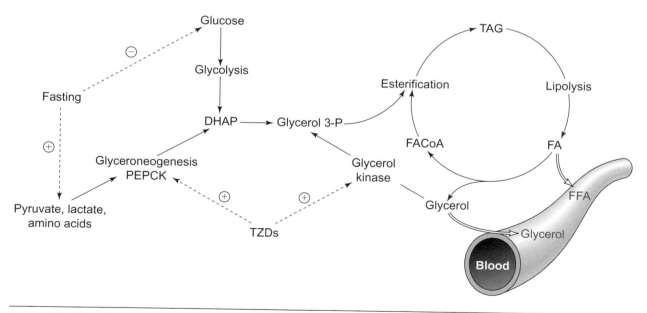

Figure 6.12 Triacylglycerol (TAG)-fatty acid cycling in adipose tissue and the role played by glyceroneogenesis and the antidiabetic drug class, the thiazolidinediones (TZDs). Fasting increases the provision of glycerol 3-P from glyceroneogenesis while decreasing that from glucose and glycolysis. The TZDs upregulate genes for glycerol kinase and phosphoenolpyruvate carboxykinase (PEPCK), in effect increasing the proportion of fatty acids released by lipolysis that will be re-esterified back to TAG. This helps to reduce the concentration of free fatty acids (FFA) in the blood. DHAP is dihydroxyacetone phosphate.

fatty acids stored in TAG. Capillaries in skeletal muscle also contain lipoprotein lipase, so that the fatty acids in the capillaries surrounding muscle fibers are those bound to albumin as FFA and those immediately obtained through hydrolysis of VLDL from liver by lipoprotein lipase. In order for the free energy stored in FFA molecules to be obtained, they must be transported from the blood, across the cell membrane of the utilizing cell, through the cytosol of the cell, across the inner mitochondrial membrane, and into the matrix. There, the fatty acids are broken down into two-carbon acetyl groups attached to CoA by beta-oxidation. We will look at these steps in detail. In skeletal muscle, fatty acids derived from intracellular TAG lipolysis must also be transported into mitochondria for oxidation.

Intracellular Transport of Fatty Acids

Because fatty LCFA are poorly soluble in an aqueous environment, it was assumed in the past that they could cross cell membranes by simple diffusion. This is a mechanism for transport, but we know now that there are also fatty acid transporters that allow the fatty acids to be moved down their concentration gradient into the cytosol of fat-utilizing cells. With its ability to increase its metabolic rate to high levels, muscle is of primary importance for oxidizing fatty acids. Several transporters have been identified in skeletal muscle, including a plasma membrane fatty acid-binding protein ($FABP_{pm}$) and a fatty acid translocase known as FAT/CD36. A diet high in fat induces an increase in both $FABP_{pm}$ and FAT/CD36, while endurance training was shown to increase $FABP_{pm}$ only in male subjects (Roepstorff et al. 2004). Recent studies suggest that there is an intracellular store of FAT/CD36 transporters that can be mobilized to the muscle sarcolemma with the onset of exercise (Chabowski et al. 2004). In this sense, fatty acid uptake responds to exercise in much the same way as glucose uptake, that is, with an increase in transporters in the sarcolemma.

Because of their insolubility, fatty acids in the cell cytoplasm of many tissues are attached to a cytosolic fatty acid-binding protein ($FABP_c$). In muscle, fatty acids attached to $FABP_c$ can be those originating in adipose tissue and taken up from the blood or those released when IMTG is hydrolyzed by skeletal muscle HSL. To be oxidized in the mitochondria, fatty acids must first be activated, forming a fatty acyl CoA, which we saw earlier in this chapter.

This reaction, catalyzed by *acyl CoA synthetase*, is as follows:

$$\text{fatty acid} + \text{ATP} + \text{CoA} \rightarrow \text{fatty acyl CoA} + \text{AMP} + \text{PPi}$$

Acyl CoA synthetase is an outer mitochondrial membrane enzyme. The reaction it catalyzes is essentially irreversible because the PPi is hydrolyzed by inorganic pyrophosphatase to two Pi (not shown), which drives the reaction to the right. The fatty acyl CoA that is formed is an energy-rich molecule, much like acetyl CoA. Formation of a fatty acyl CoA costs two ATP because two phosphates are removed from ATP.

Transport As Acylcarnitine

Fatty acyl CoA (often called simply acyl CoA) must be transported across the inner mitochondrial membrane to the matrix where it will be broken down into acetyl CoA units by the process of beta-oxidation. However, the mitochondrial inner membrane is impermeable to CoA and its derivatives, which permits separate regulation of CoA compounds in mitochondrial and cytosolic compartments. Transport of fatty acyl CoA into the mitochondrial matrix occurs using three different proteins and the small molecule **carnitine** (see figure 6.13). Only fatty acids attached to carnitine are able to cross the inner mitochondrial membrane. Therefore, the first step is to exchange a CoA for carnitine to create a fatty acylcarnitine. This is catalyzed by an enzyme in the outer membrane known as *carnitine palmitoyl transferase I* (CPT I). Carnitine, formed from the amino acids lysine and methionine, can cross the inner membrane, and a fatty acyl form of carnitine can also cross in the opposite direction. The *carnitine-acylcarnitine translocase* is also known as the carnitine-acylcarnitine antiport because this membrane protein transfers two different substances across the inner membrane in opposite directions.

On the matrix side of the inner mitochondrial membrane, a reverse exchange occurs in which the carnitine is exchanged for CoA using *carnitine palmitoyl transferase II* (CPT II), located on the matrix side of the inner membrane. Like acyl CoA, acylcarnitine is an energy-rich molecule, so the two reactions catalyzed by CPT I and CPT II are equilibrium reactions with equilibrium constants near 1. The net transfer of acyl groups attached to CoA in the matrix is given by the fact that the acyl CoA undergoes the process of beta-oxidation. Transport of fatty acyl units into

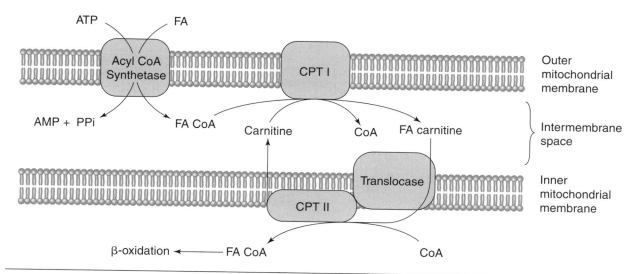

Figure 6.13 Transport of fatty acyl CoA into the mitochondrial matrix. Fatty acids are activated through attachment of a CoA using an outer mitochondrial membrane acyl CoA synthetase. The CoA is then exchanged for a carnitine using an outer membrane protein, carnitine palmitoyl transferase I (CPT I). The fatty acylcarnitine is conveyed across the intermembrane space in one direction, while a carnitine moves the opposite way using a carnitine-acylcarnitine translocase. The carnitine attached to fatty acylcarnitine is exchanged for CoA using an inner membrane carnitine palmitoyl transferase II (CPT II).

the mitochondria for oxidation is regulated at the CPT I step. This enzyme is inhibited by malonyl CoA, which can place a limit on fat oxidation in skeletal muscle. We will discuss this in more detail later in this chapter.

Carnitine deficiency, due to the body's inability to convert lysine into carnitine, is not an unusual metabolic disease. Patients with this deficiency have muscle weakness and poor exercise tolerance due to accumulation of triacylglycerol in muscle and the inability to oxidize fatty acids. Carnitine supplementation by endurance athletes has been common in the belief that it will enhance the ability to oxidize fatty

acids. Carnitine is now commonly taken by many in the general population because it may reduce oxidative stress.

Beta-Oxidation of Saturated Fatty Acids

The initial process in the oxidation of fatty acids is known as beta-oxidation, which occurs in the matrix of the mitochondrion. Beta-oxidation begins as soon as the fatty acyl CoA appears in the matrix, using repeated cycles of four steps, so that with each cycle the fatty acyl CoA is broken down to form a new fatty acyl CoA, shortened by two carbon atoms, plus an acetyl CoA. Figure 6.14 summarizes the four steps in detail. Reactions 1 to 3 are designed to change carbon 3 (the β-carbon) from a methylene (CH_2) group to a carbonyl group (C=O). Reaction 4 introduces a CoA group to carbon 3, cleaving off an acetyl CoA and leaving an acyl CoA shortened by two carbon atoms. This process is repeated until all the carbon atoms in the original fatty acyl CoA are in the form of acetyl CoA. For example, starting with palmitoyl CoA (a 16-carbon fatty acyl group attached to CoA), the four reactions of beta-oxidation would be repeated seven times to generate a total of eight acetyl CoA.

The first reaction is catalyzed by the enzyme acyl CoA dehydrogenase, located on the matrix side of the inner membrane of the mitochondrion. Using FAD as a coenzyme, acyl CoA dehydrogenase removes two

▶ K E Y   P O I N T ◀

The carnitine transport system is necessary for the long chain fatty acids (16-20 carbon atoms) that are typically stored as triacylglycerol in adipose tissue and skeletal muscle. **Medium chain triacylglycerols** (MCT) contain fatty acids that have a length of 6 to 12 carbon atoms. Compared to triacylglycerols with long chain fatty acids, the MCT are more rapidly digested and their fatty acids absorbed because these are not incorporated into chylomicrons. In addition, the MCT can enter mitochondria for oxidation without the need for carnitine.

Figure 6.14 Beta-oxidation of fatty acids, a four-step process that removes acetyl CoA units one at a time from a fatty acyl CoA. The starting material is a fatty acyl group in which carbon 1 is attached to the CoA; carbon 2 (or α-carbon) is the next carbon, and the one beyond that is carbon atom 3 or β-carbon. Steps 1 to 3 are designed to change carbon 3 from a methylene (CH_2) to a carbonyl (C=O). The fourth step splits the original fatty acyl CoA between carbons 2 and 3 by introducing a CoA, resulting in an acetyl CoA and a new acyl CoA shortened by two carbons. This process is repeated until the original fatty acid is completely changed to acetyl CoA groups.

electrons as hydrogen atoms, one from carbon 2 and one from carbon 3. In the process, FAD is reduced to $FADH_2$. The electrons are then passed through an electron transfer flavoprotein to coenzyme Q, resulting in the formation of coenzyme QH_2. The removal of the two hydrogen atoms in step 1 results in the formation of a carbon-to-carbon double bond between carbons 2 and 3 of the fatty acyl group. The double bond is trans because the two hydrogens are on opposite sides, although this is not shown. The free energy released in the electron transfer from acyl

CoA to FAD to coenzyme Q is insufficient to transport six protons from the matrix to the cytosolic side of the inner membrane. In this regard, the reaction is like that of succinate dehydrogenase or mitochondrial glycerol phosphate dehydrogenase.

In the second reaction, trans enoyl CoA is hydrated, accepting a water molecule. The enzyme catalyzing this reaction, *enoyl CoA hydratase*, is located in the matrix. The OH part of the water is added to carbon 3 (β-carbon) while the other hydrogen atom of water is added to carbon atom 2. The product is a 3-hydroxyacyl CoA (or β-hydroxyacyl CoA). The enzyme, enoyl CoA hydratase, is located in the matrix. In the third reaction, catalyzed by the matrix enzyme *3-hydroxyacyl CoA dehydrogenase*, the β-carbon (carbon 3) of the 3-hydroxyacyl CoA is oxidized, losing a hydride ion and a proton. Such oxidation reactions always use the nicotinamide coenzymes, NAD⁺ in this case. This reaction changes the OH group to a carbonyl. In the last step, CoA attaches to carbon 3 (the β-carbon), which allows carbons 1 and 2 to come off as an acetyl CoA, leaving a new acyl CoA shortened by two carbon atoms. The enzyme responsible for the fourth reaction is *thiolase* (called thiolase because the CoA contains a terminal SH [thiol] group).

> ### ▶ K E Y   P O I N T ◀
>
> The steps in the citric acid cycle from succinate to oxaloacetate are remarkably similar to the first three steps in beta-oxidation. For both, there is a dehydrogenation, generating a carbon-to-carbon double bond (fumarate in the citric acid cycle); a hydration (forming malate); and then a further dehydrogenation generating the keto group in oxaloacetate.

The reactions in figure 6.14 are shown as if they are irreversible. Actually, the free energy changes overall are modest, and the reactions could rightfully be described as equilibrium (reversible). Direction is given to beta-oxidation by the fact that the acetyl groups feed into the citric acid cycle. There is a considerable energy yield from beta-oxidation of fatty acids. For example, from the 16-carbon palmitic acid, the yield is 8 acetyl CoA, 7 FADH₂, and 7 NADH. If you do an ATP yield analysis from the oxidation of a fatty acid, remember to account for the fact that formation of fatty acyl CoA costs the equivalent of two ATP.

Oxidation of Unsaturated Fatty Acids

Many fatty acids stored in body fat are unsaturated. These fatty acids are also sources of energy in the form of acetyl CoA units and reduced coenzymes during beta-oxidation. Recall that the naturally occurring unsaturated fatty acids in foods and those stored in our bodies are in the cis configuration. However,

step 2, catalyzed by enoyl CoA hydratase, needs a trans double bond for its substrate. Therefore, an additional step is required that is catalyzed by a new enzyme, *enoyl CoA isomerase*, which converts the cis to a trans carbon-to-carbon double bond.

Other unsaturated fatty acids, such as linoleic acid, have their double bonds in the wrong position as well as the wrong (cis) configuration for beta-oxidation. For these, *enoyl CoA isomerase* converts the cis to a trans double bond, but this is not enough. An additional enzyme known as a *reductase* converts the carbon-to-carbon double bond in the wrong position into a carbon-to-carbon single bond by the addition of hydrogen in the form of NADPH and H⁺. As we have seen, NADPH is closely related to NADH and serves to reduce double bonds during synthetic reactions.

Oxidation of Odd-Carbon Fatty Acids

Most of the fatty acids found in land animals contain an even number of carbon atoms. However, plants and marine animals contain fatty acids with an odd number of carbon atoms. Oxidation of these odd-carbon fatty acids occurs through the normal route of beta-oxidation. In the last cycle of beta-oxidation, the starting substrate is a five-carbon acyl CoA. When this is split by the thiolase enzyme, the result is acetyl CoA and propionyl CoA, which contains the three-carbon propionyl group attached to CoA. Propionyl CoA is not treated as a fat. In fact, it is converted into the citric acid intermediate succinyl CoA in a three-step process. This means that propionyl CoA generates a gluconeogenic precursor (see chapter 5).

OXIDATION OF KETONE BODIES

Ketone bodies are sometimes described as the water-soluble energy-providing lipids. Figure 6.15 shows the structure of the three ketone bodies. D-3-hydroxybutyrate and acetoacetate are formed primarily in the matrix of liver mitochondria and to a minor extent in the kidneys. These are soluble in aqueous medium and are the major representatives of the ketone body family. Acetoacetate undergoes slow, spontaneous decarboxylation to yield acetone and carbon dioxide. The acetone is formed in only very small amounts, but because it is volatile, its presence, like that of alcohol, can be smelled in the breath.

Figure 6.15   The three members of the ketone body family. D-3-hydroxybutyrate and acetoacetate are water-soluble lipids that can be oxidized in cells. Acetone, formed by spontaneous decarboxylation of acetoacetate, is present in only trace amounts.

Formation of Ketone Bodies

Ketone body formation accelerates in normal individuals when the body carbohydrate content is extremely low, for example during starvation or self-controlled fasting, when extremely low-carbohydrate diets are eaten, and during prolonged exercise with insufficient carbohydrate ingestion. In all of these conditions, carbohydrate content in the body is low, and thus carbohydrate utilization and blood insulin concentration are low. Accelerated ketone body formation also occurs during uncontrolled diabetes mellitus. Although blood glucose concentration is high because insulin is lacking, carbohydrate utilization by insulin-dependent tissues is low.

With starvation or fasting, low-carbohydrate diets, prolonged exercise, or uncontrolled diabetes mellitus, the adipose tissue releases large quantities of fatty acids due to an imbalance between triacylglycerol formation and lipolysis caused by low blood insulin concentration. Recall that insulin promotes triacylglycerol formation in fat cells and inhibits lipolysis. Thus the net effect of low blood insulin is that lipolysis greatly exceeds triacylglycerol formation, resulting in a large increase in blood FFA concentration. For most healthy people, blood FFA concentrations range between 0.25 to 0.50 mM over the day. In uncontrolled diabetes mellitus, for

example, blood FFA concentrations can reach toxic levels of 4 mM.

Under normal conditions, the liver is able to extract about 30% of the FFA that pass through it. With high blood FFA concentration, the liver extracts even more. The fate of the extracted fatty acids by liver is formation of fatty acyl CoA, then subsequent formation of triacylglycerol or phospholipid, or entry into the mitochondrial matrix. During conditions favoring ketone body formation, entry of fatty acyl CoA into liver mitochondria is accelerated. Beta-oxidation of the fatty acyl CoA is greatly augmented, forming acetyl CoA at a rate that far exceeds the capacity of the liver mitochondria to oxidize it by the citric acid cycle. Moreover, conditions favoring ketone body formation are characterized by low matrix concentrations of oxaloacetate in the liver. Accordingly, much of the acetyl CoA is directed to the formation of acetoacetate. Some acetoacetate is reduced to D-3-hydroxybutyrate. Figure 6.16 summarizes ketone body formation. The two major ketone bodies formed are acetoacetate and D-3-hydroxybutyrate. The ratio of 3-hydroxybutyrate to acetoacetate depends on the [NADH]/[NAD+] ratio in the liver.

Figure 6.17 shows in more detail the steps between acetyl CoA and acetoacetate. In the first step, two acetyl CoA combine to form acetoacetyl CoA catalyzed by *thiolase*, the same enzyme catalyzing the fourth reac-

Figure 6.16   The steps in the formation of ketone bodies, starting with increased blood FFA (free fatty acid) concentrations and leading to the formation of acetoacetate in liver mitochondria. Acetoacetate can be reduced to 3-hydroxybutyrate; thus liver produces both acetoacetate and 3-hydroxybutyrate.

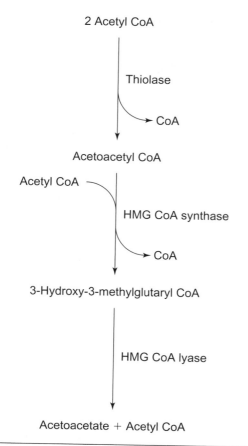

2 Acetyl CoA

↓ Thiolase

→ CoA

Acetoacetyl CoA

Acetyl CoA ↘

HMG CoA synthase

→ CoA

3-Hydroxy-3-methylglutaryl CoA

↓

HMG CoA lyase

↓

Acetoacetate + Acetyl CoA

Figure 6.17 The formation of acetoacetate from acetyl CoA in three steps, each catalyzed by a separate enzyme. HMG represents 3-hydroxy-3-methylglutaryl.

tion in beta-oxidation. This reaction is freely reversible. Acetoacetyl CoA combines with another acetyl CoA to form 3-hydroxy-3-methylglutaryl (HMG) CoA, catalyzed by *HMG CoA synthase*. The last step is splitting off an acetyl CoA, leaving acetoacetate. This reaction is catalyzed by *HMG CoA lyase*.

The Fate of Ketone Bodies

Ketone bodies are used as a fuel for mitochondria in extrahepatic tissues (i.e., nonliver tissues), chiefly skeletal muscle, heart, and the brain. Normally, the brain uses glucose as its primary fuel because FFA cannot pass the blood–brain barrier. However, as blood glucose concentration falls and blood ketone body concentration rises, the brain can extract and use 3-hydroxybutyrate and acetoacetate. Ketone body oxidation in tissues (see figure 6.18) occurs as follows:

1. Ketone bodies are taken up by extrahepatic cells and transported into the mitochondrial matrix.
2. D-3-hydroxybutyrate is oxidized to acetoacetate using the enzyme *3-hydroxybutyrate dehydrogenase*. This reaction generates acetoacetate and NADH.
3. Acetoacetate takes a CoA from succinyl CoA and forms acetoacetyl CoA.
4. Acetoacetyl CoA is split into two acetyl CoA using the last enzyme of beta-oxidation of fatty acids, *thiolase*.
5. The acetyl CoA is oxidized in the citric acid cycle.

Ketosis

Prolonged depletion of body carbohydrate stores or uncontrolled diabetes mellitus leads to a condition known as ketosis. This results from accelerated ketone body formation and is characterized by the following:

1. **Ketonemia**—an increase in ketone body concentration in the blood
2. **Ketonuria**—loss of ketone bodies in the urine
3. Acetone breath—loss of volatile acetone in expired air
4. Elevated blood FFA concentration, resulting in accelerated formation of ketone bodies and low blood insulin

3-Hydroxybutyrate + NAD⁺ ← 3-Hydroxybutyrate dehydrogenase → Acetoacetate

Acetoacetate + Succinyl CoA ← CoA transferase → Acetoacetyl CoA + Succinate

Acetoacetyl CoA + CoA ← Thiolase → 2 Acetyl CoA

Figure 6.18 Acetoacetate and 3-hydroxybutyrate as fuels for mitochondria. The 3-hydroxybutyrate must first be oxidized to acetoacetate.

5. Acidosis (or ketoacidosis)—a fall in blood pH due to the formation of H$^+$ when acetoacetate is formed and the loss of cations (e.g., Na$^+$) when ketone bodies are excreted (this can be fatal in uncontrolled diabetes mellitus)

6. Hypoglycemia (in normal persons) or hyperglycemia (in uncontrolled diabetes mellitus)

> ### ▶ K E Y P O I N T ◀
>
> Ketones can be a useful fuel during submaximal exercise, sparing the use of muscle glycogen and blood glucose. Use of ketone bodies to fuel exercise performance is enhanced with a week or more of a ketogenic diet. This increases the activity of enzymes needed to form acetyl CoA in mitochondria from ketone bodies. The net effect is to reduce the need to provide the citric acid cycle with acetyl CoA from pyruvate.

SYNTHESIS OF FATTY ACIDS

Most fatty acids used by humans come from dietary fat. However, humans can synthesize fatty acids from acetyl CoA in liver, mammary gland, and adipose tissue, although this is not a major process in people with a relatively high-fat diet. The acetyl CoA comes from amino acids, carbohydrates, and alcohol; but carbohydrate normally provides the bulk of the carbon atoms for fatty acid synthesis. The synthesis of fatty acids is often described as de novo lipogenesis (DNL), which means synthesis of fatty acids starting from acetyl CoA. The 16-carbon palmitic acid is synthesized first and can be extended in length or desaturated to make some unsaturated fatty acids (but not the essential fatty acids needed in the diet). Synthesis of palmitic acid occurs in the cytosol, whereas oxidation occurs in the mitochondria. Synthesis and degradation of fatty acids also use

different enzymes, permitting separate regulation of these two opposing processes.

Pathway to Palmitic Acid

Synthesis starts with the carboxylation of acetyl CoA to make a three-carbon molecule known as malonyl CoA (see figure 6.19). The cytosolic enzyme *acetyl CoA carboxylase* catalyzes this reaction, which involves the carboxylation of acetyl CoA. The actual substrate is the bicarbonate ion (HCO$_3^-$), which is transferred to acetyl CoA by the B vitamin biotin. This reaction is the committed step in fatty acid synthesis and is positively affected by insulin. We have also seen biotin involved in the pyruvate carboxylase reaction in gluconeogenesis. The malonyl CoA units are used to make fatty acids in the cytosol.

To make malonyl CoA in the cytosol, a continuous supply of acetyl CoA is necessary. Recall that acetyl CoA is formed in the mitochondrial matrix, primarily from pyruvate in the pyruvate dehydrogenase reaction and beta-oxidation of fatty acids. Thus there must be a mechanism to get acetyl CoA from the mitochondrial matrix to the cytosol. Moreover, there is a need for NADPH in fatty acid synthesis, so there must be a way of producing this in the cytosol. NADPH can arise from the pentose phosphate pathway (see chapter 5). It can also be produced by *malic enzyme* as shown in the following reaction.

malate + NADP$^+$ ↔ pyruvate + CO$_2$ + NADPH + H$^+$

Figure 6.20 outlines how an excess of glucose can be used to make malonyl CoA, the precursor for fatty acid synthesis. Glycolysis converts glucose to pyruvate. It is translocated into the matrix and converted to acetyl CoA by pyruvate dehydrogenase. Acetyl CoA combines with oxaloacetate to produce citrate, using the first enzyme of the citric acid cycle, *citrate synthase*. The citrate is transported into the cytosol by an antiport, with malate simultaneously going from the cytosol to the matrix. In the cytosol, citrate is cleaved into oxaloacetate and acetyl CoA by

Figure 6.19 Creation of malonyl CoA, the precursor for the synthesis of fatty acids, by the carboxylation of acetyl CoA.

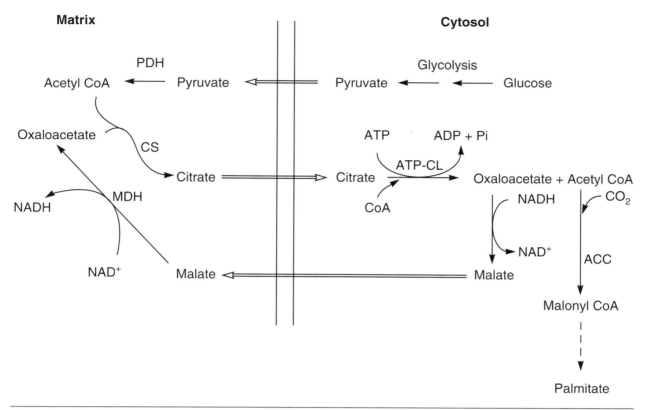

Figure 6.20 Use of excess glucose to make malonyl CoA. Conversion of glucose into fatty acids starts in the cytosol with the conversion of glucose to pyruvate in glycolysis. Pyruvate is translocated into the mitochondrial matrix and converted to acetyl CoA by pyruvate dehydrogenase (PDH). Acetyl CoA combines with oxaloacetate to form citrate using citrate synthase (CS). Citrate crosses the inner mitochondrial membrane to the cytosol using an antiport, with malate going the other way. In the cytosol, citrate is split into acetyl CoA and oxaloacetate by ATP-citrate lyase (ATP-CL). The acetyl CoA is carboxylated to make malonyl CoA by acetyl CoA carboxylase (ACC). The oxaloacetate is reduced to malate by a cytosolic malate dehydrogenase, and malate returns to the matrix using the citrate-malate antiport. Open arrows refer to transport across a membrane.

ATP-citrate lyase. This enzyme uses the free energy of hydrolysis of ATP not only to split citrate, but also to attach a CoA. The acetyl CoA in the cytosol is now a substrate for acetyl CoA carboxylase. The oxaloacetate is reduced to malate and transported back into the matrix by the citrate-malate antiport.

Palmitic acid synthesis occurs using a large enzyme, *fatty acid synthase* (FAS), which is composed of two very long, multifunctional polypeptide chains. The FAS has seven distinct enzyme activities. We will describe fatty acid synthesis as it proceeds in stages. Refer to figure 6.21 for the details. In the loading stage, an acetyl CoA and a malonyl CoA are each transferred by separate *transacetylase* enzymes to the thiol end (–SH) of an *acyl carrier protein* (ACP) part of the FAS. These transfers result in an acetyl-ACP and malonyl-ACP. In the condensation stage, the acetyl group attaches to the end of the malonyl-ACP, and a CO_2 is eliminated. This creates a four-carbon acetoacetyl group attached to one of the ACP. This contains

a carbonyl group (C=O) that must be converted to a methylene (CH_2), which is accomplished in three stages. In the first stage, the carbonyl is reduced to an alcohol group by reduction using NADPH. The next stage involves a dehydration, which eliminates the alcohol OH group and another H, creating a carbon-to-carbon double bond. Finally, another reduction, again using NADPH, eliminates the double bond. The reduction, dehydration, and reduction are exactly opposite to the first three reactions in beta-oxidation, where instead of elimination of a carbonyl group as occurs in fatty acid synthesis, the carbonyl is created by dehydrogenation, hydration, and then dehydrogenation.

Figure 6.21 shows the formation of a four-carbon acyl group only. The FAS complex can form a 16-carbon palmitoyl group, which is removed as palmitate. To go from the four-carbon group attached to an ACP as shown in figure 6.21 to a palmitoyl ACP, we need to add 12 more carbons. The first stage in elon-

Figure 6.21 Synthesis of fatty acids on a large, multiunit enzyme, fatty acid synthase (FAS), containing acyl carrier protein (ACP) subunits with a free thiol (SH) group. The process begins as an acetyl CoA and a malonyl CoA each attach to an ACP; CoA is released. The acetyl group leaves its ACP and attaches to the malonyl group with the elimination of a CO_2. This forms a four-carbon acetoacetyl ACP. The carbonyl group (C=O) is eliminated in three steps involving reduction using NADPH, a dehydration, and another reduction using NADPH. The resulting product is butyryl ACP. The process continues when another malonyl CoA is attached to the free ACP. Ultimately, a 16-carbon palmitoyl ACP is produced and then a palmitic acid is released.

gating our butyryl ACP in figure 6.21 would involve, first, attachment of a malonyl CoA to the free ACP group. This malonyl group would condense with our butyryl ACP with the elimination of a CO_2, and we would create a new six-carbon ACP with a carbonyl group that we would need to eliminate by reduction, dehydration, and reduction. This process continues until a palmitoyl ACP is formed. At this point, the palmitoyl ACP is split by a *thioesterase* enzyme, releasing palmitate (palmitic acid).

To make one palmitate requires the following precursors: one acetyl CoA, seven malonyl CoA, and 14 NADPH + 14 H^+. It is obvious that a supply of reducing equivalents in the form of NADPH is necessary, emphasizing the importance of the pentose phosphate pathway and the malic enzyme as suppliers of NADPH. The need for malonyl CoA is obvious, and as we will see in the next section, the formation of malonyl CoA is a crucial point for regulation.

Regulation of Fatty Acid Synthesis

As we have seen, control of a number of enzymes in liver is dependent upon our diets. For example, the genes coding for enzymes of gluconeogenesis are stimulated by fasting or a low-carbohydrate diet, but repressed if the diet is high in carbohydrate. On the other hand, genes for the enzymes of glycolysis are repressed by fasting or a low-carbohydrate diet and activated if the diet is rich in carbohydrate. Accordingly, we might suspect that genes coding for enzymes of de novo lipogenesis (DNL) should depend on the energy and macronutrient content of the diet, with maximal expression if there is an excess of food energy in the form of carbohydrate. This is the case, as a carbohydrate-rich diet leads to an increased expression of the genes for acetyl CoA carboxylase, which makes the key precursor, and FAS, which produces palmitate. Since the concentration of insulin in blood is also influenced by feeding, especially carbohydrate, it should not be a surprise to learn that insulin plays a key role in promoting DNL. On the other hand, demonstrating DNL has been difficult in humans because of the size of our fat depots compared to the amount of palmitate actually formed by DNL each day. For example, a person with 15% body fat and a body weight of 70 kg (154 lb) has approximately 10 kg (22 lb) of stored TAG. Assuming that the person mobilizes each day a maximum of 100 g of fat (a large amount), this represents only 1% of the total stored fat. In addition, if the person ingests about 100 g of fat, one can easily understand how difficult it is to demonstrate that 20 g of palmitate is formed each

day through DNL by the liver and adipose tissue. The reality is that DNL is of minor importance to the overall energy balance of the average person on a typical mixed diet.

With its role in providing malonyl CoA, the enzyme acetyl CoA carboxylase (ACC) is a prime site for regulation of DNL. Formation of malonyl CoA provides the precursor to synthesize palmitate. Moreover, malonyl CoA is a potent inhibitor of CPT I, which regulates the oxidation of LCFAs. Thus the activation of ACC both promotes palmitoyl synthesis and inhibits fatty acid oxidation. Acetyl CoA carboxylase is controlled by allosteric and phosphorylation mechanisms. Citrate is a positive allosteric effector for ACC, whereas fatty acyl CoA and malonyl CoA allosterically inhibit ACC. The potency of the inhibition by the fatty acyl CoA molecules is proportional to the carbon length of the fatty acyl units. An abundance of fatty acyl CoA and malonyl CoA signals a surfeit of products and therefore no need to make more fatty acids. On the other hand, an increase in citrate concentration indicates an overabundance of acetyl CoA and a need to synthesize fatty acids. An abundance of carbohydrate means that fatty acid synthesis from carbohydrate should take place. An accumulation of fatty acyl CoA reflects a lack of carbohydrate substrate, and thus fatty acid oxidation should predominate.

Acetyl CoA carboxylase activity is inhibited by covalent phosphorylation, directed by a cAMP-dependent protein kinase (protein kinase A-PKA) and AMP-activated protein kinase (AMPK). We learned in the previous chapter that AMPK is active under stressful conditions in which ATP is increasingly needed. Dephosphorylation by protein phosphatases activates ACC. Phosphorylation and dephosphorylation of ACC depend on the relative levels of insulin and glucagon, with insulin promoting the unphosphorylated (active) form of ACC and glucagon promoting its phosphorylation. The latter results in inhibition of ACC. A relative rise in insulin concentration leads to a decrease in cAMP concentration by activation of cAMP phosphodiesterase; insulin also activates phosphoprotein phosphatases, which dephosphorylate ACC.

Malonyl CoA is also a potent inhibitor of the oxidation of fatty acids. In tissues where fatty acids can be oxidized to acetyl CoA in the mitochondria and where fatty acids can be simultaneously synthesized in the cytosol (primarily liver), malonyl CoA can inhibit the initial transfer of the fatty acyl group from CoA to carnitine at the cytosolic side of the inner membrane using the enzyme carnitine acyl transferase I (see figure 6.13 on p. 159). This process prevents futile cycling in which fatty acids are actively broken down in the mitochondria while at the same time being synthesized in the cytosol. Such regulation is active in the liver and adipose tissue, the main sites for fatty acid synthesis in humans. On another level, a relative increase in the intake of carbohydrate in the diet induces the transcription of genes coding for lipogenic enzymes. A high-fat diet, on the other hand, results in a decrease in the transcription of genes for lipogenic enzymes.

> ### ▶ KEY POINT ◀
>
> If de novo lipogenesis is not a prominent feature of human metabolism, one might ask how people can become obese on a high-carbohydrate and relatively low-fat diet. People become obese whenever they have a prolonged excess of food energy intake, whether it is carbohydrate or fat. If the diet provides excess calories as carbohydrates, the body uses carbohydrates more as an energy source so that fat stores and dietary fat are oxidized less. The excess carbohydrate, beyond what is needed to restore liver and muscle glycogen, is converted to fatty acids or used as a source of glycerol 3-phosphate to help store even the small amount of dietary fat.

FAT AS A FUEL FOR EXERCISE

Human fat stores are huge compared to carbohydrate stores. Moreover, considering the importance of glucose as fuel for the brain, it is important to use fat instead of carbohydrate during exercise, if this is possible. The lipid used to fuel muscular work comes from fatty acids released from adipose tissue, traveling to muscle as FFA. Intramuscular triacylglycerol is also a significant source of fatty acids. A third possibility is fatty acids released from plasma triacylglycerols by lipoprotein lipase, although the evidence suggests that this can provide at most 10% of fat oxidized during exercise lasting an hour or more.

Free Fatty Acids and Exercise

Blood glucose levels are well maintained during exercise lasting up to 60 min. This is due to the careful match between uptake of glucose from the blood and

release of glucose to the blood from the liver. The relative constancy of blood glucose concentration during exercise is not observed with the concentration of FFA. With exercise, there is an increase in lipolysis, brought about mainly by the increase in epinephrine in the blood, increased sympathetic nervous activation of adipocytes through norepinephrine, and the decrease in plasma insulin concentration. In addition, re-esterification of fatty acids in fat cells decreases, so more are released to the blood. Aiding this is a small increase in adipose tissue blood flow that allows a greater proportion of fatty acids released from fat cells to enter the general circulation (Jeukendrup 2003). Figure 6.22 illustrates the effect of 90 min of exercise at a workload of 60% of $\dot{V}O_2$max on arterial FFA concentration for a subject in the postabsorptive state, and then for the same subject, on a separate occasion, 2 h after eating a meal containing 50% of the food energy as carbohydrate. Overall, the shapes of two curves are similar, but the curve for the fasted state is displaced higher. The arterial FFA concentration reflects the relative balance between release of fatty acids from adipose tissue (R_a, rate of appearance) and FFA uptake by exercising muscle (R_d, rate of disappearance).

At the start of submaximal exercise, there is an immediate increase in the rate of FFA uptake by the exercising muscle (R_d). This exceeds the more slowly responding increase in adipose tissue lipolysis (R_a), such that the plasma FFA initially falls. With the gradual increase in lipolysis, the rate of release of FFA from adipose tissue meets, then exceeds, the rate of

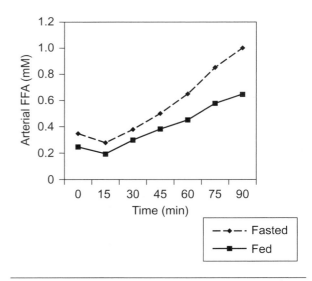

Figure 6.22 The arterial free fatty acid (FFA) concentration during 90 min of exercise at 60% of $\dot{V}O_2$max when a subject had fasted before the exercise and then 2 h following a meal containing 50% of the energy from carbohydrate.

uptake of plasma FFA ($R_a > R_d$). As a result, the concentration of plasma FFA tends to rise over the course of the exercise period. In the fed state, the adrenergic effect to stimulate lipolysis is blunted by the effect of the previous meal, especially the concentration of insulin. Thus plasma FFA concentration begins at a lower level and remains lower throughout the 90 min exercise period. As we will show in a subsequent section, doing the same exercise protocol in the fed compared to the fasted state is characterized by a higher respiratory exchange ratio (RER), reflecting greater oxidation of carbohydrate.

During more intense exercise, release of fatty acids from adipose tissue is reduced even though there is a greater adrenergic stimulus for lipolysis. Two mechanisms have been proposed to explain this. At higher intensities of exercise, there is a higher concentration of lactate in blood. Researchers have proposed that lactate inhibits lipolysis, but this is a controversial mechanism. It is possible that lactate increases re-esterification of fatty acids into adipose tissue triacylglycerols through a process of glyceroneogenesis, discussed earlier. This means that rather than being released from the fat cell to enter the blood, the fatty acid is esterified to a new triacylglycerol molecule. A more plausible explanation is that adipose tissue blood flow is reduced at higher intensities of exercise, since blood flow regulation will aim at directing flow to the active muscles by decreasing flow to the liver, kidney, and adipose tissues. This effect is related to the relative intensity of exercise. Therefore, fatty acids released by lipolysis are less likely to enter the general circulation due to a lower adipose tissue blood flow (Jeukendrup et al. 1998b).

To focus only on changes in plasma FFA concentration as an indication of fat oxidation during exercise can be very misleading, as this does not tell us what is happening with these fatty acids. We could measure the RER to determine the relative use of fat, since we know that the lower the value, the more that fat is being oxidized. Again, this does not indicate what tissues are using the fat, although with a large increase in metabolic rate because of the exercise it is safe to assume that exercising skeletal muscle is the major tissue contributing to the RER value. We could measure the concentration of FFA in arterial and venous blood across an exercising limb to determine the relative use of plasma FFA. However, accurate measures of FFA uptake into skeletal muscle require good values for blood flow, and this is not a simple technique. Moreover, subcutaneous adipose tissue can be an active source of FFA to the blood of the exercising limb, again leading to complications in the interpretation of the arterial and venous FFA concentration differences. A technique that is proving quite helpful is to infuse a labeled fatty acid into a vein at a precise rate. Sampling blood before and during exercise for changes in the concentration of the labeled fatty acid allows one to make predictions about the rate of disappearance of fatty acids from the blood (Bülow 2003).

Use of Intramuscular Triacylglycerol

As we discussed previously, skeletal muscle contains intracellular lipid drops, called intramuscular triacylglycerol (IMTG) or intramyocellular lipid (IMCL). There is also a skeletal muscle HSL. These facts are indisputable. However, the extent to which the IMTG is oxidized to support metabolism of muscle during prolonged low- to moderate-intensity exercise has generated considerable controversy (Watt et al. 2002). Differences in IMTG used during exercise in a variety of studies can be related to how this is measured. One technique is to chemically determine the amount of triacylglycerol extracted from muscle biopsy samples obtained before and after exercise. This technique has been used in numerous studies, but significant decreases in IMTG concentration over an exercise bout may not be observed, and there can be marked intraindividual differences. Nonetheless, Watt and colleagues (2002) suggested that the difference in IMTG after versus before exercise in a cross section of studies could provide about two-thirds of the energy obtained from glycogen oxidation.

A second technique is to use ^1H magnetic resonance spectroscopy (MRS). Although this technique provides reliable data and is noninvasive, the equipment is expensive, beyond the means of most exercise laboratories (van Loon et al. 2003). A third technique to determine IMTG utilization during exercise employs a labeled fatty acid (e.g., ^{13}C palmitate bound to albumin) infused into a vein. The R_d of the labeled fatty acid from the blood would reflect FFA disappearance and presumably oxidation. Because some of the labeled fatty acid that disappears from the blood could be incorporated into IMTG, this technique can be misleading unless one also determines the oxidation of the labeled fatty acid by determining labeled CO_2. Properly performed, this technique assesses IMTG oxidation as the difference between total fat oxidation, based on total oxygen consumption and RER values, and the FFA oxidation based on expiration of labeled CO_2.

A technique that was used previously was to measure the release of glycerol from exercising muscle based on the assumption that the glycerol released during lipolysis of TAG in skeletal muscle could not be metabolized. There is now convincing evidence that skeletal muscle can convert glycerol to glycerol 3-phosphate because of the presence of the enzyme *glycerol kinase*. This means that earlier measures of IMTG lipolysis based on glycerol release underestimated its importance. One of the confounding issues when one is attempting to understand the usage of lipid as a fuel is the fact that its use can be so dramatically influenced by the availability of carbohydrate stores in the form of muscle glycogen. Moreover, feeding before or during exercise, or the time from the last meal, can play a major role in determining fuel utilization during exercise. Depressing the release of fatty acids from adipose tissue by the use of high concentrations of nicotinic acid results in elevations in IMTG utilization during exercise (Watt, Holmes, et al. 2004). However, increased IMTG use cannot compensate for the reduction of FFA available to muscle, so the overall carbohydrate contribution to exercise is increased. The next section outlines some of the factors that contribute to our current understanding of exercise metabolism.

> ### ▶ K E Y P O I N T ◀
>
> While the body contains limited carbohydrate stores, there is an abundance of stored triacylglycerol. Carbohydrate is a better fuel for exercising muscle for three reasons: (1) It can generate acetyl CoA for the citric acid cycle at a much higher rate than can fatty acids from adipose tissue and intramuscular triacylglycerols; (2) more ATP per unit of oxygen is consumed with carbohydrate; and (3) carbohydrate can generate ATP in the absence of oxygen via anaerobic glycolysis.

METABOLISM DURING EXERCISE: FAT VERSUS CARBOHYDRATE

At rest, during the postabsorptive state, the RER is about 0.82, indicating that lipid is the predominant fuel being oxidized by the body. The respiratory quotient (RQ) across rested muscle is even lower, demonstrating that inactive muscle uses fat as its primary fuel. If a person in the rested state eats a carbohydrate-rich meal, blood glucose concentration is elevated, blood insulin concentration is increased, blood FFA concentration is reduced, whole-body RER is increased, the RQ across rested muscle increases, and there is an increase in lactate release from the muscle, even though it is at rest (Didier et al. 2000). In a prolonged fasted state, when body carbohydrate stores are severely reduced, blood FFA levels are elevated and overall body RER is reduced. Observations like these were reported by Randle and his colleagues in the 1960s; and from these studies the authors proposed a *glucose-fatty acid cycle*. The proposition regarding this cycle is that there is a reciprocal relationship between carbohydrate and fat in terms of oxidation. As already mentioned, an increase in carbohydrate increases its oxidation and depresses that of fat. If fatty acid concentration in the blood is elevated, fat oxidation is promoted and carbohydrate oxidation is depressed (Roden 2004). As we will see, the use of carbohydrate and fat during exercise follows a similar trend, but there are limitations as we will discuss.

Role of Exercise Intensity

As noted in the previous chapter, it has been convincingly demonstrated that in well-nourished individuals during exercise, the RER increases in proportion to the increase in exercise intensity. This means that as exercise intensity increases, the contribution of carbohydrate oxidation to ATP formation increases whereas that of lipid oxidation decreases. This is illustrated in figure 6.23, which shows the relative release of FFA and glucose into the blood and the utilization of glycogen during exercise at different intensities. Several points are immediately clear. The release of fatty acids into the blood from adipose tissue stores rises in parallel with exercise intensity to approximately 50% of $\dot{V}O_2$max, then gradually declines. On the other hand, release of glucose into the blood from the liver increases with exercise intensity. As we have noted, glycogen utilization increases exponentially with increasing exercise intensity.

George Brooks has proposed a **"crossover concept"** to explain fuel utilization during exercise in terms of the balance between carbohydrate and fat. The crossover point is that relative exercise intensity at which ATP formation from the use of carbohydrate exceeds that of lipid. Exercise at power outputs beyond this point will rely more and more on carbohydrate oxidation and less and less on fat. Brooks illustrated this with data obtained from a

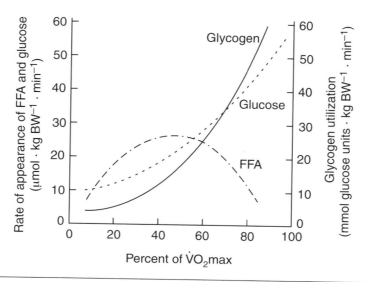

Figure 6.23 Use of glucose, glycogen, and free fatty acids (FFA) during exercise at various relative intensities, expressed as a percentage of maximum oxygen uptake (VO₂max). The use of FFA and glucose is based on their rates of appearance in the blood. Glycogen use is determined by the rate of its loss from skeletal muscle.

number of studies, as shown in figure 6.24. Data were obtained from a variety of experiments using four different species of animals, varying widely in terms of their VO₂max expressed as milliliters per kilogram per minute. In all cases, the data generally

fell along the lines shown in figure 6.24, indicating that the relative use of fuels is closely tied to relative intensity of exercise.

Role of Diet

In the previous chapter, we showed (see figure 5.14 on p. 125) that pre-exercise diet, and hence muscle glycogen concentration, could play an important role in fuel utilization during submaximal exercise. With elevated muscle glycogen stores, more glycogen is utilized and more lactate is produced compared to values from the identical exercise task performed with low glycogen stores. Elevated muscle glycogen stores are even associated with lower blood FFA levels at rest (Didier et al. 2000). Although not directly addressed in chapter 5, any reduction in carbohydrate use during exercise is made up by lipid oxidation because the only other possible fuel, protein, provides at most only about 5% or less of total energy needs. Thus it is clear that the reciprocal relationship between carbohydrate and lipid oxidation at rest extends to exercise up to approximately 90% of VO₂max. Beyond this intensity, exercise relies almost exclusively on carbohydrate. At rest, carbohydrate ingestion raises blood glucose and insulin levels. Insulin not only promotes glucose uptake by skeletal and cardiac muscle as well as adipose tissue; it is a potent inhibitor of lipolysis in adipose tissue, thereby depressing blood FFA concentrations.

Figure 6.24 Relative contribution of lipid (fat) and carbohydrate (CHO) oxidation to energy needs during exercise at different intensities. The crossover point is the exercise intensity beyond which energy produced by carbohydrate oxidation exceeds that of lipid. The actual lines represent means from a variety of studies using four different species (goats, dogs, rats, and humans).

A diet enriched in fat may alter metabolism during exercise. It is important to distinguish acutely elevating plasma FFA concentration and its effects on exercise metabolism and performance from the effects that accrue as a result of seven days or more of a high-fat diet. Acutely increasing plasma FFA levels can be accomplished by infusion of triacylglycerol into a vein along with heparin, which is known to increase lipoprotein lipase activity. Such a protocol depresses blood glucose utilization at rest and during submaximal exercise and increases the use of fat during exercise as evidenced by a lower RER. A number of studies have shown that a high-fat diet for seven or more days results in an increase in IMTG and an increased reliance on fat oxidation during submaximal exercise (Vogt et al. 2003; Zderic et al. 2004). During exercise, blood FFA and glycerol concentrations are higher, revealing a higher rate of adipose tissue lipolysis. Surprisingly, some studies show no or only slightly reduced muscle glycogen concentrations with a high-fat diet.

Medium chain triacylglycerols containing fatty acids with 6, 8, or 10 carbon atoms have been used as a lipid supplement by some athletes. These are reputed to reduce carbohydrate oxidation because they are not metabolized in the same way typical ingested lipid is. Because of the short length of the fatty acid chain, they are not converted into triacylglycerols and then into chylomicrons in intestinal cells the way typical lipids are. Rather, they exit the gut in the blood and pass through the liver. From the general circulation, they enter cells and can pass into mitochondria without the need for the carnitine transport system. Therefore, unlike LCFAs, their oxidation is not depressed by carbohydrate through a malonyl CoA mechanism. In a variety of exercise studies, MCT use has not been shown to have a significant impact on endurance performance, since

> ### ▷ K E Y   P O I N T ◁
>
> A wonderful theoretical rationale for MCT can be made, but the proof of this effect during exercise studies in the lab is not convincing. This is just one example of many in which hype about certain supplements and reality diverge. Although the effect is not prominently mentioned by proponents, MCT use can produce gastrointestinal distress that is really counter to exercise performance (Vistisen et al. 2003).

MCT ingestion spares neither muscle glycogen nor blood glucose use.

Effects of Feeding During Exercise

We noted that if a person performs a given amount and intensity of exercise with full muscle and liver glycogen stores, he or she will use more glycogen and produce a higher level of muscle and blood lactate compared to levels with the exact same exercise performed when glycogen stores are low. This tells us that if all things are equal, muscle will choose to use more muscle glycogen and blood glucose if available. If a person doing prolonged submaximal exercise does not ingest glucose, the RER during exercise will gradually decline; and after about 1 h, blood glucose levels will decline. If the exercising subject takes in glucose during prolonged exercise, blood glucose levels are better maintained and the RER does not decline to the same extent. In addition, insulin concentrations are elevated, blood FFA levels are depressed, and the total oxidation of carbohydrate is increased and that of fat decreased (Watt, Krustrup, et al. 2004). This is another example of the effect of carbohydrate on promoting its own oxidation at the expense of fat.

Role of Training

Regularly performed endurance exercise can play a significant role in fuel utilization during exercise. The trained muscle has a much larger capacity for oxidative metabolism than untrained muscle, as we have already addressed. During exercise at the same absolute intensity (e.g., 200 W) after compared to before training, fat oxidation is increased, carbohydrate oxidation is decreased, and the ability to exercise to exhaustion is vastly increased. If a person increases his or her $\dot{V}O_2max$ by 20% following an endurance training program, he or she is obviously capable of exercising at a much higher intensity. If we compare the oxidation of carbohydrate and fat during exercise at the same relative intensity posttraining versus pretraining, we see a different metabolic response. During exercise at, say, 60% of posttraining $\dot{V}O_2max$, the workload may be 180 W, while at 60% of pretraining $\dot{V}O_2max$ it may be 144 W. Therefore, we can say with absolute certainty that after training the exercise $\dot{V}O_2$ will be higher at the same relative workload, and more total fuel will be oxidized during exercise. The relative use of carbohydrate and fat during exercise with the greater energy demand is more controversial. Some studies show that the relative use of that fuel is in about the same proportion as before training. Other studies show that the RER is lower, suggesting that relatively more fat is being oxidized to support the higher $\dot{V}O_2$.

Trained individuals have larger IMTG stores and use these stores to a greater extent than the untrained. Trained muscle has a higher lipoprotein lipase activity, so the ability to hydrolyze and use fatty acids derived from VLDLs is enhanced (Morio et al. 2004). One study showed that the genes coding for key enzymes involved in fatty acid transport (FAT/CD36 and CPT I) were increased with just nine days of training. As well, fat oxidation during 60 min of submaximal cycling was increased by almost 25% with this short training program (Tunstall et al. 2002). Training adaptations favoring lipid metabolism are also experienced by older subjects (Pruchnic et al. 2004). As we will emphasize in a later section on mechanisms regulating carbohydrate and lipid metabolism in muscle, AMPK plays a major role in skeletal muscle fuel selection. Endurance training increases total AMPK activity in skeletal muscle and the basal activity of this enzyme (Frøsig et al. 2004).

Role of Body Composition and Gender

Despite the fact that overweight and obese individuals store a much greater proportion of triacylglycerol than their lean counterparts, the extra fat does not predispose overweight individuals to its use, compared to carbohydrate, during exercise (Mittendorfer et al. 2004). Indeed, overweight and obese individuals demonstrated lower levels of adipose tissue lipolysis and oxidation of FFA during exercise. This suggests a blunted response to catecholamines with increasing adiposity. Given gender differences in total body fat and body fat distribution, we might expect this as well. However, whether there are gender differences in fuel oxidation during exercise is a controversial issue (Bülow 2003). Nielsen and colleagues (2003) reported that women released 40% more FFA than men when subjects were compared on the basis of their resting energy expenditure. However, there were no differences between men and women for resting values for RER or FFA concentrations, suggesting that the extra FFA released by women was re-esterified. Venables and colleagues (2005) tested a cross section of 300 trained and untrained men and women in terms of fuel utilization, using indirect calorimetry during a graded exercise test to

exhaustion. Compared to men, women had higher rates of fat oxidation and a later shift to carbohydrate oxidation as exercise intensity increased.

Mechanisms for Glucose–Fatty Acid Cycle

We have discussed the effects of fuel availability on metabolism. In this section, we will look at mechanisms to account for these observations.

Increased Fat Availability and Carbohydrate Utilization

Randle and his colleagues (1963, 1964) offered a mechanism to account for the effects of increased fatty acids and ketone bodies on depressing carbohydrate oxidation (figure 6.25). They proposed an inhibition of glucose utilization in the muscle cell, blocked at two sites. The first step is at the pyruvate dehydrogenase reaction due to an increased concentration of acetyl CoA and NADH in the mitochondrial matrix (see "Regulation of Pyruvate Oxidation" on p. 81 in chapter 4). Second, the elevated matrix acetyl CoA concentration would lead to an increase in citrate in the matrix and therefore an increase in cytosolic citrate. As discussed in chapter 5, phosphofructokinase, the prime regulatory enzyme of glycolysis, is inhibited by citrate, leading to a rise in the concentration of its substrate fructose 6-phosphate and thus glucose 6-phosphate. Since glucose 6-phosphate inhibits hexokinase, this would reduce the gradient for glucose transport into the muscle cell. Moreover, glycogen utilization would be depressed since glycogenolysis is depressed if glucose 6-phosphate concentration is elevated. Overall, muscle carbohydrate oxidation would be depressed, and the difference would be made up by fatty acid oxidation. Ketone bodies, if elevated by very low-carbohydrate diets or prolonged fasting, are able to inhibit glucose uptake and oxidation by the same mechanisms.

While the original observations of Randle and his colleagues have been supported by numerous studies since, the mechanisms they proposed have not been supported (Roden 2004). For example, intracellular citrate concentrations have not been shown to be elevated to the extent needed to account for phosphofructokinase inhibition. While glucose uptake is depressed if blood FFA concentrations are increased sharply, this is not associated with marked increases in glucose 6-phosphate concentrations. Finally, the PDH activity is decreased in the postabsorptive state, but a dramatic increase in blood FFA concentration

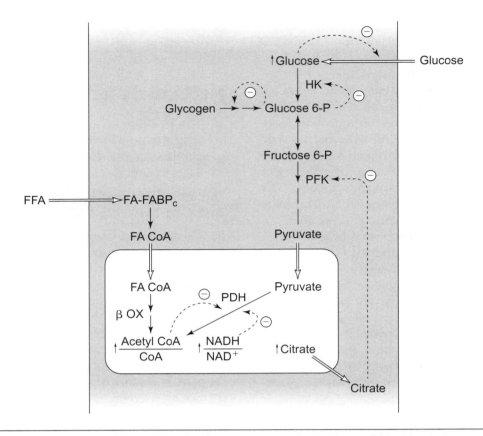

Figure 6.25 The mechanism proposed by Randle and colleagues to account for the glucose–fatty acid cycle. Increased availability of blood FFA leads to increased FFA uptake by the cell, increased intracellular fatty acids (shown as FA-FABP$_c$), increased transport into mitochondria, and increased beta-oxidation (β OX), raising both the acetyl CoA/CoA and NADH/NAD$^+$ ratios, inhibiting pyruvate dehydrogenase (PDH) and pyruvate oxidation. Increased fatty acid oxidation leads to increased citrate concentration, which is transported to the cytosol where it inhibits phosphofructokinase (PFK), and subsequently an increased intracellular glucose concentration, which reduces the transportation gradient and thus glucose uptake. An increase in glucose 6-phosphate inhibits glycogenolysis. Open arrows refer to transport across a membrane.

may not depress PDH activity further. In summary, the glucose-fatty acid cycle operates without doubt, but the mechanisms proposed by Randle and colleagues were incomplete and failed to account in total for their observations.

We know that a high-fat diet can raise blood FFA concentrations but, surprisingly, can increase blood glucose concentrations. This is characteristic of insulin resistance, and as discussed in the previous chapter, is a marker for the early type 2 diabetes condition. Researchers have also abruptly increased blood FFA levels by infusing triacylglycerol into a vein and then adding heparin, a known activator of the enzyme lipoprotein lipase. This results in a sharp rise in blood FFA but also produces hyperglycemia, even in healthy individuals. The blood glucose concentration in the postabsorptive state reflects the balance between glucose output from the liver, derived from liver glycogenolysis and gluconeogenesis, and glu-

cose disposal by peripheral tissues; skeletal muscle is responsible for more than 70% of whole-body glucose disposal. Therefore, the FFA-induced hyperglycemia could arise from an increased glucose output from the liver, a suppression of glucose disposal by skeletal muscle, or both. We know that an elevation in FFA can counteract the normal suppressive effect of insulin on liver glucose output with the result that too much glucose is released from the liver. It is also known that the FFA or some intracellular metabolite from FFA interferes with glucose uptake or its phosphorylation (or both), which could also lead to hyperglycemia. It is now clear that elevated blood FFA plays a major role in decreasing glucose uptake. In the short term, this effect is mediated through the insulin receptor signal cascade system that regulates the number of GLUT-4 transporters in the muscle cell membrane (see chapter 5). For prolonged elevations in FFA, the genes for fatty acid transport are increasingly expressed and the gene for GLUT-4 transporter is reduced.

Regulation of Fatty Acid Oxidation in Muscle

In an earlier section of this chapter, we discussed the formation of malonyl CoA by the ATP-dependent carboxylation of acetyl CoA, catalyzed by the enzyme ACC (see figure 6.19 on p. 164). This reaction is important in liver, adipose tissue, and mammary gland, which use the malonyl CoA to synthesize palmitate. We also noted that malonyl CoA can inhibit the activity of the enzyme CPT I, which is responsible for the exchange of carnitine for long chain fatty acyl CoA at the outer mitochondrial membrane (see figure 6.13 on p. 159). Inhibition of CPT I blocks entry of LCFAs into the matrix, thereby preventing their oxidation.

Skeletal and cardiac muscle are important consumers of fatty acids, but they do not make fatty acids; that is, they are not lipogenic tissues. Why then do they synthesize malonyl CoA? The answer now seems clear: the malonyl CoA is there to regulate fatty acid metabolism by muscle. We know that the oxidation of fatty acids increases during fasting and light exercise. Under both conditions, the concentration of malonyl CoA in muscle decreases, which would help facilitate the transfer of fatty acids into the mitochondria. On the other hand, if glucose and insulin levels are rapidly and acutely raised, muscle malonyl CoA levels increase within minutes. This would block entry of fatty acids into muscle mitochondria, sharply reducing the oxidation of fatty acids.

Malonyl CoA is synthesized in muscle by an isozyme of ACC, known as ACC_β or ACC_2. This isozyme of ACC is different from ACC_α, also known as ACC_1,

which is found in the lipogenic tissues (liver and adipose tissue). Unlike that of the liver isozyme, the ACC_β content in muscle does not depend on the composition of the diet (e.g., high or low in carbohydrate), nor is it sensitive to insulin and glucagon concentrations. In this way it is distinct from the liver isoform. Moreover, ACC_β is not phosphorylated by the cAMP-dependent protein kinase. It is, however, phosphorylated by an AMPK that inactivates ACC_β. Citrate is a positive allosteric effector for ACC_β. Regulation of the activity of ACC_β is shown in figure 6.26. An increase in AMP concentration increases the activity of AMPK, which phosphorylates and inactivates ACC_β. This leads to decreased formation of malonyl CoA. Removal of the phosphate is catalyzed by a phosphoprotein phosphatase. We discussed this protein kinase in detail in the previous chapter. AMPK can be activated by an increase in the concentration of AMP, which comes about when adenylate kinase converts two ADP to AMP and ATP. The AMP both allosterically activates AMPK and activates an upstream kinase AMPK kinase (AMPKK) that activates AMPK.

If malonyl CoA is synthesized in skeletal and cardiac muscles, it must also be degraded, because its only use in skeletal and cardiac muscle is regulatory. An enzyme known as malonyl CoA decarboxylase (MCD) breaks down malonyl CoA to acetyl CoA (Sambandam et al. 2004). Figure 6.27 shows how malonyl CoA is created by ACC_β and degraded by MCD, as well as how some physiological conditions

Figure 6.26 Regulation of acetyl CoA carboxylase beta (ACC_β) in skeletal and cardiac muscle through reversible phosphorylation and dephosphorylation. In the unphosphorylated state, ACC_β is active. The activity can be further increased allosterically by citrate. When phosphorylated (shown by the circled P), the enzyme is inactive. AMP-activated protein kinase (AMPK) phosphorylates ACC_β while a phosphoprotein phosphatase removes the phosphate, leading to the active enzyme.

influence the concentration of malonyl CoA. The steady state concentration of malonyl CoA in cytosol is therefore based on the relative activities of ACC_β and MDC. It is apparent that exercise decreases ACC_β activity and at the same time increases the activity of MDC. The net effect of this is to decrease malonyl CoA concentrations, to relieve the inhibition of malonyl CoA at CPT I, and to increase the oxidation of LCFAs by mitochondria. On the other hand, an abundance of the alternate fuel glucose leads to activation of ACC_β and an increase in the formation of malonyl CoA, thus decreasing the oxidation of LCFAs. The mechanism for the glucose effect is based on an increase in cytosolic citrate, which acts as a positive allosteric effector for ACC_β. It has been shown in rat and human skeletal muscle that an increase in glucose availability leads to an increase in cytosolic citrate.

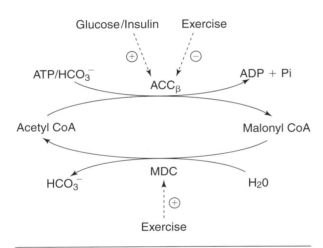

Figure 6.27   Creation and degradation of malonyl CoA. In skeletal and cardiac muscle, the concentration of the regulatory molecule malonyl CoA is based on the relative activities of acetyl CoA carboxylase beta (ACC_β), which makes malonyl CoA, and malonyl CoA decarboxylase (MDC), which breaks it down. Exercise and the attendant increase in the concentration of AMP lead to inhibition of ACC_β but activation of MDC. Citrate is a positive allosteric activator for ACC_β.

Figure 6.28 summarizes how an increased availability of glucose can lead to an increase in malonyl CoA concentration and a block in the oxidation of LCFAs, whether they are derived from the breakdown of IMTG or are taken up as FFA from the blood. There is evidence that an elevated skeletal muscle glycogen concentration acts to reduce AMPK activity by binding to one of its subunits. Such an effect would lead to less phosphorylation of ACC_β and thus greater activity. This would lead to a higher concentration of malonyl CoA and thus greater inhibition of LCFA

oxidation. On the other hand, low muscle glycogen would be associated with greater AMPK activity, more phosphorylation of ACC_β, decreased formation of malonyl CoA, and more LCFA oxidation.

Convincing evidence that changing concentrations of malonyl CoA can account for all differences in fuel oxidation in human skeletal muscle during exercise is scarce. Instead, there are data suggesting that a deficit in the availability of carnitine may act to limit fat oxidation during exercise with elevated levels of muscle glycogen or glucose (or both) and insulin (Roepstorff et al. 2005). Figure 6.29 summarizes how a deficit in carnitine could limit fat oxidation. During steady state exercise, ATP needs are met by oxidation of carbohydate and fatty acids. If one fuel is more dominant, the other is used less. An excess capacity to oxidize carbohydrate—brought about by high blood glucose concentrations, elevated muscle glycogen stores, or both—results in high rates of glycolysis and therefore pyruvate formation. The major fate of pyruvate is to enter the mitochondria and be converted to acetyl CoA. When there is a surfeit of carbohydrate, pyruvate dehydrogenase (PDH) activity is elevated and acetyl CoA is formed at a rapid rate. If the concentration of acetyl CoA exceeds the ability to form citrate in the citric acid cycle, an enzyme, carnitine acetyl transferase (CAT), can reversibly exchange a CoA for carnitine, producing acetyl carnitine. It has been shown that under carbohydrate-loaded exercise conditions, considerable carnitine can be tied up as acetyl carnitine (Roepstorff et al. 2005). This could lead to a limitation of transport of long chain fatty acyl groups from the cytosol to the matrix, limiting the rate of beta-oxidation and the production of fat-derived acetyl CoA. Thus with high carbohydrate stores, the increase in malonyl CoA concentration and its effect to limit long chain fatty acyl group entry into the mitochondrion could be supplemented by an inadequate amount of carnitine.

> ### ▶ K E Y   P O I N T ◀
>
> Clearly the body regulates its use of carbohydrate very carefully when fat will do the trick, especially during exercise. Blood glucose is critical to the function of the central nervous system and other tissues. However, it is curious how these carbohydrate-sparing mechanisms are overridden and carbohydrate use is increased if there is a surfeit of blood glucose or muscle glycogen.

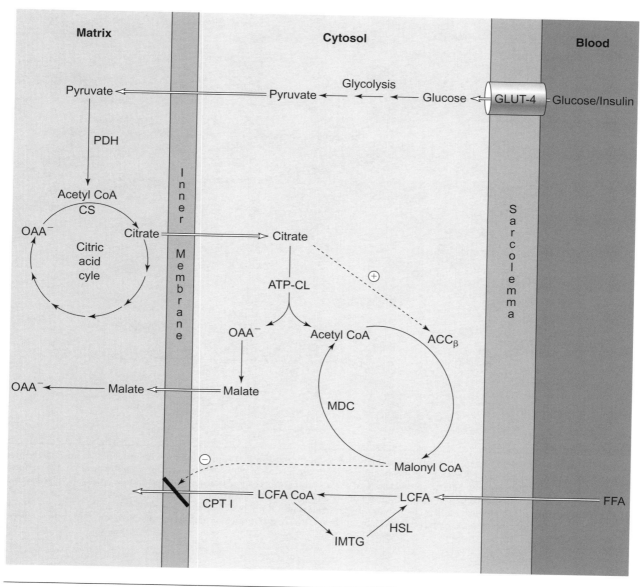

Figure 6.28 An increase in the availability of glucose to a skeletal or cardiac muscle cell leading to a shift in fuel oxidation from fatty acids to carbohydrates. The increased breakdown of glucose by glycolysis and the increase in the conversion of pyruvate to acetyl CoA in mitochondria through the action of pyruvate dehydrogenase (PDH) result in overproduction of citrate. Some of the citrate is transported to the cytosol, where it is broken down to acetyl CoA and oxaloacetate (OAA⁻). Acetyl CoA is converted to malonyl CoA in the cytosol by acetyl CoA carboxylase beta (ACCᵦ). Citrate is a positive allosteric effector for ACCᵦ. Malonyl CoA can block the entry of long chain fatty acyl CoA into the mitochondria by inhibiting the first key enzyme involved in LCFA transport, carnitine palmitoyl transferase I (CPT I). Blockage of LCFA entry to mitochondria means that the entry of LCFAs as FFA (free fatty acids) in blood is decreased. It also leads to a reduction in lipolysis of intramuscular triacylglycerol (IMTG) by the muscle hormone-sensitive lipase (HSL). Open arrows refer to transport across a membrane.

ADIPOSE TISSUE AS AN ENDOCRINE TISSUE

Lyon and colleagues (2003, p. 2195) stated, "Fat is both a dynamic endocrine organ, as well as a highly active metabolic tissue." So far, our emphasis has been on the role of adipose tissue as a critical store of triacylglycerol. New evidence, however, reveals that adipose tissue secretes a number of important peptide regulators, known as cytokines. These substances may be released to the blood where they may act on other cells in an endocrine fashion. Alternatively, they may be released locally, acting on cells in their immediate environment in a paracrine way. Expression of the genes for many of these adipose-secreted cytokines

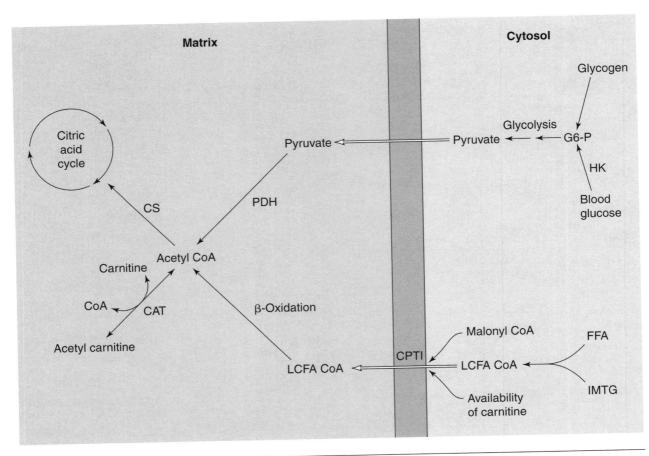

Figure 6.29   How a deficit in carnitine can limit fat oxidation. An increase in pyruvate formation via glycolysis because of elevated muscle glycogen or blood glucose (or both) results in an increase in conversion of pyruvate to acetyl CoA in mitochondria. When acetyl CoA is formed beyond the capacity of the citric acid cycle to use it, acetyl groups can be transferred to carnitine, forming acetyl carnitine, catalyzed by carnitine acetyl transferase (CAT). A decrease in carnitine needed to allow long chain fatty acyl groups (LCFA) to cross into the mitochondrion can supplement the blockade caused by elevated levels of malonyl CoA at the level of CPT I. Abbreviations: G 6-P, glucose 6-phosphate; HK, hexokinase; PDH, pyruvate dehydrogenase; IMTG, intramuscular triacylglycerol. Open arrows refer to transport across a membrane.

(also known as adipocytokines or adipokines) and their secretion from the fat cell are regulated by feeding and fasting. **Adipokines** interact with a variety of cell types, regulating metabolic, neural, or secretory actions with those cells. Inappropriate regulation (dysregulation) of some of these adipokines is associated with obesity and the metabolic syndrome (discussed in chapter 5). Although a discussion of all of these is beyond the scope of this book, we will mention a few critical adipokines.

Leptin is one of the first adipokines that was discovered. Its formation and release are tied to the amount of adipose tissue in the body. As adipose tissue mass increases as a result of a prolonged excess of food energy, leptin synthesis and release are increased. Thus circulating levels of leptin reflect the size of the adipose tissue mass for the individual. One of the main functions of leptin is to signal to

the hypothalamus that energy stores are too large and thus food energy intake should decrease and energy expenditure should increase.

Some inflammatory cytokines are released from adipocytes as well as from other tissues. These cytokines mediate inflammatory responses in the body. It is becoming increasingly clear that many of the problems associated with obesity, the metabolic syndrome, and type 2 diabetes, as well as atherosclerosis, are characterized and perhaps caused by inflammation. The obesity-mediated inflammation is likely secondary to local oxidative stress as a sequel to increased adipose tissue mass (Furukawa et al. 2004). (We discussed oxidative stress in chapter 4.) *Tumor necrosis factor alpha* (TNFα) is an inflammatory mediator secreted by adipose tissue that promotes insulin resistance within the adipose tissue cells. With loss of body fat, the synthesis and release of TNFα decrease. *Interleukin-6* (IL-6) is also a proinflammatory peptide secreted by adipocytes. Interleukin-6 is secreted by muscle tissue too, especially when it is stressed by prolonged exercise, and is also released by white blood cells. Release of IL-6 from adipose tissue can be about one-third of the total release in the body. The synthesis and release of IL-6 reflect the size of the adipose tissue mass in the body, and TNFα may play a major role in determining IL-6 release from adipocytes. Release of IL-6 is exaggerated with cancer and may promote the loss of adipocyte triacylglycerol during the cachexia that accompanies many cancers. *Plasminogen activator inhibitor-1* (PAI-1) is another proinflammatory adipokine whose blood concentration is elevated in obesity. This adipokine is thought to play a role in obesity-promoted type 2 diabetes (Venugopal et al. 2004). *Resistin* is a cytokine produced in and released from adipose tissue to the blood where it promotes peripheral insulin resistance. Its production and release parallel the change in adipose tissue mass (Guerre-Millo 2003). The concentration of resistin is low during food deprivation but increased greatly following feeding; it is especially high in people who are obese (Wolf 2004).

As already mentioned, many of the harmful adipokines are produced and released in response to a prolonged positive-calorie balance and oxidative stress and underlie many of the problems of obesity. On the other hand, *adiponectin* is a polypeptide released from adipose cells that opposes the effects of the proinflammatory adipokines. From a health perspective, adiponectin provides beneficial effects such as increasing insulin sensitivity and reducing atherosclerosis in arteries by reducing vascular inflammation. Serum adiponectin levels tend to be lower in obese subjects, but weight loss is associated with increased concentrations.

Besides secreting important cytokines, adipose tissue cells express receptors that recognize other cytokines in addition to the β (β₁, β₂)- and α₂-adrenergic and insulin receptors. **Ghrelin** is a polypeptide secreted by the gastric cells. In adipose tissue, ghrelin signals the formation of new fat cells, increasedTAG formation, and decreased lipolysis. Its effect is to increase the storage of fat in the body, including the liver (Barazzoni et al. 2005).

CHOLESTEROL

Cholesterol is an important lipid, but we seem to hear only about its bad properties. Because of its hydrocarbon content (figure 6.30 illustrates the chemical

structure), cholesterol is not soluble in water. In the blood, cholesterol either has a fatty acid attached to the cholesterol OH group, forming a cholesterol ester (about 70%), or is simple cholesterol (about 30%). Both cholesterol and cholesterol esters are in lipid-protein complexes called lipoproteins. The two main cholesterol-containing lipoproteins from a health perspective are the low-density lipoproteins (LDL) and the high-density lipoproteins (HDL). Chylomicrons represent another important lipoprotein fraction in the blood, transporting dietary fat. These are formed in intestinal cells, secreted into the lymphatic system, which empties into the bloodstream, and circulated throughout the body. Lipoprotein lipase hydrolyzes chylomicron and VLDL triacylglycerols, releasing the fatty acids.

Figure 6.30 Shorthand structure for cholesterol. The molecular formula is $C_{27}H_{47}O$. Can you find all of the carbon atoms? The numbers illustrate the location of the hydroxyl group, the double bond, and the alkyl side chain.

Determinations of the concentrations of cholesterol and cholesterol-rich lipoproteins in the blood are important clinical tests. Total blood cholesterol concentration is reported in three different ways: milligrams cholesterol per deciliter, milligrams cholesterol per liter, and millimolar. In the United States, the first method predominates (mg/dL); many other parts of the world use the SI system (mM units). If the molecular weight of cholesterol is 387, you should be able to convert a blood value of 150 mg/dL to the value of 3.88 mM. More important in determining a health profile is to ascertain the cholesterol concentration in the LDL and HDL fractions. From this perspective, it is best if the LDL cholesterol level is low while the HDL cholesterol is high.

Cholesterol is a component of membranes, a precursor for the synthesis of steroid hormones (e.g., cortisol, testosterone, estrogen, etc.), bile salts, and vitamin D, and a major component of myelin in nerves. Cholesterol is synthesized from acetyl CoA units. Although present in all cell types, cholesterol synthesis is most important in the liver, intestines, and adrenal and reproductive glands. For most people, about 60% to 70% of the body's cholesterol is synthesized; the remainder comes from the diet.

Although no clear evidence exists that regular physical exercise decreases blood cholesterol levels, it may alter the type of lipoprotein carrying cholesterol in the blood, raising the level of HDL. New drugs have been shown to markedly decrease total blood cholesterol levels.

S U M M A R Y

In the body, lipids exist primarily as fatty acids, triacylglycerols (triglycerides), phospholipids, and cholesterol. Triacylglycerols consist of three long chain fatty acids containing saturated or unsaturated carbon-to-carbon bonds, attached by ester bonds to the three-carbon alcohol glycerol. Triacylglycerols are mainly stored in specialized cells called fat cells or adipocytes. In these cells, triacylglycerols are made as fatty acids are joined to glycerol. The reverse reaction, lipolysis or lipid mobilization to yield fatty acids and glycerol, is regulated by hormone-sensitive lipase. Triacylglycerol formation is favored and lipolysis inhibited by insulin, whereas epinephrine from the adrenal medulla and norepinephrine from the sympathetic nervous system play a major role in enhancing the rate of lipolysis.

Fatty acids released from the fat cell during lipolysis travel in the bloodstream attached to the protein albumin. The fatty acids can be used by other cells as fuel. For this to happen, the fatty acid is transported across the cell membrane using fatty acid transport proteins, converted to a fatty acyl CoA, then transported into the mitochondrial matrix through attachment to carnitine. In a four-step process known as beta-oxidation, long chain fatty acids are broken down into two-carbon acetyl units attached to CoA. These acetyl CoA units can then feed into the citric acid cycle. When lipolysis is increased because body carbohydrate stores are low or during poorly controlled diabetes mellitus, fatty acids can be a source of carbon for the liver to make ketone bodies. Ketone bodies are also a source of energy, but unlike fatty acids, they can be used by the brain as a fuel. Many of the problems associated with diabetes mellitus relate to uncontrolled formation of ketone bodies. Acetyl CoA units produced by the partial breakdown of excess carbohydrate or amino acids are used to make fatty acids.

We obtain most of our fatty acids from our diet, but the liver and adipose tissue have a limited ability to make new fatty acids. Unlike lipolysis, the synthesis of new fatty acids (de novo lipogenesis) occurs in the cell cytosol and involves an intermediate known as malonyl CoA. Malonyl CoA is formed by acetyl CoA carboxylase, whose activity is regulated by allosteric and phosphorylation mechanisms. The oxidation of fat can play a major role in ATP formation by an active muscle. However, because of a limited rate at which fatty acids can be oxidized, their use is more important to a rested muscle or a muscle that is moderately active. The relative use of carbohydrate and fatty acid oxidation by muscle depends on a variety of factors, including exercise intensity, exercise training, and diet. There is a reciprocal relationship between the availability of fatty acids and carbohydrate: Fatty acids are able to reduce glucose utilization, and elevated glucose or muscle glycogen is able to reduce the use of fatty acids. Thus the same exercise task performed with elevated muscle glycogen stores is performed with greater use of carbohydrate than would be the case if muscle glycogen stores were lower. Malonyl CoA, and potentially the availability of carnitine, play roles in determining fuel oxidation by muscle.

Adipose tissue is more than a collection of cells that make and store triacylglycerol. New research shows that adipocytes produce and secrete a variety of biologically active peptides or cytokines. A number of these adipokines are produced in response to increased storage of fat due to a calorie imbalance. The increased fat content of adipocytes may lead to oxidative stress and the upregulation of expression of adipokine genes. A number of the adipokines (interleukin-6, tumor necrosis factor α, resistin, plasminogen activator inhibitor-1) are released from enlarged fat cells and affect both the adipose tissue and other cells, contributing to such conditions as type 2 diabetes and atherosclerosis.

REVIEW QUESTIONS

1. Draw the chemical structures for arachidonic acid [20:4 (5, 8, 11, 14)] and elaidic acid, which is the trans form of oleic acid.

2. In the blood draining an adipose tissue site, the ratio of FFA to glycerol is 1.5 to 1. What proportion of triacylglycerol molecules being broken down are resynthesized?

3. During steady state exercise, the average non-protein RER is 0.87 over 60 min. According to tables, such an RER means that 58.5% of the energy comes from carbohydrate and 41.5% from fat, and that each liter of oxygen has the energy equivalent value of 5.043 kcal (21.09 kJ). If the average $\dot{V}O_2$ during 60 min is 2.4 L/min, what is the energy expended by the body during this time? Assuming an energy value of 4 kcal per gram (16.7 kJ/g) and 9 kcal per gram (37.6 kJ/g) for carbohydrate and fat, respectively, how many grams of carbohydrate and fat are oxidized over this period of time? Are we overlooking any other substance that may contribute to the energy expended?

4. It is reported that ingestion of caffeine can increase the concentration of FFA compared to an identical protocol in which a placebo is given. If caffeine does not have a direct influence to increase lipolysis, by what other mechanisms might it act? You will need to do some research on this.

5. Assuming that the average molecular mass of stored triacylglycerol is 860 g/mol, what percentage of this is glycerol?

6. During rhythmic leg kicking exercise, it is determined that 3 kg of quadriceps muscle is active over a 1 h period. Based on arterial and venous blood sampling across the active leg, it is determined that the average $\dot{V}O_2$ for this exercise is 1.0 L per minute. Before the exercise there are 12 mmol per kilogram wet muscle weight of stored intramuscular triacylglycerol, and after 60 min of exercise the value is 10 mmol per kilogram wet weight. Assume that the metabolic response during this exercise, and due entirely to the active muscle, was three times that of the rest of the body. Approximately how many kilocalories (kcal) of energy generated were due to the exercise itself? What is the approximate percentage of energy devoted to kicking that was derived from intramuscular triacylglycerol? What is the total mass of fat oxidized? What would be the other sources of energy? Assume that the RER was the same as in question 3 and that the average molecular weight of the stored triacylglycerol was the same as in question 5.

7

AMINO ACID METABOLISM

We first looked at the chemistry of amino acids in chapter 1. In the present chapter we view amino acids from a different perspective. Amino acids are obtained in dietary protein and are considered macronutrients. Protein provides approximately 17 kJ/g (4 kcal/g) of physiological energy, compared with the same value for carbohydrates and 38 kJ/g (9 kcal/g) for lipids. Humans typically ingest about 10% to 15% of their calories in the form of dietary protein. We eat protein to obtain the amino acids that are used to make body protein and other specialized substances. Amino acids are also used as a source of energy. We begin the chapter with an overview of amino acid metabolism. This is followed by sections on the breakdown of amino acids, the body's means of disposal of amino acid nitrogen groups through formation of urea, and the fate of the major portion of amino acids—their carbon skeletons—when amino acids are catabolized. Amino acids can be fuels, and the ways in which they are used during exercise is important. We complete the chapter by reviewing roles for amino acids beyond their roles as fuels or components of proteins.

OVERVIEW OF AMINO ACID METABOLISM

Figure 7.1 provides a summary of amino acid metabolism. Our principal source of amino acids is in the form of food protein. During digestion, protein is broken down to free amino acids that are absorbed into the blood. The amino acids in the blood and extracellular fluids represent an **amino acid pool.** Although the size of this pool is very small compared

to the mass of body protein, the amino acids turn over very rapidly. Amino acids from dietary proteins enter this pool from the gut. They also enter the pool following release from cells. Amino acids are taken up from the pool when they are transported into cells to make cell protein. Each cell makes its own specific proteins, using amino acids released when proteins break down and taking up the rest from the pool. The body of a normal 60 kg (132 lb) woman has approximately 10 kg of protein and about 170 g of free amino acids. As mentioned in chapter 1, there is a constant protein turnover, and protein synthesis balances protein breakdown for much of our lives. When put into numbers, the pace of protein synthesis in the human body is truly staggering. An average-sized adult needs to synthesize about 1 million billion hemoglobin molecules each second just to replace those degraded as red blood cells are broken down. In fact, there are other proteins, such as myosin, in greater abundance and with shorter half-lives than hemoglobin whose rates of synthesis far outpace that of hemoglobin.

The liver and, to a smaller extent, the kidney are responsible for much of the amino acid metabolism in the body. More than half of the amino acids absorbed in the intestines following digestion of a protein meal are taken up by the *liver* (Gropper et al. 2005). It is believed that the liver regulates the composition of the amino acid pool, ensuring that there is a balanced mix of amino acids. The liver is capable of synthesizing some of the 20 amino acids needed to make our body proteins. These are described as *nonessential amino acids* (dispensable amino acids), as opposed to the **essential amino acids** (indispensable amino acids) that cannot be

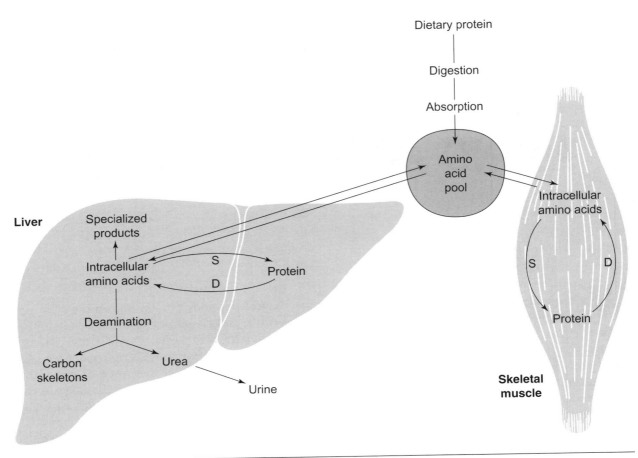

Figure 7.1   An overview of amino acid metabolism in the body showing two representative tissues—skeletal muscle and liver. Our principal source of amino acids is ingested protein. After digestion and absorption, amino acids enter an amino acid pool, which is in the blood and extracellular fluids. Cells take up amino acids from the pool to synthesize proteins (S). Degradation (D) releases amino acids that may be used again in protein synthesis or released to the amino acid pool. The liver makes a variety of specialized products from amino acids and breaks down most amino acids. The nitrogen on amino acids is converted to urea; the remaining carbon skeletons can be a source of glucose, be oxidized, or be converted into lipid.

synthesized and must be obtained in the diet from a variety of food sources.

- **Nutritionally nonessential:** alanine, asparagine, aspartate, cysteine, glutamate, glutamine, glycine, proline, serine, and tyrosine
- **Nutritionally essential:** arginine, histidine, isoleucine, leucine, lysine, methionine, phenylalanine, threonine, tryptophan, and valine

The categorization of amino acids as essential or indispensable and nonessential or dispensable really applies to healthy adults eating a sound diet. Some of the nonessential amino acids can become indispensable if they cannot be produced at a rapid enough rate. Special conditions such as infant growth, disease, organ failure, or the presence of drugs can make a dietary source of the amino acids arginine, cysteine, glutamine, and proline more important. Amino acids

that are synthesized in the liver are used within the liver or released to the blood. The *kidneys* also play a role in the synthesis of some amino acids, as well as in the exchange of amino acids to and from different tissues and organs. Thus the blood and extracellular fluids contain a pool of amino acids resulting from dietary intake, catabolism of cellular protein, and amino acid synthesis in and release from the liver.

Skeletal muscle is the largest repository of free and protein-bound amino acids in the body; but in terms of the scope of metabolism of amino acids, muscle is rather limited. About 45% of the body weight of an adult is composed of skeletal muscle. This proportion is higher in lean, well-trained individuals. This can represent an important nitrogen reserve during stressful conditions such as trauma, infection, and starvation. A variety of hormones and growth factors such as insulin, insulin-like growth factor-1 (IGF-1), growth hormone, and testosterone promote skeletal muscle growth.

We have no ability to store amino acids as we can carbohydrate (glycogen) and fatty acids (triacylglycerol). Moreover, the protein content of our adult bodies is remarkably constant, and so we might expect that we would oxidize amino acids at a rate commensurate with our intake, that is, about 10% to 15% of our daily energy expenditure. In other words, those amino acids not used in protein synthesis or converted to other substances (e.g., heme groups; hormones such as serotonin, adrenaline, and noradrenaline; nucleotides; glutathione; carnitine; creatine) are simply used as fuels. As mentioned in chapter 5, amino acids first lose their amino group(s) or other N atoms; and the resulting carbon skeleton can be oxidized directly, used to make glucose (gluconeogenesis), or converted into fat for storage.

> ### ▶ K E Y P O I N T ◀
>
> Although the carbon skeletons from excess amino acids can be degraded to acetyl CoA, converted to malonyl CoA, and used to make palmitate in the liver (see chapter 6), this is not very significant compared to fatty acid synthesis from excess carbohydrate. Immediate oxidation of amino acid carbon skeletons or conversion to glucose (gluconeogenesis) is the principal fate for excess amino acids.

Not shown in figure 7.1 is the transport of amino acids from the amino acid pool into cells. Because amino acids have charged groups, they need protein transporters to move them from the extracellular to the intracellular compartment or from the intracellular to the extracellular compartment. There are a number of amino acid transporters, which fall into two broad categories. The transporters can be sodium ion dependent or sodium independent. If the transporter is sodium dependent, the amino acid moves into a cell down a Na^+ concentration gradient. Therefore, this is a symport transport system, and the amino acids can be moved against their concentration gradient. Amino acid transporters may have broad specificity, recognizing a number of amino acids, or narrow specificity, recognizing only one or two closely related amino acids. (For a review of the various categories of amino acids, refer to figure 1.4 on p. 8.) Amino acid transporters are also subject to

regulation by hormones and growth factors. Insulin, in particular, is important to stimulate amino acid uptake, particularly into skeletal muscle fibers.

DEGRADATION OF AMINO ACIDS

For a well-nourished individual, the supply of amino acids arising from the degradation of intracellular proteins and those provided through the diet are normally sufficient to meet the immediate needs for protein synthesis. However, alterations in the balance of amino acids can be achieved through conversion reactions in which one amino acid is changed into another by transfer of an amino group. In addition, excess amino acids are catabolized, but the amino groups must be removed beforehand. In chapter 5 we talked about the potential for 18 of the 20 amino acids to provide all or part of their carbon atoms to the formation of glucose in the process of gluconeogenesis; these are sometimes described as glucogenic amino acids—the other two are described as ketogenic amino acids. As noted in our discussion of gluconeogenesis, amino acids consist of carbon skeleton and amino groups. Humans cannot use the nitrogen in the amino groups as an energy source, although it is needed to form some specialized products; most amino nitrogen is excreted. The carbon skeletons are very valuable, and as pointed out in chapter 5, after removal of the amino groups, many of the carbon skeletons are intermediates or can be easily converted into intermediates in the citric acid cycle or pyruvate.

Amino acids undergo constant oxidative degradation under the following metabolic circumstances:

- During normal synthesis and degradation of proteins in the body, amino acids released in the constant breakdown process may not be immediately reused in synthesis. Since we cannot store amino acids, they will be degraded.

- When we ingest more amino acids than our bodies can use to make proteins or convert to other substances, these amino acids are degraded. In North America, most people eat more protein than they need.

- During starvation, fasting, dieting, or uncontrolled diabetes mellitus, when carbohydrates are either not available or not used due to an absence of insulin, amino acid catabolism accelerates.

- When we overtrain so that there is an imbalance in protein turnover favoring protein degradation, amino acids are degraded.

Catabolism of individual amino acids has two major stages. First, the amino acids must lose their nitrogen atoms because we cannot obtain usable energy from nitrogen groups. Second, the resulting carbon skeletons are fed into specific energy-yielding pathways so that their chemical energy can be retrieved. The liver removes the amino groups from most amino acids, although skeletal muscle has a significant capacity for **deamination** of the branched chain amino acids (BCAAs).

Transamination Reactions

Transfer of amino groups takes place with all of the amino acids except threonine and lysine. Even the other essential amino acids can undergo amino group transfer, but only in the direction of removing the amino groups because the body cannot synthesize their carbon skeletons. Transamination is carried out by *aminotransferase enzymes* (formerly known as transaminases), which contain vitamin B6. Most of the aminotransferase enzymes transfer the amino group of an amino acid to α-ketoglutarate (also known to many as 2-oxoglutarate), making glutamate and a new carbon skeleton in the form of a keto acid (see figure 7.2). The vitamin B6 is a precursor for the coenzyme pyridoxal phosphate (PLP; see table 2.1 on p. 25). Notice in figure 7.2 that the ketone group in α-ketoglutarate is next to the carboxylate; thus it is in the 2 or α position, making it an α-keto acid. Additional aminotransferase enzymes are specific for amino acids other than glutamate. For example, *alanine aminotransferase* (see figure 5.20 on p. 133) involves the reversible transamination between the amino acid alanine and its α-keto acid, pyruvate, along with α-ketoglutarate and glutamate. The key word is reversible, for an excess of alanine is converted to glutamate or an excess of glutamate is converted to alanine. The actual concentrations are based on the equilibrium for each aminotransferase. *Aspartate aminotransferase* is involved in the reversible transamination between the amino acid aspartate and its α-keto acid, oxaloacetate, as well as α-ketoglutarate and glutamate. Alanine aminotransferase and aspartate aminotransferase are prominent enzymes that are known clinically as ALT (formerly known as glutamate pyruvate transaminase, GPT) and AST (formerly glutamate oxaloacetate transaminase, GOT), respectively. Serum ALT and AST activities are commonly used to determine if there has been acute trauma or disease of the liver.

Transamination reactions are freely reversible, with equilibrium constants of about 1 and standard free energy change values near zero. Their net direction depends on the relative concentrations of the four reactants. The overall effect is to transfer amino groups from a variety of amino acids to α-ketoglutarate (2-oxoglutarate), making glutamate. We will discuss the BCAAs as a special case, for their metabolism is primarily in skeletal muscle.

Deamination

Amino groups can be shuttled from one amino acid to α-ketoglutarate to make glutamate or from glutamate to another keto acid to create a new amino acid, as we have seen. However, most of us excrete an amount of amino nitrogen each day that equals our intake. Therefore, the body must rid itself of amino groups by forming urea. This means that there must be processes to remove amino groups

Figure 7.2 Aminotransferases, a class of enzyme that transfers amino groups reversibly between the amino acid glutamate and its α-keto acid (α-ketoglutarate, also known as 2-oxoglutarate). The other participants are specific to the other individual amino acid and its carbon skeleton, an α-keto acid.

and then turn them into urea molecules in the liver. Some key reactions are used to remove amino groups from amino acids. Therefore, the amino group on glutamate (which comes from amino acids) must be transferred to the liver (if it is not already there) and into the urea molecule. Nitrogen from amino groups in the liver in the form of glutamate can be released as ammonia (principally the ammonium ion NH_4^+) in the *glutamate dehydrogenase* reaction.

$$\text{glutamate} + H_2O + NAD^+ \leftrightarrow \alpha\text{-ketoglutarate} + NADH + H^+ + NH_4^+$$

As shown, the coenzyme is NAD^+, but $NADP^+$ can also be used. The freely reversible glutamate dehydrogenase reaction takes place only in the mitochondrial matrix, whereas most aminotransferase enzymes exist in both the mitochondrial matrix and the cell cytosol. The glutamate dehydrogenase reaction, as it proceeds to the right, is a deamination reaction since it removes the amino group. Because NADH is also formed, it is an oxidative deamination reaction. The glutamate dehydrogenase reaction removes nitrogen from glutamate, which can be used to make urea. As we will discuss later, exercising muscle releases ammonia (as the NH_4^+ ion). The glutamate dehydrogenase reaction in muscle is one source of this ammonia. As well, ammonia is produced by the deamination of AMP in the AMP deaminase (adenylate deaminase) reaction we discussed in chapter 3. This latter reaction is generally not significant unless muscle is working at a high intensity. Overall, the production of ammonium ion and its release from muscle are proportional to the intensity of the exercise.

Glutamine

Glutamine is a special amino acid even though it is not essential for most of us. The content of the free

KEY POINT

Ammonia is a toxic substance, but it is produced in small quantities by some tissues (kidney and muscle), and some is excreted in the urine. We have been using the term *ammonium ion*—the principal form we would find in the blood—rather than *ammonia*, although the former is simply protonated ammonia. Ammonia (ammonium ion) released from muscle is taken up by liver (primarily) and kidney. In the liver, ammonia can be used to make urea.

amino acid glutamine is high in a variety of cells and in the blood, in concentrations beyond that of any other amino acid. It represents about 60% of the amino acid pool. We obtain some glutamine in our diets, but most is synthesized in the body from glutamate using the enzyme *glutamine synthetase* (see figure 7.3). Glutamine is an important fuel for the gut and immune system. Notice that the additional nitrogen in glutamine is an amide of the side chain carboxylate of glutamate. Skeletal muscle is an important site for glutamine synthesis, but it is also made in brain, adipose tissue, and the heart. As already mentioned, ammonia is toxic, so converting it to glutamine is a good strategy. Glutamine content in skeletal muscle and other tissues appears to have a regulatory role in whole-body protein synthesis. During a variety of catabolic conditions, its content is decreased; and during anabolic states, intracellular glutamine content is elevated. Glutamine is also an important fuel for macrophages and lymphocytes, cells important to our health and well-being.

The last reaction to consider is the removal of the amino group from the side chain of glutamine, which

Figure 7.3 The glutamine synthetase reaction. This reaction is very active in skeletal muscle, transferring ammonium ions to glutamate to create glutamine.

is the reverse of the glutamine synthetase reaction. *Glutaminase*, found mainly in liver mitochondria, catalyzes this deamination reaction:

$$\text{glutamine} + H_2O \rightarrow \text{glutamate} + NH_4^+$$

The Branched Chain Amino Acids

Branched chain amino acids leucine, isoleucine, and valine are the most common essential amino acids in proteins. Unlike the other 17 amino acids, the BCAAs are metabolized mainly in skeletal muscle, although they can also be metabolized in liver. These essential amino acids are treated as a group because the first three reactions in their catabolism are catalyzed by the same enzymes, *branched chain amino acid aminotransferase*, *branched chain ketoacid dehydrogenase*, and *acyl CoA dehydrogenase* (see figure 7.4). After these first three reactions, carbon skeletons remaining from the three amino acids take their separate reaction directions. Both valine and isoleucine produce a *succinyl CoA* as a product. Since this is a citric acid cycle intermediate, valine and leucine are glucogenic amino acids. The products from leucine catabolism, acetoacetate and acetyl CoA, cannot be converted to glucose, so leucine, like lysine, is considered a ketogenic amino acid. Recall that acetoacetate is a ketone body (see chapter 6).

The first reaction in the catabolism of the BCAAs is a transamination in which the amino group on each BCAA is transferred to α-ketoglutarate, making glutamate. While transamination reactions are reversible, the branched chain amino acid aminotransferase goes only one way because the products undergo an oxidative decarboxylation catalyzed by branched chain ketoacid dehydrogenase (BCKAD). This mitochondrial enzyme is a multienzyme complex, regulated by phosphorylation and dephosphorylation mechanisms; this is reminiscent of the regulation of pyruvate dehydrogenase (PDH) by phosphorylation and dephosphorylation. Indeed, PDH and BCKAD have an identical subunit. Phosphorylation of BCKAD by BCKAD kinase renders the enzyme inactive. A BCKAD phosphatase removes the phosphate group, leading to an active BCKAD.

In chapter 5, figure 5.21 (p. 133) showed that the amino groups on the BCAAs, transferred to α-ketoglutarate making glutamate, can be transferred from glutamate to pyruvate by the action of alanine aminotransferase. Alanine is released from muscle to the blood, where its amino group is available for urea synthesis. In this way, the amino groups of the BCAAs are removed from muscle to be disposed of by the liver. Leucine and its α-ketoacid, α-ketoisocaproate, can suppress protein degradation in skeletal muscle. Leucine can also enhance protein synthesis by stimulating the initiation of translation (Rennie and Tipton 2000). Leucine is available as a supplement at many nutrition retail suppliers.

THE UREA CYCLE

Ammonia is quite toxic, especially to the brain. However, two safe forms of ammonia exist: the amino group of glutamate and the side chain amide nitrogen in glutamine. Although we can temporarily store ammonia in these innocuous forms, we must eliminate the nitrogen we cannot use. Mammals convert the nitrogen to urea, whereas birds and reptiles eliminate amino groups by converting nitrogen to uric acid. Urea, whose structure is shown next, is a simple molecule formed in the liver and excreted from the kidney when urine is formed:

$$\begin{array}{c} O \\ \parallel \\ H_2N - C - NH_2 \end{array}$$

The two amino groups in urea allow the body to rid itself of nitrogen. The carbonyl group comes from carbon dioxide. The nitrogen comes from the ammo-

Figure 7.4 Catabolism of the three branched chain amino acids leucine, isoleucine, and valine. The process starts with the same three enzymes: BCAAAT (branched chain amino acid aminotransferase), BCKAD (branched chain ketoacid dehydrogenase), and ACDH (acyl CoA dehydrogenase). Isoleucine and valine produce succinyl CoA as a final product, which can be converted to glucose in the liver by the process of gluconeogenesis.

nium ion and from the amino group of aspartate. Figure 7.5 summarizes the path taken by nitrogen from amino acids in the body, using only three organs as examples. Muscle contains more protein than any other tissue. Amino acids released when muscle protein is degraded can be reused to make new protein, partially catabolized in muscle, as in the case of the BCAAs or released to the blood. Amino groups from the BCAAs are first passed to α-ketoglutarate to make glutamate. Some glutamate transfers its amino group to pyruvate to make alanine. Some glutamate combines with ammonia to create glutamine. We have

seen that ammonia production increases in muscle during exercise, especially intense exercise.

Muscle releases alanine and glutamine, which contain much of the muscle nitrogen. As we saw in figure 5.21 (p. 133), the source of the nitrogen on alanine is BCAAs. In the liver, the amino group on alanine is transferred to α-ketoglutarate to make glutamate, catalyzed by alanine aminotransferase. The glutamine released from muscle results from the amination of glutamate in the glutamine synthetase reaction. The liver is the major site for nitrogen removal from most of the amino acids (except the BCAAs). The amino

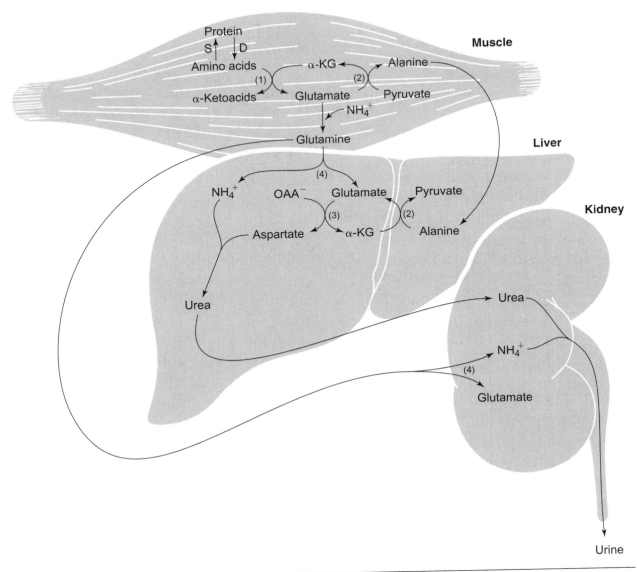

Figure 7.5   The interorgan transport of nitrogen, involving skeletal muscle, the liver, and kidneys. Branched chain and other amino acids generated by the degradation (D) of protein and not used in protein synthesis (S) can lose their amino groups by transfer to α-ketoglutarate (α-KG) to make glutamate. Amino groups on glutamate can be transferred to pyruvate to make alanine. Glutamate can also be converted to glutamine using ammonia (NH_4^+). Muscle releases alanine and glutamine. Alanine amino groups can end up on aspartate, and glutamine can be converted into glutamate and ammonia. Urea is synthesized in liver using amino groups from ammonia and aspartate. The kidney releases most nitrogen as urea and a smaller amount as ammonia. Reactions: (1) branched chain amino acid aminotransferase or other aminotransferase; (2) alanine aminotransferase; (3) aspartate aminotransferase; (4) glutaminase.

groups end up on glutamate. Muscle also releases ammonium ions, especially during exercise. Urea synthesis requires the ammonium ion that comes from ammonia taken up by liver from the blood, from glutamine via the glutaminase reaction, or from glutamate via the glutamate dehydrogenase reaction. Aspartate, which provides the other urea nitrogen, comes from a transamination reaction in which oxaloacetate undergoes transamination from glutamate to make aspartate and α-ketoglutarate.

The enzyme *aspartate aminotransferase* catalyzes the following reaction.

glutamate + oxaloacetate ↔ aspartate
+ α-ketoglutarate

The urea cycle consists of four enzymatic steps that take place in the liver cells, in both the mitochondrial matrix and cytosol. Before the actual cycle begins, carbamoyl phosphate must be synthesized from ammonia (as ammonium ion) and carbon dioxide (as bicarbonate), in a reaction catalyzed by *carbamoyl phosphate synthetase* (see figure 7.6) This reaction takes place in the mitochondrial matrix; it is driven by the simultaneous hydrolysis of two ATP and involves the formation of carbamoyl phosphate, catalyzed by carbamoyl phosphate synthetase. The carbamoyl phosphate synthetase reaction, the committed step in urea synthesis, controls the overall rate of the urea cycle. Carbamoyl phosphate synthetase activity is allosterically activated by *N-acetyl glutamate,* formed from acetyl CoA and glutamate by the enzyme *N-acetylglutamate synthase.* When amino acid degradation is accelerated, glutamate increases in the liver. By its conversion to N-acetylglutamate and subsequent allosteric activation of carbamoyl phosphate synthetase, it is signaling a need to speed up the urea cycle to get rid of the amino groups of the degraded amino acids.

The actual urea cycle is shown in figure 7.7. The carbamoyl phosphate enters the urea cycle in the mitochondrial matrix, joining with *ornithine* to form

citrulline in a reaction catalyzed by *ornithine transcarbamoylase.* The citrulline exits the mitochondrial matrix. The second nitrogen, in the form of the *aspartate* amino group, enters the cycle when aspartate combines with citrulline to form *argininosuccinate.* This step involves ATP hydrolysis and is catalyzed by *argininosuccinate synthetase.* The argininosuccinate is then cleaved to arginine and fumarate in a reaction catalyzed by *argininosuccinate lyase.* In the final step, arginine is cleaved by *arginase* to yield ornithine and urea. The ornithine is now regenerated and enters the mitochondrion to combine again with carbamoyl phosphate.

Urea is extremely water soluble. Thus it leaves the liver via the blood and enters the kidneys, where it is filtered out. The urea cycle is metabolically expensive because the ATP cost to make one urea molecule is four. Two are used to make carbamoyl phosphate. Can you think where the other two ATP come from? (Hint: How many phosphate groups are removed from ATP during the condensation of citrulline and aspartate?)

$$NH_4^+ + HCO_3^- + 2\ ATP \xrightarrow{\text{Carbamoyl phosphate synthetase}} \underset{\underset{O^-}{|}}{H_2N - \overset{\overset{O}{\|}}{C} - O - \overset{\overset{O}{\|}}{P} - O^-} + 2\ ADP + P_i$$

Figure 7.6 The reaction controlling the rate of urea synthesis, catalyzed by carbamoyl phosphate synthetase in the matrix of liver mitochondria.

Figure 7.7 The urea cycle. This cycle is a sequence of four enzyme-catalyzed reactions that produce urea from nitrogen on carbamoyl phosphate and aspartate. There is no net consumption or formation of ornithine or other intermediates in the cycle, shown in bold letters.

FATE OF AMINO ACID CARBON SKELETONS

After the amino groups are removed from the amino acids, carbon skeletons remain, in many cases in α-keto acids such as pyruvate, oxaloacetate, or α-ketoglutarate (see figure 7.8). These carbon skeletons have a variety of fates, such as gluconeogenesis; 18 of the 20 amino acids can be a source of glucose and are described as glucogenic amino acids. Note in figure 7.8 that leucine and lysine do not produce either pyruvate or citric acid (TCA) cycle intermediates. Leucine and lysine are described as ketogenic amino acids because their carbon skeletons produce only acetoacetyl CoA and acetyl CoA. Note that leucine degradation initially produces acetoacetate and acetyl CoA (figure 7.4). However, as with the ketone body acetoacetate, a

CoA is transferred from succinyl CoA, making acetoacetyl CoA and succinate. The carbon skeletons of all amino acids can also be used for immediate oxidation because they form citric acid cycle intermediates or acetyl CoA. Finally, all are potential sources of carbon to make new fatty acids because all are also potential sources of acetyl CoA. As already mentioned, amino acid carbon skeletons are not a very significant source for de novo fatty acid synthesis.

Figure 7.8 illustrates what the carbon skeletons of the amino acids have in common with the citric acid cycle or substances directly related to this cycle. The amino acids shown can generate the citric acid cycle intermediates or related molecules by simple removal of their amino groups via transamination (alanine, glutamate, and aspartate) or through a number of steps not shown. As we discovered in chapter 5, the formation of citric acid cycle intermediates through

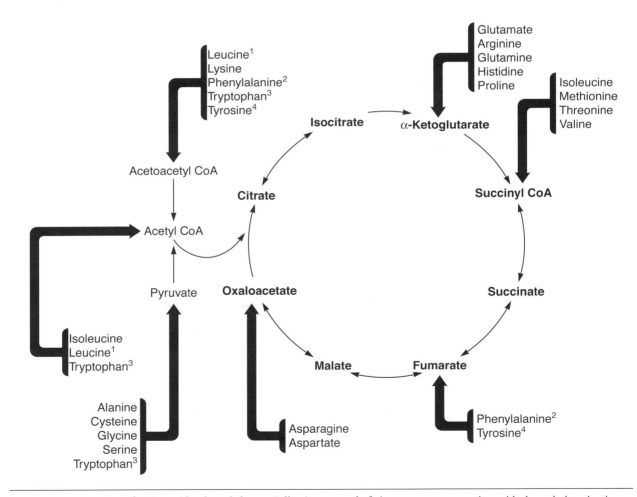

Figure 7.8 Conversion of amino acid carbon skeletons. Following removal of nitrogen groups on amino acids through deamination or transamination, the amino acid carbon skeletons are converted into citric acid cycle intermediates (shown in bold) or related molecules. The large arrows show the major paths taken by the carbon skeletons. The carbon atoms of some large amino acids, identified with superscript numbers, give rise to two or three different molecules.

removal of the amino groups from amino acids is known as anaplerosis. It is an important mechanism to maintain citric acid cycle intermediate concentration, especially during exercise when demand on the citric acid cycle is high.

AMINO ACID METABOLISM DURING EXERCISE

During exercise, carbohydrate and lipid supply most of the energy needs of the body as already discussed. However, metabolism of amino acids also takes place at an accelerated rate in a manner designed both to transfer amino acid nitrogen (amino groups) out of muscle and to move their carbon skeletons from muscle to liver. To get a proper perspective on the effects of exercise, we will look at low- to moderate-intensity exercise and high-intensity exercise, because the effects of the two are quite different. First, it is important to understand how we learn about muscle amino acid metabolism.

The study of metabolism of amino acids during exercise is complex. In humans it is carried out through measurement of the concentrations of amino acids in the arterial and venous blood across an exercising muscle. From concentration differences and muscle blood flow measures, one can determine what amino acids are being taken up and released from muscle and the rate at which this is taking place. One can then describe the release of an amino acid from muscle in terms of the concentration difference in arterial and venous blood times the blood flow, producing such numbers as micromoles of amino acid released per minute. In addition, muscle biopsies can be taken to determine the concentrations of amino acids and a variety of other metabolic intermediates in the muscle samples obtained before, during, and after exercise. For example, the lack of release of an amino acid or metabolite from muscle may mean that nothing is happening with it compared to what occurs in the inactive condition, or that its concentration is increasing within the muscle and that is why it is not being released. Thus it is necessary to measure both changes within muscle and changes in exchange across the muscle.

Even when good measures of muscle metabolites, arterial and venous concentration differences, and blood flow can be obtained, it may still be difficult to get an accurate picture of muscle amino acid metabolism during exercise. The challenge is to estimate with reasonable accuracy the mass of muscle involved in the actual exercise. In other words, we need to be able to describe changes taking place in muscle amino acids by describing releases or uptakes in terms of micromoles per minute per unit mass of muscle (i.e., micromoles per minute per kilogram of muscle). Although a detailed analysis of this type of work is beyond the scope of the present discussion, it is worthwhile pointing out that much of the published data in current literature on amino acid metabolism in human muscle during exercise probably represent reasonable estimates and not precise values.

Moderate-Intensity Exercise

Figure 7.9 summarizes some of the important changes in amino acid metabolism that take place in exercising skeletal muscle. During exercise in the postabsorptive state, skeletal muscle is in a net protein catabolic state, in which protein breakdown exceeds the rate of protein synthesis. Most of the amino acids produced by the net protein catabolism are released from the muscle. The major exception is glutamate, because there is a net uptake of glutamate at rest and even more so during exercise. Glutamate is needed as a precursor for glutamine, which, along with alanine, is released in increasing amounts with exercise intensity and duration. The release of glutamine and alanine is far out of proportion to their content in skeletal muscle proteins, suggesting that both amino acids are being formed in skeletal muscle at an accelerated rate during exercise. Figure 7.9 illustrates the derivation of the ammonia, alanine, and glutamine.

We have already discussed the fact that during glycolysis, approximately 1% of the pyruvate produced is converted to alanine. Transamination of pyruvate with glutamate by alanine aminotransferase produces alanine and α-ketoglutarate. As shown previously, the alanine leaves the muscle and travels in the blood to the liver (see figure 5.21, p. 133). In the liver, alanine aminotransferase converts alanine to pyruvate and α-ketoglutarate. The pyruvate is a source of carbon to make glucose by gluconeogenesis. If the glucose produced in liver is released to the blood and catabolized to pyruvate in muscle, a cycle is produced that is known as the *glucose-alanine cycle*.

The principal source of the amino group that is released from muscle in alanine is from the BCAAs. As mentioned earlier, these amino acids are preferentially transaminated in skeletal muscle by branched chain amino acid aminotransferase. This reaction converts α-ketoglutarate to glutamate and the deaminated products of the BCAAs, known as branched

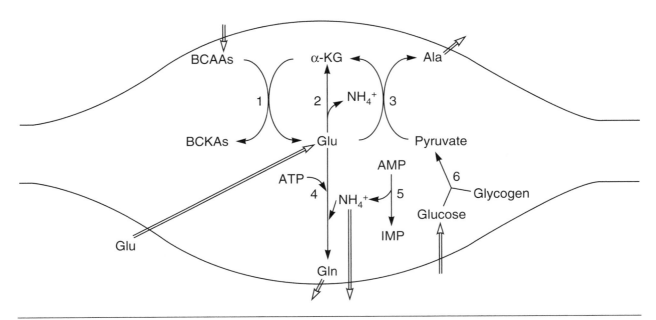

Figure 7.9   Net release of amino acids from skeletal muscle during exercise, which occurs with most amino acids. The exceptions are glutamate and the branched chain amino acids (BCAAs), for which there is a net uptake. Alanine (Ala) and glutamine (Gln), formed by reactions 3 and 4, respectively, are released from muscle at rates far beyond those for the other amino acids. Alanine is synthesized from pyruvate and an amino group donated by the BCAAs. The source of the pyruvate is glycolysis (6), using glucose or glycogen as a precursor. Glutamine is synthesized from glutamate (Glu) using glutamine synthetase (4). Glutamate may also be deaminated by glutamate dehydrogenase (2), forming α-ketoglutarate (α-KG). Active in muscle during vigorous exercise is AMP deaminase (5), which produces ammonium ion (NH_4^+) and IMP (inosine monophosphate). Exercising muscle releases NH_4^+. Reaction 1 is catalyzed by branched chain amino acid aminotransferase. Reaction 3 is catalyzed by alanine aminotransferase. BCKAs are branched chain keto acids.

KEY POINT

If one considers the net effect of the glucose-alanine cycle, the carbon atoms of glucose are simply recycled from liver to muscle (as glucose) and muscle to liver (in alanine). However, this cycle also allows the muscle to transfer nitrogen from branched chain amino acids to liver for urea synthesis.

chain keto acids. These may undergo further oxidative metabolism within the muscle or be released to the blood for metabolism in the liver. The carbon skeleton of leucine can be completely oxidized in the citric acid cycle since one product is acetyl CoA and the other, acetoacetate, gives rise to two acetyl CoA. Degradation of isoleucine and valine generates succinyl CoA, which can help to maintain citric acid cycle intermediates.

During low- to moderate-intensity exercise, muscle releases alanine at a rate far in excess of the rate at rest. As shown in figure 7.9, the source of the pyruvate for alanine formation is that produced during glycolysis from glucose and, more importantly, stored glycogen.

We also know from chapter 5 that the contribution of glycogen to ATP production during prolonged exercise declines as the glycogen stores are gradually reduced. Although this is partially compensated by an increased use of blood glucose, it is typically noted that the respiratory exchange ratio during constant-intensity exercise declines over time. With glucose and glycogen as the sources of pyruvate to make alanine, we would expect that the release of alanine declines with exercise duration, and this is what is observed.

As shown in figure 7.9, skeletal muscle also releases glutamine, especially during exercise. Amination of glutamate, using the glutamine synthetase reaction (shown in figure 7.3), is the major source of glutamine, because muscle proteins are not rich in the amino acid glutamine. The ammonium groups needed to form glutamine from glutamate can be derived from glutamate via the glutamate dehydrogenase reaction and also from the deamination of AMP by the enzyme AMP deaminase. With its net uptake from the blood, glutamate is the source for glutamine synthesis and some of the ammonia produced. Indeed, so much glutamate is needed to make glutamine that the concentration of glutamate actually declines during exercise, unlike that of the other amino acids.

Interestingly, when subjects were given large doses of monosodium glutamate before exercise so that plasma levels of glutamate were greatly elevated, there were increases in the release of both alanine and glutamine, but release of ammonia was attenuated during the exercise (Mourtzakis and Graham 2003).

The changes just noted for moderate-intensity exercise can be modulated depending on the nutritional state of the exercising person as well as the concentration of muscle glycogen. In a fasted, low-glycogen state, less pyruvate is formed in glycolysis to produce alanine. In this state, protein breakdown is likely much greater, with a corresponding increase in the release of most amino acids from muscle. Early exercise studies also suggested that the rate of catabolism of amino acids by the liver increases in a fasted, low muscle glycogen state as evidenced by increased formation of urea and larger concentrations of urea in sweat. Some studies have also noted an increase in muscle IMP and a decrease in **total adenine nucleotide (TAN)** content in untrained subjects during prolonged moderate-intensity exercise when muscle glycogen concentration is severely reduced. Apparently the steep decline in muscle glycogen leads to an increase in ADP concentration and, thereby, AMP concentration. This can result in a modest flux through AMP deaminase by a mass action effect, producing IMP even though conditions favoring complete activation of AMP deaminase are not present.

▶ K E Y   P O I N T ◀

Many athletes ingest carbohydrate-containing fluids during training. While they may be thinking that this is to provide fuel for their training, it will also help to maintain protein balance in the muscle by modulating the rate of protein breakdown, a process normally activated by strenuous exercise.

The Purine Nucleotide Cycle

During exercise, there are increases in cytosolic ADP concentration even though the level of ATP is well preserved by the three energy systems we discussed in chapter 3. The ADP can be converted into ATP and AMP in a freely reversible reaction catalyzed by the enzyme *adenylate kinase*. As shown in table 3.3 (p. 53), the increase in AMP concentration during exercise is directly related to exercise intensity.

$$2\ ADP \leftrightarrow ATP + AMP$$

In chapter 3 we also discussed the deamination of AMP, catalyzed by the enzyme *adenylate deaminase*. This enzyme converts AMP to IMP and ammonia as shown in the following:

$$AMP + H_2O \rightarrow IMP + NH_3$$

Proton addition to the base NH_3 generates ammonium ion (NH_4^+). As discussed in chapter 3, the adenylate deaminase reaction results in an actual decrease in the TAN content of the muscle.

Adenylate deaminase is more active in type II (fast-twitch) than type I (slow-twitch) muscle fibers. Therefore, it is unlikely to be activated significantly in low-intensity exercise since the force production will be handled to a greater extent by motor units containing type I muscle fibers. Furthermore, adenylate deaminase activity is low in rested or slowly contracting muscle, increasing only if the pH is decreased and the concentrations of AMP and ADP increase. Not only do type I fibers have less adenylate deaminase, they are unlikely to develop the rise in AMP and decrease in pH that are needed to activate the enzyme, since these fibers have a higher capacity for oxidative metabolism and lower glycolytic activity. Therefore, we can anticipate that adenylate deaminase is unlikely to be significant during low to moderate levels of exercise.

Since the adenylate deaminase can reduce TAN levels by up to 50% in very severe exercise, there must be a way to regenerate the adenine nucleotides. The process for doing this is known as the purine nucleotide cycle and is shown in figure 7.10. In reaction 1, which is catalyzed by adenylate deaminase, IMP and NH_3 are produced. As already noted, this reaction is active during intense muscle activity. Adenosine monophosphate is regenerated by two additional reactions. In the first, aspartate combines with IMP at the cost of a guanosine triphosphate (GTP), forming adenylosuccinate. This is catalyzed by *adenylosuccinate synthetase*. In the final step, adenylosuccinate is split into AMP and fumarate, catalyzed by *adenylosuccinate lyase*. The net reaction shows that it costs energy to regenerate AMP in the form of a GTP. In the process, an aspartate is deaminated, producing fumarate, a TCA cycle intermediate. The regeneration of AMP by adenylosuccinate synthetase and adenylosuccinate lyase takes place predominantly after exercise because the activities of these enzymes are too low to produce appreciable AMP during exercise.

Figure 7.10   The purine nucleotide cycle. In this simple pathway, AMP is deaminated during intense muscle activity, catalyzed by adenylate deaminase (1). The next two steps regenerate AMP. First, IMP combines with aspartate, catalyzed by adenylosuccinate synthetase (2). Secondly, adenylosuccinate is split into AMP and fumarate, catalyzed by adenylosuccinate lyase. The net reaction shows that the regeneration of AMP costs a GTP, and the amino acid aspartate is deaminated and a fumarate is produced.

High-Intensity Exercise

When the intensity of exercise is high, there are likely to be only modest increases in the release of glutamine and alanine from muscle. Glutamate uptake by muscle also appears to be little influenced by increased exercise intensity. Glutamine synthesis consumes ATP, and its release represents a loss of glutamate from muscle. Since glutamate is a source for the citric acid cycle intermediate α-ketoglutarate, accelerated formation and release of glutamine may not be a productive reaction when ATP demand is high.

The key reaction that is accelerated by an increase in exercise intensity is the adenylate deaminase reaction. In particular, the metabolic changes in an intensely contracting fast-twitch muscle fiber (increase in AMP and H+) favor activation of adenylate deaminase, with the attendant formation of IMP and ammonia. As a phosphorylated molecule, IMP is trapped within the fiber, but ammonia is released at an accelerated rate. Studies generally show a positive relationship between lactate formation and release from skeletal muscle and release of ammonia.

ADDITIONAL ROLES FOR AMINO ACIDS

We have seen in previous chapters that amino acids have purposes beyond simply serving as the building blocks of proteins. Table 7.1 summarizes roles for amino acids other than as components of body proteins.

Creatine is created in a two-step process from the amino acid *glycine* and part of the amino acid *arginine*. A methyl group must be transferred, and this is accomplished through the use of *S-adenosylmethionine* (SAM), a derivative of the essential amino acid *methionine. Carnitine,* the essential molecule used in the transfer of long chain fatty acyl groups across the mitochondrial inner membrane, begins as a *lysine* residue that is methylated on its side chain amino group with SAM. We have also discussed the essential role played by glutathione in maintaining cellular redox status. Glutathione is a tripeptide formed from the amino acids *glutamate, cysteine,* and *glycine.* Biosynthesis of pyrimidines, such as **uracil, thymine,** and cytosine, depends on aspartate and glutamine. For purine biosynthesis, including the bases adenine and guanine, the amino acids glutamine, glycine, and aspartate are needed. Heme proteins, such as hemoglobin, myoglobin, and the cytochromes, have a **heme** group with a central iron ion. The heme group is built from a structure class known as **porphyrins,** which are synthesized in the body using carbons from succinyl CoA and the amino acid glycine.

The amino acid *tyrosine* can be created by hydroxylation of the essential amino acid phenylalanine. Tyrosine is the precursor for three important molecules, the neurotransmitters *dopamine* and **norepinephrine** and the adrenal medulla hormone epinephrine. Tyrosine is first converted to dopamine, a neurotransmitter in the brain. Hydroxylation of dopamine creates norepinephrine, and methylation of norepinephrine produces epinephrine. *Histamine,*

Table 7.1 Roles for Amino Acids Other Than As Components of Proteins

Precursor amino acid	Special product	Role of product
Alanine and histidine	Carnosine	Dipeptide involved as an antioxidant
Arginine	Nitric oxide	Inter- and intracellular signaling
Arginine, glycine, and methionine	Creatine/Creatine phosphate	Energy transfer reactions
Aspartate, glutamine, and glycine	Purines and pyrimidines	Nucleic acids, nucleoside phosphates
Aspartate	Aspartate	Excitatory neurotransmitter
Cysteine, glutamate, and glycine	Glutathione	Maintain cellular redox
Glutamate	Glutamate	Excitatory neurotransmitter
Glutamate	Tetrahydrofolic acid (THFA)	One-carbon transfer reactions
Glycine	Porphyrins	Heme groups in proteins
Glycine	Glycine	Inhibitory neurotransmitter
Histidine	Histamine	Neurotransmitter
Lysine and methionine	Carnitine	Fatty acyl transfer
Methionine	S-adenosylmethionine	Methyl group transfers
Tryptophan	Serotonin	Neurotransmitter
Tyrosine	Dopamine, norepinephrine, and epinephrine	Neurotransmitters, hormone

another neurotransmitter, is synthesized from the essential amino acid histidine by decarboxylation of its carboxyl group. In a simple two-step process, the essential amino acid tryptophan is converted to the brain neurotransmitter *serotonin*.

There are some special amino acids, found in specific tissues. *Carnosine* (β-alanylhistidine) is a dipep-

tide formed from a modified form of the amino acid *alanine* (β-alanine) and *histidine*. It and two related forms, *anserine* and *homocarnosine*, are found in skeletal and cardiac muscle and brain in particular. These novel amino acid derivatives are thought to play roles as antioxidants. Carnosine has been tested as a supplement to reduce symptoms of autism.

SUMMARY

The content of protein in the adult body remains remarkably constant. Because most adults take in about 10% to 15% of their dietary energy in the form of protein, an equivalent amount of amino acids must be lost each day. The body cannot store excess amino acids. Those not needed to make protein lose their nitrogen groups, and the carbon skeletons are used to make glucose (gluconeogenesis) or are converted to acetyl CoA, which can feed into the citric acid cycle or be used to make fatty acids. The first step in disposing of excess amino acids is to remove the amino groups, which are primarily transferred to α-ketoglutarate by aminotransferase enzymes to produce glutamate and α-keto acids. This process occurs in the liver for most amino acids, but the branched chain amino acids (leucine, isoleucine, and valine) lose their amino groups primarily in muscle through the action of branched chain amino acid aminotransferase. Skeletal muscle contains a high concentration

of glutamine, synthesized by glutamine synthetase from glutamate and ammonium ions. During exercise, muscle releases most amino acids, but particularly alanine, ammonium ions, and glutamine, at accelerated rates. During exercise, branched chain amino acids and glutamate are taken up by muscle. Amino groups on amino acids are removed from the body in the form of urea, a molecule synthesized in liver using the urea cycle and excreted in the urine. The simple urea cycle eliminates amino groups that cannot be oxidized. The purine nucleotide cycle is active in skeletal muscle. During vigorous exercise, AMP deaminase is activated, changing AMP into IMP and ammonia. Regeneration of AMP, and thus the other adenine nucleotides, is accomplished by two reactions of this cycle. Many amino acids have functions beyond their role as building blocks of proteins. A number of amino acids can be converted into special molecules that function as hormones.

R E V I E W Q U E S T I O N S

1. What do the terms *energy balance* and *protein balance* mean?

2. The nitrogen content of proteins is considered to average 16% by mass of the body's proteins. The excretion of nitrogen can be measured in a simple manner. If a young adult excretes 16 g of nitrogen on average each day for a week, what is the average daily intake of protein if we assume that he or she is in nitrogen balance?

3. You determine that the content of urea in the urine averages 40 g per day. What is the approximate intake of protein assuming that the person is in energy balance?

4. The measurement of the excretion of 3-methylhistidine has been taken as an index of skeletal muscle protein breakdown. If you want to estimate muscle protein degradation, you need to determine the concentration of 3-methylhistidine in the urine. What else do you need to measure?

5. A young woman in energy and nitrogen balance takes in 8,000 kJ of food energy per day. She excretes 30 g per day of urea. Assuming that each gram of amino acids when oxidized yields 18 kJ, what percentage of her daily food intake comes from proteins?

6. The data in the following table were obtained from subjects at rest and during exercise on a cycle ergometer at 70% of $\dot{V}O_2$max for 1 h. Blood flow was measured, and exchange of amino acids across the leg was determined from blood samples obtained from the femoral artery and femoral vein. Leg blood flow averaged 0.3 L/min during rest and 8.0 L/min during exercise. The average concentration difference between arterial and venous blood for glutamate, glutamine, alanine, and leucine is shown for rest and exercise conditions. The molecular weight (MW) for each amino acid is also shown.

Amino acid	MW	Arterial minus venous blood concentration difference per minute (μmol/min)	
		Rest	Exercise
Glutamate	146	12	40
Glutamine	146	−25	−50
Alanine	89	−30	−100
Leucine	133	−1	20

a. What does a negative arterial minus venous concentration difference mean?

b. What was the total volume of blood flow through the leg during the 60 min of exercise?

c. What was the fold increase in blood flow from rest to exercise?

d. How many millimoles of glutamate were exchanged by the leg during the exercise period?

e. What are the two major sources of carbon for the alanine released from the muscle?

f. How many grams of alanine were released from the muscle?

g. Assuming that all of the alanine released during the exercise period is taken up by the liver, how many grams of pyruvate would be produced?

h. If the alanine released from muscle during the 60 min of exercise was entirely converted to glucose in the liver, how many millimoles of glucose would this produce?

III

TRANSCRIPTION, PROTEIN SYNTHESIS, AND DEGRADATION

We have already discussed the essential role of proteins in everything that takes place in an organism. In part III we learn how proteins are made and placed in their most appropriate locations within and outside of cells. As mentioned in the first part of the book, the information to make a protein is coded in the sequence of bases in DNA in cell nuclei. This DNA information is copied to make a sequence of bases in an RNA molecule, known as messenger RNA, in a process known as transcription. The information in the base sequence of messenger RNA is translated into a precise sequence of amino acids in a protein. In chapter 8, we review DNA, then focus on the process of transcription, learning how it is controlled in an appropriate way. Chapter 9 deals with the translation of messenger RNA information into the amino acid sequence of a protein. In addition, we will learn how newly synthesized proteins are modified to their functional forms and moved to the precise location where they will carry out their role. Since proteins are continuously turned over, we discuss processes that break down old proteins so that newly synthesized proteins can take their place.

Before we begin, a review of some terms is in order. The phenotype of an organism represents its physical, observable characteristics, whereas the genotype represents the genetic factors responsible for creating the phenotype. Our genome is the same in every cell nucleus, consisting of 3.1647×10^9 nucleotide bases organized into 46 chromosomes. Although our genome represents all the genes in all the chromosomes in the cell nucleus, not all of the genome may be expressed. The word **gene** is familiar to most people, yet it can have a variety of definitions. From a classic biology perspective it may be characterized as the basic unit of inheritance. From a modern perspective it may be described as the segment of DNA that provides the information for the amino acid sequence of a polypeptide chain or RNA molecule needed in the formation of a polypeptide. The majority of our 30,000 genes code for the amino sequence of proteins, yet this amounts to less than 2% of our genome. Finally, despite all of our unique individual characteristics, one person's genome is about 99.9% identical to that of all other people.

CHAPTER

8

GENE TRANSCRIPTION AND ITS CONTROL

In this and the next chapter, we trace the path of **gene** expression from DNA base sequence to the making of a protein. The DNA content of a nucleus is fixed. When cells divide to make new cells in the process of mitosis, the DNA must first be duplicated. Ribonucleic acid or RNA results when a section of DNA is copied in the nucleus during transcription. Ribonucleic acid plays an important intermediary role in the conversion of the DNA information into a sequence of amino acids in a protein. In this chapter, we consider the different kinds of RNA, the genetic code, gene transcription, and how transcription is regulated. We pay particular attention to the regulation of transcription, especially by way of signaling pathways that play a critical role in matching the rates of gene expression in individual cells to the overall needs of the organism. This chapter begins with a review of DNA, a topic many have already studied in biology courses.

DNA

You are no doubt familiar with the double-helix structure of DNA and aware that DNA in each nucleus provided the information that allowed us to grow from a tiny embryo into adulthood. With the exception of human eggs and sperm, each nucleus in every cell contains 46 **chromosomes.** Each chromosome consists of a single, but very long, DNA molecule plus many different kinds of proteins. Human males

contain two copies of chromosomes 1 through 22 plus an X and a Y chromosome. Human females also have duplicate copies of chromosomes 1 through 22, but differ in that they have no Y chromosome, but instead two X chromosomes. Our **genome** consists of 3.1647×10^9 bases, and the information contained in it codes for an estimated 30,000 genes. While each of us thinks we are unique and indeed we are, in terms of both our fingerprints and DNA, 99.9% of the sequence of bases in anyone's DNA is exactly the same as in everyone else's.

DNA Structure

We will talk more about the proteins associated with DNA, but for now our focus is on the structure of DNA. The components of DNA are a sugar (deoxyribose), phosphate, and four different bases. There are two *pyrimidine bases (cytosine* and *thymine)* and two *purine bases (adenine* and *guanine)* (figure 8.1). Deoxyribonucleic acid uses the sugar deoxyribose, whereas the other nucleotides we discussed in previous chapters (ATP, GTP, etc.) have ribose as the sugar. The difference between deoxyribose and ribose can be seen in figure 8.1, where a hydrogen atom at the 2' position in deoxyribose replaces an OH group as found in ribose. In DNA, a base is attached to the 5' carbon of deoxyribose. The combination of a purine or pyrimidine base to ribose or deoxyribose generates a *nucleoside* or *deoxynucleoside.* Besides containing the sugar ribose, instead of deoxyribose, RNA has

Figure 8.1 *(a)* Purine and *(b)* pyrimidine bases and *(c)* the structures for ribose and deoxyribose sugars.

the pyrimidine base uracil instead of thymine (see figure 8.1). Examples of nucleosides are adenine plus ribose, forming adenosine, and adenine plus deoxyribose, producing a deoxyadenosine. As mentioned in chapter 3, the addition of one or more phosphate groups to a nucleoside produces a nucleotide. Figure 8.2 shows the structures for deoxyadenosine, a nucleoside, and adenosine 5'-monophosphate (AMP), a nucleotide.

A single-stranded polynucleotide chain consists of sugars (ribose or deoxyribose, shown by S) joined together by a phosphate group (shown by P), attached to the 5' carbon of one sugar and to the 3' carbon of the next sugar. The sugar-phosphate bonds are actually ester bonds or *phosphoester* bonds.

Attached to each sugar is a purine or pyrimidine base. This is shown in a shorthand way in figure 8.3. If the sugar is deoxyribose and the base is thymine, we call it single-stranded DNA; if the sugar is ribose and the base uracil replaces thymine, it is a single strand of RNA.

Deoxyribonucleic acid is found as a double-stranded molecule. The two strands are held together by hydrogen bonding between bases as shown on the right side of figure 8.4. In all cases, the base adenine (A) in one strand is hydrogen-bonded to a thymine (T) in the opposite strand. Guanine (G) in one strand is hydrogen-bonded to a cytosine (C) in the other strand. There are two possible hydrogen bonds between each A and T and three hydrogen

Deoxyadenosine   Adenosine 5'-monophosphate

Figure 8.2   Structures for deoxyadenosine and adenosine 5'-monophosphate. The base adenine attached to the 1' carbon of deoxyribose creates the nucleoside deoxyadenosine. Addition of a phosphate group to the 5' carbon of the sugar ribose, itself attached to adenine, produces the nucleotide adenosine 5'-monophosphate (5'-AMP or simply AMP).

5' End contains a terminal phosphate group bonded only to 5' carbon of sugar.

Phosphodiester bond joins 3' carbon of upper sugar to 5' carbon of sugar below.

3' End contains a terminal deoxyribose without a phosphate group attached.

Figure 8.3   Representation of a single-stranded polynucleotide, shown with only five bases, identified by A, T, G, and C. S refers to the sugar deoxyribose, and P represents the phosphate group. The sugar-phosphate backbone is maintained by phosphodiester bonds that join the 3' carbon of one deoxyribose to the 5' carbon of the next deoxyribose.

bonds between each G and C. Note that the backbone of each strand in a DNA molecule is made up of deoxyribose sugar and phosphate groups. Each strand in DNA has a polarity. As shown in figure 8.4, one strand begins with a free phosphate group attached to the 5' carbon of deoxyribose. The end of the strand contains a free 3' OH group on the terminal deoxyribose. Note that the opposite strand has reverse polarity, with a free 3' OH at the top and the free 5' phosphate group at the bottom. The polarity of DNA is very important. You should also understand that the content of G in a DNA molecule is equal to that of C; as well, A equals T.

You are more familiar with the three-dimensional picture of DNA in which the two strands are wound around a central axis forming a right-handed double

▶ K E Y P O I N T ◀

The capital letters A, G, C, and T can represent the four bases in DNA. These letters are also used to designate the nucleotides (base + sugar + phosphate) that form the basic units of polynucleotides such as DNA and RNA.

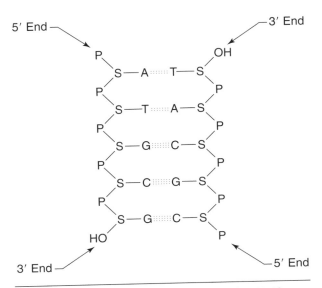

Figure 8.4 Representation of double-stranded DNA. The two strands are antiparallel, with one strand running from 5' to 3' and the other strand from 3' to 5' (top to bottom). Double-stranded DNA is held together by hydrogen bonds between complementary bases (A and T; G and C) shown by dotted lines.

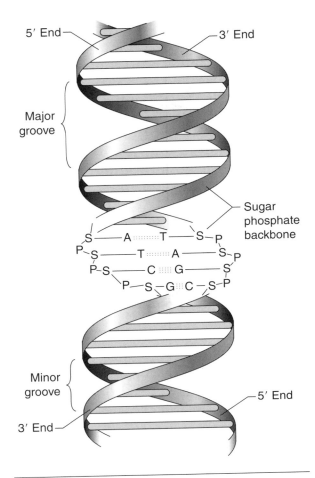

Figure 8.5 The double-helix structure of a portion of a DNA molecule, showing the two antiparallel polynucleotide strands. The strands are antiparallel in that the sugar-phosphate backbones (S-P-S-P . . .) in each strand run in opposite directions (3' to 5' vs. 5' to 3' direction). Attached to each sugar molecule is a purine base (adenine or A, guanine or G) or a pyrimidine base (cytosine or C, thymine or T). The two strands are complementary in that an A in one strand is always adjacent to a T, and a C is always adjacent to a G. The two strands are held together by hydrogen bonding (dotted lines) between adjacent bases, with two hydrogen bonds between each A and T and three hydrogen bonds between each C and G. The arrangement of the helix creates both major and minor grooves. One complete turn of the helix spans a distance of 3.4 nm.

helix (figure 8.5). The two strands are said to be **antiparallel;** that is, one runs from 5' to 3' and the other from 3' to 5'. We say the two strands are **complementary** in that a G in one strand is opposed by a C in the other, while an A and a T match each other. Because of the nature of the helical structure, there are both major and minor grooves in the helix. A complete turn of the helix spans a distance of 3.4 nm and contains 10 stacked pairs of bases. The most common form of DNA in our cells is known as B-DNA. The hydrogen bonds that maintain the three-dimensional, complementary double-stranded structure of DNA can be broken by heat, which separates the two strands. However, the ability of the bases in each strand to recognize and hydrogen-bind to complementary bases in the other is powerful, so that cooling allows the two strands to spontaneously re-form to exactly the way they were before, much as a zipper operates. This process, known as **hydridization,** forms the basis for detecting particular single-stranded polynucleotide molecules.

> ➤ **K E Y   P O I N T** ◄
>
> With a common sugar-phosphate backbone, the message carried by DNA lies in the order of the four bases. Therefore we can say that the language of DNA is based on only four letters.

Chromatin and Nucleosomes

Deoxyribonucleic acid molecules are very large, containing millions of **base pairs** (bp). The DNA in a single cell nucleus would reach about 2 m (6.6 ft) in length if extended. Nuclear DNA is compacted more than 100,000 times by folding and coiling (Chakravarthy et al. 2005). Each chromosome has roughly an equal mass of protein and DNA. Many of these proteins are needed to control the copying of sections of DNA. Histones are an important class of proteins closely associated with DNA because they contain many amino acid residues with positively charged side chains that can interact by electrical attraction with the negatively charged phosphate groups on the sugar-phosphate backbone. The combination of DNA and its associated proteins in a cell is known as **chromatin.** Two general classes of proteins are involved with DNA in chromatin: histone and nonhistone proteins. Chromosomes are not visible as discrete structures except during cell division.

The DNA-protein complex known as chromatin is densely packed and distributed within the nucleus. There is a structural organization to chromatin that is based on a class of proteins known as **histones,** which are intricately arranged with DNA to produce repeated structures known as **nucleosomes.** Each nucleosome consists of a disc-like core containing two pairs each of the histones H2A, H2B, H3, and H4. Tightly wrapped around each histone disc is a single DNA strand with 147 bp of DNA making almost two complete turns (figure 8.6). Nucleosomes are joined by a linker DNA strand that is variable in length between 15 and 55 bp. A histone H1 also plays a role in linking adjacent histones. Histone proteins are rich in the basic amino acids lysine and arginine, so they carry a net positive charge that complements the negatively charged sugar-phosphate backbone of DNA. Because of nucleosomes, chromatin can look like a beaded necklace, with each nucleosome core particle making up a bead and the linker as the string that holds together the beads. Considering that there are roughly 200 bp of DNA per nucleosome core particle plus one section of linker DNA, we would expect there to be roughly 30 million nucleosomes in each human diploid cell, consisting of 6.33×10^9 bp of DNA (Alberts et al. 2002).

Chromatin has other levels of structural organization in which nucleosomes are packed together in even more condensed forms, generating higher levels of structure. Regions of chromatin with low levels of transcription activity are very tightly packed into a form referred to as *heterochromatin.* Such dense DNA packing in transcriptionally silent areas contrasts with more open chromatin packing in regions with higher rates of gene expression, known as *euchromatin.* The structure of the histone core shown in figure 8.6 does not reveal that parts of each histone protein in the core have tails that extend outside; these are described as histone tails. Lysine residues in these tails have positively charged side chains that allow interaction with neighboring DNA to produce more densely packed structures. We will see how

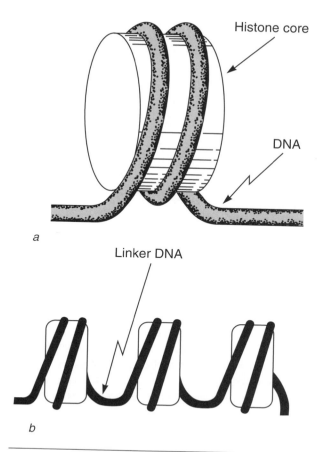

a

b

Figure 8.6 Nucleosomes. *(a)* A single nucleosome, showing 147 bp of DNA wrapped almost twice around a core of histone proteins, consisting of two of each of the histones H2A, H2B, H3, and H4. *(b)* Three nucleosomes joined together by a strand of linker DNA.

modification of these histone tails can influence the transcription of genes.

DNA Replication

Before a cell divides to form two new (daughter) cells, the DNA must first be duplicated, so that two exact copies of the DNA are formed. This process, known as *DNA replication,* is carried out by a special complex called a *replisome,* which contains a specific enzyme **(DNA polymerase)** and precursor molecules called deoxynucleoside triphosphates (dATP, dGTP, dGTP, and dTTP). Each strand of each molecule of DNA in the nucleus is copied to make an exact **complementary DNA (cDNA)** strand in a process described as **semiconservative** replication. As a result of DNA replication, the parent cell prior to cell division has two exact copies of its DNA. With mitosis, the duplicated DNA is partitioned such that each daughter cell contains the exact complement of DNA of the parent.

Figure 8.7   A schematic of DNA replication showing the template strand (though both strands are copied). Dotted lines represent hydrogen bonds between complementary bases on the template and newly synthesized strand. The next deoxyribonucleotide to be added contains the base adenine (A) because the template strand has the base thymine (T). The 3' hydroxyl group of the last nucleotide will bind to the incoming dATP, adding a new nucleotide to the growing new strand.

Figure 8.7 summarizes the process of DNA replication. The two DNA strands of each DNA molecule are unwound, exposing the single strands that act as a template on which a complementary copy is formed. Note that deoxyribonucleotides are added one at a time to the 3' end of a growing polynucleotide chain. Since only one phosphate group from precursor deoxynucleoside triphosphates is needed, a pyrophosphate (PPi) is released. Because of base pairing between A and T and G and C, fidelity in copying is assured. Since DNA molecules are so long, replication takes place simultaneously at multiple sites on the same DNA molecule.

> ➤ **K E Y P O I N T** ◄
>
> The small circular mitochondrial DNA molecule we discussed in chapter 4 also undergoes replication. Indeed, we have multiple copies of mitochondrial DNA in each mitochondrion. Our mitochondrial DNA is obtained from the egg and not the sperm, and the distinctive characteristics of our mitochondrial DNA can provide a history of the mothers in our families.

TYPES OF RNA

Ribonucleic acid is a single-stranded polynucleotide composed of the bases adenine, guanine, cytosine, and uracil; the sugar ribose; and phosphate groups. We could represent a single-stranded ribonucleotide

chain as in figure 8.3 if the sugar (S) was ribose and if the base uracil (U) replaced thymine (T). The ends of RNA molecules are designated as 5' if there is a free phosphate on the 5' carbon of ribose, or 3' if the ribose sugar has a free 3' OH group. Thus, like each DNA strand, an RNA molecule has a direction. Ribonucleic acid molecules are created when a section of a strand of DNA (a gene) is copied in a process known as transcription. The length of the RNA molecule and the sequence of bases (nucleotides) dictate what role the RNA molecule will perform. As discussed in chapter 1 in the context of the sequence of amino acids in a protein, the sequence of bases in an RNA molecule is also known as its primary structure.

Messenger RNA (mRNA) is the actual template for protein synthesis in the cytosol. This means that the base sequence on mRNA specifies the sequence of amino acids in a polypeptide chain. Most of our genes generate specific mRNA, which has a short lifetime—usually minutes to several hours or more. Messenger RNA is also the least abundant of the three types of RNA, representing about 2% of total cellular RNA.

Transfer RNA (tRNA) is the smallest of the RNA molecules, usually between 73 and 93 nucleotides (Nt) in length. Transfer RNA attaches to specific amino acids and brings them to the complex of mRNA and ribosomes on which a polypeptide is formed. There is at least one tRNA molecule for each of the 20 amino acids involved in protein synthesis. All the tRNA molecules represent approximately 15% of the total cellular RNA.

Ribosomal RNA (rRNA) is the most abundant RNA, representing more than 80% of all the RNA in a cell. A ribosome is a complex of protein and ribosomal RNA where proteins are synthesized. There is one other type of RNA, known as **small nuclear RNA (snRNA)**. The snRNAs are found in the nucleus associated with protein in particles described as small nuclear ribonucleoprotein particles or snRNPs. These are involved in the processing of the primary RNA transcripts during their conversion to mRNA molecules. We will mention them later.

THE GENETIC CODE

The sequence of four bases in a gene in DNA is transcribed to form a sequence of four bases in mRNA that must specify the sequence of the 20 different amino acids used to make proteins. We may ask ourselves how 20 different amino acids can be uniquely described by only four different bases. The only way that four different bases in mRNA (identified as A, G, C, and U) can specify 20 different amino acids in a polypeptide chain is for the bases to be read in groups of three, known as **codons.** Four different bases read three at a time gives rise to 4^3 different possibilities, or 64 codons. This process is called triplet coding, that is, three bases read together to give a message. Since mRNA has two ends, one of them—the 5' end—is

considered to be at the beginning, and the message is read in groups of three toward the 3' end.

The genetic code is the relationship between the base sequence of DNA (A,T,G,C), transcribed to mRNA (U,A,C,G), and the sequence of amino acids in a polypeptide. It has a fixed starting point, an initiation codon, and the bases are then read from this point in groups of three nonoverlapping bases at a time (table 8.1). We usually consider the genetic code from the perspective of the codons of mRNA that spell out amino acids. Of the 64 possible codons, 61 specify amino acids, and the remaining three are stop or termination signals. Because there are 61 codons for only 20 amino acids, most amino acids have two or more codons. We say, therefore, that the genetic code is **degenerate.** Two amino acids (tryptophan and methionine) have a single codon, while some (leucine, serine, and arginine) have six codons. The codons for an amino acid with more than one codon are very similar. This similarity makes sense if minor errors are made. For example, the four codons for the amino acid glycine are GGA, GGG, GGC, and GGU. Notice that these all contain the same first two bases (letters). If amino acids are similar in structure, their codons are similar. For example, aspartic acid has the codons GAC and GAU, whereas glutamic acid, which is closely related in structure, has the codons GAA and GAG. If there is a base reading error, the same or a similar amino acid can still result.

Table 8.1   The Genetic Code Revealing the Three-Letter Codon for Each Amino Acid Shown by Its Three-Letter Abbreviation

First position (5' end)	Second position				Third position (3' end)
	U	C	A	G	
U	UUU Phe	UCU Ser	UAU Tyr	UGU Cys	U
	UUC Phe	UCC Ser	UAC Tyr	UGC Cys	C
	UUA Leu	UCA Ser	UAA Stop*	UGA Stop*	A
	UUG Leu	UCG Ser	UAG Stop*	UGG Trp	G
C	CUU Leu	CCU Pro	CAU His	CGU Arg	U
	CUC Leu	CCC Pro	CAC His	CGC Arg	C
	CUA Leu	CCA Pro	CAA Gln	CGA Arg	A
	CUG Leu	CCG Pro	CAG Gln	CGG Arg	G
A	AUU Ile	ACU Thr	AAU Asn	AGU Ser	U
	AUC Ile	ACC Thr	AAC Asn	AGC Ser	C
	AUA Ile	ACA Thr	AAA Lys	AGA Arg	A
	AUG Met**	ACG Thr	AAG Lys	AGG Arg	G
G	GUU Val	GCU Ala	GAU Asp	GGU Gly	U
	GUC Val	GCC Ala	GAC Asp	GGC Gly	C
	GUA Val	GCA Ala	GAA Glu	GGA Gly	A
	GUG Val	GCG Ala	GAG Glu	GGG Gly	G

*Stop codons do not have an amino acid assigned to them.

**Codes for the amino acid methionine but also is the start or initiation codon.

The codon AUG is the initiation or **start codon** that signals the start of translation. This codon also represents the amino acid methionine; thus the first amino acid used in protein synthesis is methionine. However, not all functional proteins have methionine as the first amino acid, as it can be removed after the polypeptide is completely formed. The codons UAG, UGA, and UAA are **stop (termination) codons;** they say that translation of the mRNA message is ended. The codon sequence beginning with a start codon and ending with a termination codon is known as a *reading* frame. The genetic code is universal for all organisms studied except in mitochondria, where some minor variations are noted. However, the amount of mitochondrial DNA is small (16,569 bp in humans)—just enough to code for 13 polypeptides, two kinds of rRNA, and 22 tRNA molecules.

TRANSCRIPTION

The process of transcription involves copying a gene on DNA to make an RNA molecule. A simple definition of a gene is that it is a section of DNA that provides the information to synthesize a single polypeptide or functional unit of RNA such as a tRNA or rRNA molecule. The preponderance of our approximately 30,000 genes code for polypeptide chains in proteins. We could describe these as protein-coding genes. Therefore, as shown in figure 8.8, transcription will yield an RNA molecule described as a **primary gene transcript.** This transcript will be structurally modified to produce a final RNA product destined to provide amino acid sequence information for a polypeptide (mRNA), RNA to be part of a ribosome, or one of the more than 20 different tRNA molecules.

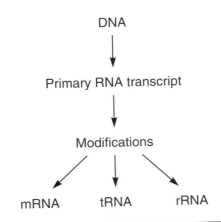

Figure 8.8 The three main kinds of RNA generated by transcription of a gene and modifications to produce the functional product.

During transcription, only one strand of the DNA in the gene is copied. This is known as the **template DNA strand,** and it is read in the 3' to 5' direction. The primary transcript RNA formed will be complementary to the template DNA strand with the exception that the RNA will contain the base uracil (U) rather than thymine (T). The primary transcript will be formed in the 5' to 3' direction, that is, antiparallel to the template strand. The precursors needed to make RNA are nucleoside triphosphates (NTP) such as CTP, GTP, ATP, and UTP.

The formation of RNA during transcription is catalyzed by a large oligomeric enzyme known as DNA-directed RNA polymerase, or simply **RNA polymerase,** of which there are three kinds. Ribonucleic acid polymerase I (abbreviated Pol I, sometimes RNAP I) transcribes three rRNA genes (5.8S, 18S, and 28S). Pol I is the most abundant RNA polymerase. Ribonucleic acid polymerase II (Pol II or RNAP II), a large protein complex with 12 subunits, transcribes genes containing the information for the amino acid sequence in a polypeptide chain. This form of RNA polymerase gives rise to mRNA. Pol II also transcribes a few small structural RNA genes. Ribonucleic acid polymerase III (Pol III or RNAP III) is the least abundant of the RNA polymerase enzymes. Its products are tRNA, another small rRNA molecule (5S rRNA), and the snRNA molecules mentioned previously, as well as some other small RNA molecules.

For genes coding for amino acid sequence information for a polypeptide, the actual gene is considered to consist of a *coding region* and a region in front of this *(regulatory region)* that controls the transcription of the coding region. Figure 8.9 shows in simplified form what happens during transcription. With two DNA strands and a regulatory region in a gene, only a portion of the template strand will be transcribed, producing a complementary copy in the form of the RNA transcript. The DNA strand in the gene that is not copied is known as the *sense strand* because it will have the same base sequence as the RNA transcript, except that U will replace T. The polarity of the sense strand and the RNA are the same and are opposite to that of the template strand. In this context, directions are often described in terms of river flow. **Upstream** refers to the 5' direction, and, by convention, the frame of reference is the sense strand of DNA. Likewise, **downstream** refers to the 3' direction of flow, that is, the direction in which the template strand is transcribed. To facilitate understanding of the genetic code, figure 8.9 also shows the amino acids spelled out by codons on the mRNA molecule.

Figure 8.10 shows the three major phases of transcription. *Initiation* begins when general transcription factors (proteins, which we will discuss next) and RNA polymerase bind to the double-stranded DNA just upstream of the **start site,** forming a preinitiation complex. The start site is where transcription actually begins, and the first DNA base copied is given the number +1 and is usually shown with a bent arrow pointing downstream. General transcription factors help the RNA polymerase find the start site

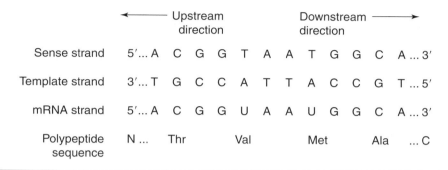

Figure 8.9   A section of a polypeptide-coding gene showing the two strands of DNA. The template strand is copied during transcription, generating a complementary copy of single-stranded mRNA. Note that the base (nucleotide) sequence of mRNA matches that of the sense strand of DNA, with U in RNA replacing T in the sense strand. To reveal the correspondence between base sequences on DNA (sense strand) and mRNA, three-base sequences on the latter are converted into an amino acid sequence determined from table 8.1. N and C refer to the amino and carboxy ends of the polypeptide, respectively.

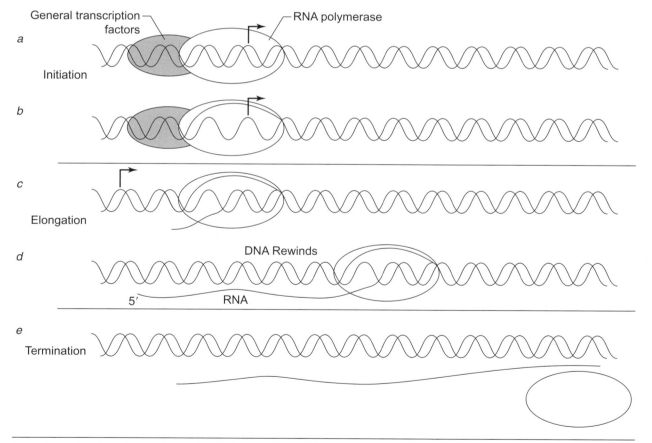

Figure 8.10   The three distinct phases of transcription. *Initiation (a, b)* occurs when RNA polymerase binds along with general transcription factors to a region encompassing the transcription start site, shown with a bent arrow. Deoxyribonucleic acid is first unwound *(b)*, exposing the template strand to be copied. During *elongation (c, d)*, RNA polymerase catalyzes the addition of nucleotides, one at a time, to the 3' end of the growing RNA molecule. *(d)* As the polymerase moves along the template strand, the DNA behind it rewinds. *(e)* When a *termination* signal is reached, transcription ceases, and the RNA polymerase and the primary RNA transcript dissociate from the DNA.

and facilitate initiation of the transcription process. Bases immediately upstream are in a region called the **promoter** and are given negative numbers. The promoter region can also include some bases just downstream from the transcription start site.

When RNA polymerase and the general transcription factors bind to the promoter region, the DNA at the start site is unwound (melted), exposing approximately 14 nucleotides in the template strand (figure 8.10b). The first nucleotide triphosphate that will become the first nucleotide on mRNA binds to the complementary base on nucleotide 1 on the template strand. Then the second NTP comes in, recognizing its complementary base, number 2 on the template strand. A phosphoester bond is formed between the 3' OH of ribonucleotide 1 and the 5' phosphate of ribonucleotide 2, and a pyrophosphate is released. During initiation, the RNA polymerase does not move along the DNA.

The elongation phase begins when the RNA polymerase dissociates from the promoter and general transcription factors and moves along the

template strand of DNA, making a complementary RNA strand (figure 8.10c). As it moves, it unwinds the DNA double helix, catalyzing the formation of the RNA strand. Behind it, the DNA double helix re-forms (figure 8.10d). For each new nucleotide added to the terminal free 3' OH of the growing RNA polynucleotide, only one of the three phosphate groups on the incoming nucleoside triphosphate is needed; the other two are released as inorganic pyrophosphate, PPi—just as occurs in DNA replication. Figure 8.11 shows the details of the addition of a new ribonucleotide to the growing RNA chain. The termination phase begins when the RNA polymerase has moved along the template strand of DNA and reaches a sequence of bases indicating that the gene message is terminated. Some call this the terminator. At this point, the RNA polymerase and the RNA strand dissociate from the DNA. The RNA polymerase is now free to transcribe the same gene again or a different gene. It is estimated that RNA polymerase catalyzes transcription at a rate of approximately 1,000 nucleotides per minute.

Figure 8.11 The RNA chain during transcription. The RNA chain grows by the addition of nucleoside triphosphates to the 3' end of the RNA chain. In the example shown, the new nucleoside triphosphate is guanosine triphosphate (GTP). On the left side of the figure, the OH group at the 3' end of the RNA chain attacks the α-phosphate of GTP, creating a new phosphoester bond and releasing inorganic pyrophosphate (PPi). The chain on the right side is now lengthened by one nucleotide.

REGULATION OF TRANSCRIPTION

Each cell nucleus contains all the genes for the given organism as base sequences in the DNA. However, cells express only the genes they need. For example, a liver cell expresses genes that are expressed by other cell types, but it also expresses a number of liver-specific genes—different genes compared to those in a nerve or heart muscle cell. The human body contains more than 250 different kinds of cells. Also, as cells differentiate during development, different genes are expressed, giving rise to new proteins, and formerly active genes are no longer expressed. Furthermore, adult cells adapt to new circumstances by expressing previously inactive genes and not expressing previously active ones. Clearly, transcription must be regulated. This means that it is necessary to regulate which genes are expressed and at what rate. Although every human cell nucleus contains 30,000 genes, only a fraction of these are transcribed at any one time, and those that are expressed depend on the cell type and its stage of development. In complex organisms such humans, regulation of transcription in a cell depends on a complex interaction of hormonal, metabolic, nutritional, and environmental signals. It is also important to emphasize that DNA is found with proteins as chromatin. This means that the substrates for transcription are actually sections of chromatin.

> ## ▶ KEY POINT ◀
>
> Proteins direct virtually all the events in a multicellular organism associated with a healthy life. The information for protein structure is based on the sequence of amino acids, dictated by the sequence of bases in a segment of DNA called a gene. Since multicellular organisms have highly specialized cells, we can assume that this specialization is due to the production and function of specific proteins. This means that specific genes must be properly transcribed in all cells.

Control of gene expression occurs primarily through directing the initiation stage of transcription. Two major types of control decisions must be described for a cell. Irreversible decisions turn specific genes on or off completely. For example, during embryo development, certain genes are initially expressed, then turned off completely. Other genes are irreversibly turned on. But the second type of decision is adjustable in terms of transient increases or decreases in the rate of transcription of an already expressed gene in response to various environmental or metabolic conditions. Consider the analogy of a light switch with a dimmer. The irreversible decision is that the switch is either on or off. Once on, however, it can be adjusted to produce a low, moderate, or high level of light intensity. Some genes in a cell are permanently turned off. One mechanism whereby this occurs is the structural organization of DNA. As we will discuss in a subsequent section, inactive genes are found in sections of chromatin that is condensed, precluding the binding of RNA polymerase and other proteins at the promoter region to initiate transcription.

Transcription Factors and Response Elements

Proteins are responsible for most of what happens in an organism, including the regulation of gene transcription. Proteins that regulate transcription are known as *transcription factors* or gene regulatory proteins. They are products of their own genes, most likely on different DNA molecules from the gene or genes they regulate. They have also been known as **trans factors** or transactivating factors. If they promote transcription, they are **activators;** and if they have a negative effect on transcription, we call them **repressors.** We can define two major classes of transcription factors: (1) general transcription factors needed for initiating transcription from most protein-coding genes and (2) those that modulate general transcription factors through activation or repression and provide specificity for cell types and differentiation. We will discuss these shortly.

In their role as regulators, transcription factors must bind to specific DNA sequences as part of the *transcription-control region;* some sources describe this as the gene control region. This can be at the promoter region (or promoter), which is the sequence of bases at the site where RNA polymerase and the general transcription factors bind to initiate transcription. A common feature of the promoter of many protein-coding genes is the **TATA box,** an A- and T-rich sequence located 25 to 35 bp upstream of the start site where the transcription initiation complex binds. The transcription-control region may also include regions close to the promoter or thousands of base pairs either upstream or downstream from the promoter. The terms *transcription-control element* or *response element* are typically applied to a DNA base sequence to which proteins can bind to regulate transcription of a gene. Another term that has been used to refer to the response element is *regulatory sequence.* Other terms are *box* (e.g., TATA box) and **enhancer.** Enhancers are response elements typically far away from the promoter, either upstream or downstream, or even within the coding region of the gene. Response elements, the actual units that make up the transcription-control regions, are normally short, usually 6 to 10 bp, although some may be larger. As parts of DNA, response elements have a negative charge because of the sugar-phosphate backbone. Some sources also identify a *proximal promoter,* a region upstream of the promoter, typically 100 to 200 bp upstream of the transcription start site. Names currently in use for these DNA response elements include the following:

Promoters

Promoter proximal elements

Response elements

Control elements

Regulatory sequences

Enhancers

Boxes

The control elements that make up a typical gene are shown in figure 8.12. The gene regulatory proteins include general transcription factors that bind at the promoter as well as several thousand other proteins that act to control the transcription of specific genes since each gene or any closely related gene in a cell appears to have its own specific regulation arrangement. Names currently in use to refer to the various transcription regulatory proteins include the following:

Transfactors or transactivating factors

General transcription factors

Activators

Repressors

Coactivators

Corepressors

Most gene regulatory proteins have a *DNA-binding domain.* This is a specific portion of the protein molecule, often described as a motif, that recognizes a particular short stretch of DNA, usually in the major groove. While the details are beyond the scope of this book, it is worthwhile identifying some common DNA-binding motifs found in eucaryotic species. The *homeodomain* contains approximately 60 amino acids arranged into short sections of alpha-helix that recognize specific response elements in the major groove of DNA. *Zinc finger* proteins contain sections of polypeptide arranged around Zn^{2+} ions that look like fingers; these can fit into the major groove of DNA. Of the 2,500 or so identified DNA-binding regulatory proteins in the human genome, zinc finger proteins are most prevalent (Roeder 2005). *Leucine zipper* proteins have the hydrophobic amino acid leucine at every seventh residue in the DNA-binding motif. A variant of the leucine zipper protein is the basic zipper or *bZip* transcription factors. *Basic helix-loop-helix proteins* (bHLH) contain a DNA-binding domain with two short sections of alpha-helix separated by a nonhelical region. The first helix in bHLH proteins contains basic amino acid residues whose net positive charge interacts with the negatively charged sugar-phosphate backbone of DNA.

Gene regulatory proteins often bind to their response elements in pairs (as dimers). The two tran-

Figure 8.12 Transcribing a typical gene. Control over transcription is exerted by the binding of a variety of transcription factors to particular DNA control (response) elements shown as boxes in this figure. Some control elements are located far upstream of the start site, shown as a bent arrow, while others may be far downstream of the coding region. The TATA box, located in the promoter, is where the transcription initiation complex binds. Immediately upstream of this, up to approximately –200 bp, are the promoter proximal elements, shown cross hatched. Other response elements are shown as open bars.

scription factors may be either the same protein or different proteins, that is, homo- and heterodimers, respectively. Therefore, we should expect that these transcription factors need dimerization domains that allow them to bind to the other protein forming the dimer complex at the response element. Transcription factors also contain a region or regions that allow them to interact with other regulatory proteins, including general transcription factors. These regions are usually called *activation domains* because most of the interactions generated promote transcription as opposed to repressing it. Activator proteins promote transcription by helping to attract and position other proteins at the promoter to facilitate transcription initiation. They may also act on the chromatin to expose the DNA to the transcription machinery. Given several thousand different gene regulatory proteins as well as the fact that a number of these are needed to initiate transcription, binding as both homo- and heterodimers, it is no wonder that there is likely a unique transcription initiation apparatus for almost every protein-coding gene.

Gene repression in eukaryotic species involves two kinds of repressor proteins. Some repressors act to keep large sections of chromatin unavailable for transcription because of the tight way it is packed in regions; this was described earlier as heterochromatin. Other repressors act locally to inhibit transcription initiation in much the same way as activators facilitate transcription. Whereas activators facilitate the assembly of the transcription initiation complex by interacting with other proteins, including those classified as general transcription factors, repressors

inhibit this process. They may do so by masking the activation domain of an activator, by blocking access to a response element of a transcription activator, or by altering the chromatin structure at the promoter to render the gene less accessible for transcription. Another class of gene regulatory proteins, the *coactivators* and *corepressors*, have activation and repression domains, respectively, but lack DNA-binding domains so they are unable to bind to response elements for the gene. Their role is primarily to facilitate or inhibit transcription initiation.

The Basal Transcription Apparatus

Ribonucleic acid polymerase II (Pol II) cannot initiate transcription by itself. Rather, it requires a number of general transcription factors to interact with it in the core promoter region and a large protein complex known as the *mediator*. The term *general transcription machinery* is used to describe Pol II, the general transcription factors, and the mediator that assembles at the promoter region of the gene. Additional proteins that will change the structure of the DNA to facilitate transcription are also necessary. Without the influence of other transcription factors bound to their cognate response elements in the promoter proximal region or of distant enhancers, the rate of transcription based only on general transcription factors is low or nil. In this section, we focus only on the basal transcription of a protein-coding gene. Figure 8.13 outlines the various protein factors found in the promoter region of a typical gene. There are general transcription factors needed for transcription

Figure 8.13 Protein factors in the promoter region of a typical gene. Transcription initiation at a typical protein-coding gene requires the presence of RNA polymerase II (Pol II) containing 12 protein subunits, a large protein complex known as the mediator with its 20 subunits, and a number of general transcription factors, including the TATA-binding protein (TBP). Common general transcription factors found at the initiation of genes transcribed by Pol II are IIB, IID, IIE, IIF, and IIH. Many promoter regions also contain general transcription factor IIA. Transcription begins at the start site, shown as a bent arrow. Chromatin remodeling, essential for transcription, is facilitated by chromatin remodeling proteins along with histone acetylase complex.

of genes by RNA polymerases I and III, but we will not discuss them.

The general transcription factors are identified as such since they are found in the promoters of all genes transcribed by RNA polymerase II. They help to position RNA polymerase II and bind in the promoter region. They also facilitate the opening of the DNA to expose the template strand. After transcription is initiated, general transcription factors help to release Pol II so it may continue on during the elongation phase. A TATA-binding protein (TBP) is first to bind at the TATA box, followed by general transcription factors IID and IIB. Transcription factor IIF along with RNA polymerase II binds next, so that Pol II is positioned over the transcription start site. Subsequent additions are at least transcription factors IIE and IIH. A large protein complex, containing at least 20 polypeptide subunits and known as the *mediator*, plays a key role in positioning Pol II and binding to activation domains of other transcription factors involved in the regulation of transcription. It is believed that to generate a basal level of transcription for most mammalian genes, at least five general transcription factors (IIB, IID, IIE, IIF, and IIH), along with Pol II and the mediator, are necessary. General transcription factor IIA is found in some promoters, but its role in all polypeptide-coding genes is unclear.

Some protein-coding genes lack a TATA box, having instead an alternate promoter element known as an *initiator*. Other protein-coding genes begin their transcription over a rather long region up to 200 bp in length. Their primary transcript can thus have multiple 5' beginnings. Genes lacking both TATA boxes and initiators may instead have *GC-rich regions* in their promoter, 100 or so bp upstream of their start site region. Such genes are generally coded at a consistent but low rate and are known as *housekeeping genes*. Their protein products are typically metabolic enzymes.

Higher Levels of Transcription Control

We have already discussed the fact that gene response elements may be located in positions far from the actual start site of the gene. In addition to the TATA box, which is so widely found in the promoter region of protein-coding genes, there are other base sequences recognized by specific proteins that can lead to higher levels of transcriptional control. These response elements may be found in the proximal promoter region (see figure 8.12), in regions imme-

diately upstream of the proximal promoter, or thousands of base pairs removed from the transcription start site. The terms *enhancers* and *repressors* have been applied to distant response elements that act to enhance or inhibit transcription initiation by the general transcription machinery. The ability of distant enhancers or repressors to influence transcription initiation at the promoter is attributed to the way higher orders of DNA bending and looping can bring distant response elements close to the promoter. Figure 8.14 illustrates how distant response elements (enhancers) with their activators bound as homo- or heterodimers can interact with the general transcription machinery via the mediator complex at the site where transcription begins. Such interactions can play an enormous role in facilitating or repressing transcription initiation (Conaway et al. 2005).

With thousands of protein-coding genes in our genome, regulation must be complex and specific. We know, for example, that the mediator plays a critical role in transducing the effects of numerous transcription factors—bound to response elements beyond the promoter—to the general transcription machinery to modulate RNA polymerase II activity to initiate transcription.

Transcription factors that augment transcription at the point of initiation may be ubiquitous, that is, found in and active in a variety of cell types and species; or they may be highly specific, expressed in a single cell type. If they are ubiquitous, the response elements they bind to will be the same in a variety of species. We say that these are conserved sequences. It is important to emphasize that transcription factors are proteins, products of their own genes, indicating that whether the gene is expressed or to what extent it is expressed can dictate the expression of possibly many other genes. Figure 8.15 illustrates a number of common gene-regulating factors in the control region of a hypothetical gene. Transcription factors and the response elements to which they bind are shown in a linear way to simplify the picture.

The transcription factor *NFκB* (nuclear factor kappa B) activates a number of genes in response to conditions such as inflammation, infection, or another immune system-activating factor. Normally, NFκB is found in the cell cytosol attached to an inhibitor, *IκB*, which prevents NFκB from functioning in the nucleus as a transcription factor. Upon the appropriate stimulation, IκB becomes phosphorylated by an IκB kinase. Phosphorylation of IκB leads to its dissociation from NFκB, which allows two subunits of NFκB (p50 and p65) to be translocated

Figure 8.14 Interactions between distant response elements and the general transcription machinery. Transcription regulation involves the components of the general transcription apparatus plus the important influence of a number of other transcription factors bound as homo- or heterodimers to the cognate response elements. Here, distant response elements/enhancers interact with the general transcription apparatus, including the TATA binding protein (TBP), through the large mediator complex. Looping and bending of DNA allows these distant response elements to be in the region of the promoter. The dashed lines are to illustrate that there are large distances along a DNA molecule between distant response element and promoter.

to the nucleus; here they bind to a response element located in more than 150 genes.

Activation protein-1 (AP-1) represents a family of transcription factors that bind as dimers. The best known of the AP-1 transcription factors are the proteins *Fos* and *Jun,* which act to regulate the transcription of many genes, especially in the immune system. Fos and Jun are the protein products of genes that are turned on very early following certain stimuli to a cell. Genes that are rapidly activated in response to some abrupt change are often called *immediate early genes* because their protein products are needed as transcription factors to turn on other genes.

Chapter 5 introduced peroxisome proliferator-activated receptor γ (PPARγ) and its coactivator,

peroxisome proliferator γ coactivator 1 (PGC-1γ). PPARγ is one member of a small family of transcription factors (the others are PPARα and PPARβ/δ) that play very significant roles in the metabolism of carbohydrates and lipids. By themselves, PPARs are not active as transcription factors, but they can assume this role when they are activated by a variety of natural ligands (e.g., unsaturated fatty acids). The PPARs are also targets of some drugs that can lead to decreases in blood lipid levels or increased insulin sensitivity, depending upon the specific drug that targets PPAR and its drug target (Chinetti-Gbaguidi et al. 2005). PPARs function as transcription factors when they bind with the *retinoic X receptor* (RXR) as heterodimers at PPAR response elements. The RXR is

Figure 8.15 The control region of a hypothetical gene, often shown as a linear DNA molecule with specific base sequences (response elements, RE) where their cognate transcription factors bind as homo- or heterodimers. Nuclear factor κB (NFκB) has two main DNA-binding subunits shown as p50 and p65. Peroxisome proliferator-activated receptor γ (PPARγ) often binds to its response element as a homodimer with the retinoid X receptor (RXR). Note that both PPARγ and RXR need to bind specific ligands (shown as small triangles) in order to function as transcription factors. PGC-1γ is a PPARγ coactivator. Sp1 is a common transcription factor, which binds as a dimer to a GC-rich response element. Fos and Jun are transcription factors often found together, known as AP-1. Ribonucleic acid polymerase II (Pol II) is found at the start site (shown as a bent arrow) near the TATA box.

itself activated as a transcription factor when it binds its ligand, a form of retinoic acid. PGC-1γ binds to PPAR and RXR heterodimers and interacts with other transcription factors as a coactivator.

Cell Signaling and Trancription

Chapters 5 and 6 described some signaling (signal transduction) pathways in which an external signal (for example, a hormone) interacts with a cell membrane receptor, generating an internal messenger (for example, cyclic AMP) that results in a change in metabolism within the cell. In complex organisms, signaling is essential to integrate the activities within individual cells so that they support the overall organism. We noted that a nerve impulse can lead to release of Ca^{2+} ions in muscle to initiate contraction. Insulin can lead to an increase in glucose uptake by cells when it binds to its receptor and signals a translocation of GLUT-4 receptors to the cell membrane to increase glucose entry into the cell. Signaling also leads to changes in gene transcription, as the following examples show.

The **steroid hormones** (glucocorticoids, testosterone, estrogen, and progesterone) as well as thyroid hormone, the active form of vitamin D, and retinoids (from vitamin A) circulate in the blood and readily enter cells because their lipophilic nature permits them to diffuse across the cell membrane. They enter the nucleus and bind to their specific protein **hor-**

mone receptors, which are also products of specific genes. The receptors are known as the *nuclear receptor superfamily* because they are all located in the cell nucleus. Each nuclear receptor has a DNA-binding domain and a ligand-binding domain. Nuclear receptors bind as homo- or heterodimers in the control region of a number of genes. In the absence of their ligands, some members of the nuclear superfamily repress transcription. When the specific hormone or active vitamin form binds to its cognate receptor in the control region of a gene, drastic alterations take place in the conformation of the receptor to activate transcription.

Estrogens are steroid hormones that play major roles in both women and men. They are involved in reproduction, as most people are aware; but they also have effects on the cardiovascular system, bones, and even our behavior. Because of its hydrophobic nature, *estradiol*, the dominant form of estrogen, can enter cells by diffusing through cell membranes. The dominant pathway involves entry of estradiol into the nucleus, where it binds to its nuclear hormone receptor, the estrogen receptor. The combination of the ligand estradiol with the estrogen receptor enhances transcription of a variety of genes in much the way we have shown for distant enhancers in figure 8.14. There is also another mechanism whereby estradiol functions in a cell to generate very rapid responses (Hewitt et al. 2005), but we will focus only on the so-called classical pathway illustrated in figure 8.16.

Figure 8.16 Regulation of transcription by a steroid hormone. The predominant form of estrogen is estradiol (shown as a small triangle), a hydrophobic molecule that can diffuse across cell membranes. In the "classical" pathway for estradiol action, estradiol enters the nucleus and binds to its specific estrogen receptor (ER), forming an estradiol-estrogen receptor complex that binds as dimers at an estrogen response element (ERE) to facilitate transcription by RNA polymerase II (Pol II) at the start site (bent arrow) of specific genes.

Protein phosphorylation and dephosphorylation play major roles in modulating the activity of a host of proteins. Therefore, we should not be surprised to learn that protein phosphorylation-dephosphorylation can influence gene transcription through changes in transcription factors that accompany addition or removal of a phosphate group. Figure 8.17 outlines a signaling pathway in which an extracellular agonist (e.g., a hormone) eventually alters the conformation of a transcription factor to modulate transcription. The agonist in figure 8.17 binds to its receptor, generating a conformational change in the receptor that is communicated to adenylyl cyclase via a stimulatory G protein (G_s). Cell membrane receptors that interact with G proteins following ligand binding are often described as G protein-coupled receptors (GPCR). Adenylyl cyclase when activated converts ATP into cyclic AMP, as we discussed in chapters 5 and 6. The cyclic AMP binds to the regulatory subunits of protein kinase A, releasing the catalytic subunits, which are then free to transfer a phosphate group from ATP

to substrates such as glycogen synthase, glycogen phosphorylase kinase, hormone-sensitive lipase, or (in the signaling system shown in figure 8.18) the cyclic AMP response element binding protein (CREB). Phosphorylated CREB transcription factors bind as dimers at the cyclic AMP response element (CRE), attracting the coactivator CBP/P300, which links CREB to the basal transcription machinery as represented by RNA polymerase II.

DNA Organization and Gene Transcription

The tight coiling of chromatin plus the nucleosome structure (see figure 8.6 on p. 205) suggests that initiating transcription must be rather difficult, considering the intricate arrangement of transcription factors and DNA response elements needed. Indeed, as they exist, nucleosomes block proper transcription initiation and make transcription elongation by the large Pol II complex difficult at best. In fact, nucleosomes

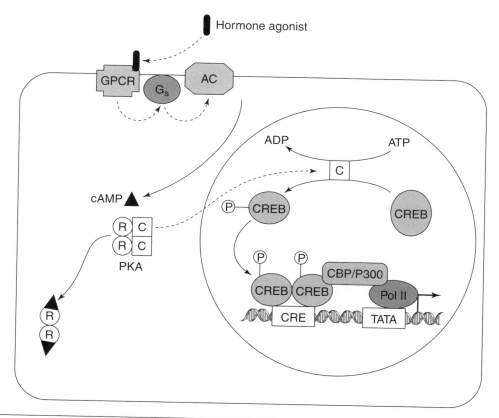

Figure 8.17 A cyclic AMP pathway leading to changes in transcription of specific genes. A hormone agonist, shown as a small oval, binds to a G protein-coupled membrane receptor (GPCR). This leads to activation of adenylate cyclase (AC) by way of a stimulatory G protein (G_s), creating cyclic AMP (cAMP, small triangle). The cAMP binds to the regulatory (R) subunits of protein kinase A (PKA), releasing the catalytic subunits (C), which translocate to the nucleus where each subunit can catalyze the phosphorylation of cAMP response element binding protein (CREB). Phosphorylated CREB binds as dimers at a cAMP response element (CRE), attracting coactivator CBP/P300 to facilitate transcription of target genes.

are regarded as general repressors by some (Boeger et al. 2005). How then are genes transcribed? There are a number of findings, including the fact that nucleosome structure is temporarily broken down at the start site to expose the gene promoter region. This suggests that the initiating event in the transcription of a particular gene is to expose the promoter region by remodeling local DNA structure.

Histone proteins in the core particles of nucleosomes have protruding short polypeptide segments, described as histone tails. Certain lysine residues in histone tails can be acetylated; that is, an acetyl group is transferred from acetyl CoA to the side chain ammonium group of lysine residues. Lysine acetylation of histones involves the changing of a small, positively charged ammonium group on the lysine side chain to a larger N-acetyl group without a charge as shown in figure 8.18. The tight binding of DNA to core histones is facilitated by attractions between the positive charge on the many lysine side chains to the negatively charged DNA backbone. Reduction in this binding facilitates the unraveling of DNA, exposing the promoter and other control regions and facilitating transcription elongation by RNA polymerase. A class of enzymes found within the nucleus carries out this acetylation process. These are known as *histone acetylases* (also known as histone acetyltransferases or HATs). Acetylation is not a permanent state, so a related enzyme known as *histone deacetylase* (HDAC) removes the acetyl group, creating a positive charge and facilitating nucleosome formation. Consistent with the idea that histone acetylation facilitates transcription, it has been discovered that transcriptionally active regions of the genome are highly acetylated. Regions with low or zero rates of gene transcription have little histone acetylation. We earlier noted the role of coactivators and corepressors as proteins without DNA-binding domains that nevertheless facilitate or inhibit transcription, respectively. Some view HAT and HDAC as coactivators and corepressors, respectively (Roeder 2005).

Histone proteins also undergo other modifications that influence transcription rate. While a detailed description of these is beyond the scope of this book, a few should be mentioned. Serine and threonine residues in histone tails can be phosphorylated by kinases. In chapters 2, 5, and 6, we discussed the addition of negatively charged phosphate groups to amino acids with side chain hydroxyls and how this alters enzyme function. Lysine residues in histone tails can also be methylated. As might be expected, modification of histone tails by whatever mechanism (acetylation, phosphorylation, or methylation) changes both the charge and the conformation of the tails, altering their ability to interact with DNA and thereby facilitating or inhibiting promoter accessibility of neighboring genes (Chakravarthy et al. 2005).

Found in the nucleus are ATP-dependent chromatin remodeling enzymes that utilize the free energy of ATP hydrolysis to alter the conformation of chromatin, exposing repressed regions due to nucleosomes and tight higher levels of chromatin packaging (Flaus and Owen-Hughes 2004). It is likely that modified histones act as targets for the ATP-dependent chromatin remodeling enzymes in promoter regions of genes. As well, modified histone tails may serve to enhance or inhibit binding of mediator, general transcription factors, and Pol II at the promoter. The complex protein–DNA and protein–protein interactions involved in transcription initiation, plus the ability of chromatin to be remodeled through histone acetylation and deacetylation, phosphorylation, and dephosphorylation as well as by ATP-dependent remodeling enzymes, underscore the complexity of gene transcription.

Skeletal Muscle

The control of gene expression in skeletal muscle has been an area of active research for many years. We have long known that endurance training increases

Figure 8.18 Lysine acetylation of histones. Lysine side chains found in the tails of core histone proteins can be acetylated by acceptance of an acetyl group donated by acetyl CoA in a reaction catalyzed by histone acetylase (also known as histone acetyl transferase). The reverse reaction is catalyzed by histone deacetylase. The addition of the acetyl group to the ammonium group in the side chain of lysine residues removes the positive charge and adds a bulky group, thus changing the conformation of the histone tail.

the content of mitochondrial proteins involved in oxidative phosphorylation, without any significant changes in contractile proteins. On the other hand, high-intensity resistance training increases the content of contractile proteins without a notable effect on mitochondrial proteins. This tells us that the stimulus induced by the specific exercise activity must selectively modulate the transcription of some muscle genes, leading to increased levels of mRNA and thus proteins. Researchers in early studies on exercise training inferred changes in the expression of genes in muscle from changes in the content of specific proteins in muscle. With the ability to measure both steady state levels and rates of formation of specific mRNA molecules, we are able to understand more precisely how activity or inactivity influences skeletal muscle genes. This is particularly significant because a change in muscle activity sufficient to influence gene expression will show up much earlier as a change in the mRNA for a protein. Furthermore, changes in mRNA are typically far in excess of the changes in the content of proteins for which they code. Indeed, the time when a change in activity results in a notable change in muscle fiber type may be weeks after a change in expression of myosin heavy chain genes. As mentioned previously, the myosin heavy chain composition of a fiber plays the dominant role in defining muscle fiber type, determining how rapidly ATP can be hydrolyzed.

Booth and Neufer (2005) summarized studies that monitored gene expression in subjects over the course of a short-term one-legged endurance training program. They described the response of genes in terms of how rapidly they responded, the duration of the response, and the peak of the response based on mRNA transcript levels. Genes that were turned on very quickly, but transiently, were described as "stress response genes"; these coded for proteins, such as transcription factors, whose content was rapidly elevated in a number of models under high-stress conditions. A second category of genes, the "metabolic priority genes," demonstrated peak expression several hours after the exercise training bout. Genes in this latter category coded for enzymes that played regulatory roles in carbohydrate metabolism. The third gene category, "metabolic/mitochondrial enzymes," was slower to respond and had a lower peak response, but the response persisted over a longer period of time. These genes coded for mitochondrial proteins, involved in oxidative phosphorylation. Studies such as those summarized by Booth and Neufer (2005) remind us that there are priorities in the transcription

of genes, the ultimate effect of which is to enhance the ability of the person to respond to further training stimuli.

Four highly specific trans factors, found only in skeletal muscle and essential for their development, are the **myogenic regulatory factors:** MyoD, myogenin, MRF4, and myf5. These proteins contain secondary and tertiary structural components, known as the basic helix-loop-helix, that recognize a six-base response element termed an *E box*. The myogenic regulatory factors form heterodimers with another class of proteins called *E proteins* (e.g., MEF2) at the E box. The E box is found in the regulatory region of some specific muscle genes, including those expressing the fast myosin light and heavy chains (see figure 1.13, showing heavy and light chains of myosin). Artificial expression of the MyoD gene family in other cell types can cause those cells to express genes transcribed only in skeletal muscle. This emphasizes the significant role of the myogenic regulatory factors as transcription factors in the development of skeletal muscle.

►KEY POINT◄

Although many exercise scientists are interested in the role of various types of exercise in modulating gene expression in muscle, some have studied more extreme forms of activity or inactivity to understand the limits of adaptation of muscle. Implanted electrodes around the nerve to a muscle can create artificial contraction situations in which a slow muscle (e.g., the soleus) is forced by high-frequency stimulation to contract in bursts. Alternatively, a fast muscle (extensor digitorum longus) is made to contract continuously with low-frequency stimulation.

Casting a limb to inhibit range of motion activity, suspending a hindlimb to unload muscles, and cutting the nerve to a muscle are other techniques used to probe the extremes of the continuum of activity and inactivity and the role played by the nerve in maintaining distinct muscle properties.

Any exercise training stress such as resistance training or endurance running leads to increases in the transcription of specific genes and later changes in the content of selected proteins. The primary question is how the training stress is linked to the activation of key genes. We have previously discussed the role of protein phosphorylation as a major regulatory mechanism in adapting metabolism to meet our needs. In this discussion, we have focused on key protein kinases such as AMP-activated protein kinase and protein kinase A. There are many more kinds of protein kinase, including a family of kinases given the name *mitogen-activated protein kinases*, MAP kinases, or MAPK. The word "mitogen" refers to something that stimulates cell proliferation (mitosis), but only some of the actions of the MAP kinases ultimately lead to the formation of new cells. Within the MAP kinase family of protein kinases are three major subfamilies: the c-Jun NH_2-terminal kinases (JNKs), the p38 MAPK, and the ERKs (extracellular signal-related kinases). A number of laboratories are addressing how an exercise stress or the absence of it, acting through one or more protein kinase pathways, ultimately leads to an adaptive response. For now, the reader is referred to a few studies that illustrate the mechanisms behind adaptations to training and inactivity (Widegren et al. 2001; Richter et al. 2004; Machida and Booth 2005).

Calcium is the signal initiating the contractile event, as we discussed in chapter 3. Released from the sarcoplasmic reticulum in response to nerve activation of a fiber, calcium is at a level in the muscle fiber that reflects the frequency and duration of fiber activation. Besides binding to the thin filament protein, troponin, to initiate contraction, calcium can bind to other cell proteins, including the ubiquitous calcium-binding protein known as calmodulin. As discussed in chapter 5, when calmodulin binds calcium, it can bind to and influence the activity of other proteins such as *calcineurin*. This is a protein phosphatase that dephosphorylates specific proteins containing one or more covalently bound phosphate groups. One of the substrates for calcineurin is a transcription factor known as **NFAT (nuclear factor of activated T cells).** NFAT is a transcription factor that was first identified in T cells. However, it is also found in skeletal muscle, where it is believed to play an important role in skeletal muscle response to altered activity (Michel et al. 2004). As shown in figure 8.19, when activated by binding calcium-bound calmodulin,

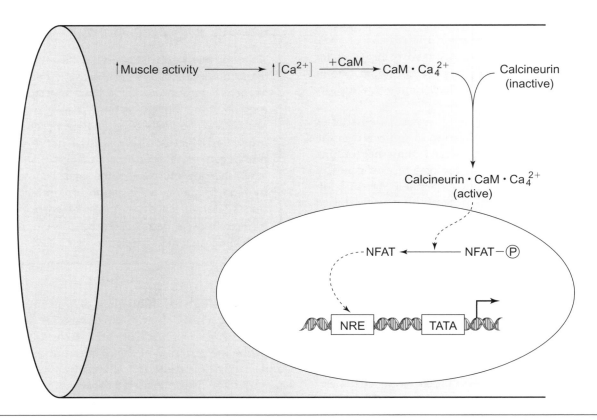

Figure 8.19 Regulation by the calcineurin NFAT pathway of gene expression in skeletal muscle in response to changes in contractile activity. An increase in muscle activity increases the concentration of calcium ion. The Ca^{2+} binds to the calcium-binding protein calmodulin (CaM), forming a complex (CaM · Ca_4^{2+}). When the protein phosphatase calcineurin binds the CaM · Ca^{2+}, it forms a calcineurin · CaM · Ca_4^{2+} complex, which is now active as a protein phosphatase. Dephosphorylation of NFAT-P produces the active transcription factor NFAT, which binds to its response element (NRE) to alter the expression of specific genes.

calcineurin can dephosphorylate NFAT, allowing it to bind to its response element in the control region of a number of its target genes in skeletal muscle. Among the effects of NFAT activation is hypertrophy.

Because the NFAT pathway responds to an increase in contractile activity, we might expect circumstances that preclude muscle activity and hence activation of the NFAT pathway to lead to loss of muscle mass.

SUMMARY

The information needed to make the thousands of different proteins in our bodies resides in our genome, the sequence of 3,164,700,000 bases that make up our DNA. A DNA molecule is composed of deoxynucleotides, joined together to make huge DNA molecules. Each deoxynucleotide consists of the sugar deoxyribose, phosphate, and one of four different bases: adenine, thymine, guanine, and cytosine. Deoxyribonucleic acid molecules are double-stranded and are arranged to form the well-known double helix. The bases in one strand are complementary to the bases in the other strand because hydrogen bonding between adenine and thymine and between guanine and cytosine is an absolute necessity. The strands in DNA are antiparallel; one strand runs from the 5' phosphate at one end to a free 3' hydroxyl group of the last deoxyribose sugar, and the complementary strand runs in the 3' to 5' direction.

Transcription of genes produces an RNA molecule called the primary RNA transcript that will be a complementary copy of the DNA strand transcribed, with the exception that the RNA transcript will have the base uracil instead of thymine and the sugar will be ribose. Four letters, U, C, A, and G, are used to designate the bases and the nucleotide sequence in RNA. Depending on the gene transcribed, the transcript is modified to produce a messenger RNA (mRNA), transfer RNA (tRNA), or ribosomal RNA (rRNA). The base sequence of the mRNA, beginning with the 5' end, is read in groups of three called codons, which constitute the words of the genetic code. Sixty-one of the possible 64 codon combinations spell out the 20 amino acids, so the genetic code is said to be degenerate. Degenerate is the word coined by molecular biologists to indicate that there is more than one codon for most of the amino acids. Three of the codons (UAA, UGA, and UAG) stop the message. Transcription is carried out by a group of multi-subunit enzymes known as RNA polymerases. These enzymes copy the genes that will become rRNA (Pol I), mRNA (Pol II), or tRNA (Pol III). During transcription, the polymerase unwinds the two DNA strands and copies part of one, reading it in the 3' to 5' direction. The nontranscribed DNA strand, called the sense strand, has the same base sequence as the primary transcript, except that thymine, in DNA, is substituted for by uracil in RNA.

The start site is the first nucleotide in the DNA strand that is copied during transcription. The region of DNA just before the transcription start site is called the promoter. The promoter, and other regions of the DNA molecule that may be far from the start site, contain sequences of bases known as response elements to which specific protein molecules called transcription factors can bind to regulate the process of transcription. General transcription factors are needed for the transcription of all genes. These bind upstream of the start site at a very common response element called the TATA box and help position RNA polymerase and start the transcription process. Virtually all genes require other tissue-specific or developmentally specific proteins to provide an additional level of transcription control so that genes are expressed when they should be and are expressed at appropriate rates to ensure the overall health of the body. These additional transcription factors may stimulate (activators) or inhibit (repressors) transcription of the gene. Circulating hormones and growth factors can influence transcription in a process termed signaling or signal transduction. Steroid hormones are hydrophobic; they can diffuse into cells and bind with their specific receptors in the nucleus, which subsequently act as transcription factors. Other hormones and growth factors bind to receptors on the cell membrane and generate signals within the cell that alter the rate of transcription. The protein-DNA complex in the nucleus known as chromatin is organized into basic structures called nucleosomes. The tight DNA loops within nucleosomes and higher orders of DNA coiling in the nucleus repress transcription. Before genes can be transcribed, the nucleosomes and other higher levels of DNA organization must be remodeled to allow access to transcription factors. An example of DNA remodeling takes place when histone tails in nucleosomes are altered by the addition of acetyl, phosphate, or methyl groups or by other ATP-dependent chromatin remodeling enzymes, adding a further level of complexity to transcription control.

REVIEW QUESTIONS

1. The base sequence of one of the strands of a 21 bp duplex of DNA is 5'-AGTCCAGCGT-TAGACCGAAGT. What is the base sequence of its complementary strand?

2. If the complementary strand of the same 21 bp duplex of DNA is part of the coding region of a gene, what is the base sequence of the mRNA it generates?

3. Using the sequence of bases in the mRNA molecule from the previous question, determine the amino acid sequence.

4. Knowing that the sequence of amino acids in a polypeptide is Glu-Lys-Met-Ala-Gly-His-Thr, design a base sequence to represent the sense strand of the DNA that generated this amino acid sequence.

PROTEIN SYNTHESIS AND DEGRADATION

Chapters 1, 5, and 7 referred to the turnover of protein. This constant cycle of protein synthesis (translation) and protein breakdown (degradation) is so important that it represents the starting point for this book (see figure 1.1). A constant cycle of protein turnover serves three important functions. First, it allows the cell to destroy modified or wrongly formed proteins, reducing the potential for these to cause harmful changes in the cell. Second, body protein is a potential source of oxidizable fuel for times when the organism is not ingesting sufficient food energy. Therefore, during starvation and fasting, rates of protein synthesis decline whereas rates of protein degradation increase, resulting in the increased appearance of endogenous amino acids for oxidation. Finally, changes in the rates of protein synthesis and degradation allow the overall organism and specific tissues to adapt to altered nutritional and environmental conditions. For example, a surfeit of carbohydrate in the diet induces increased formation of enzymes for de novo lipogenesis in liver. Exercise training can lead to changes in the content of specific proteins in the trained skeletal muscles.

In this chapter, we look at the mechanisms for creating and regulating polypeptide synthesis. In addition, we look at how newly synthesized polypeptides are combined with other polypeptides to make a functioning unit (often described as posttranslational processing). Finally, we look at the breakdown of proteins in cells, the various processes by which this occurs, and factors that control the rate at which it

occurs. Proteins are synthesized on ribosomes in the cell cytosol. The messenger RNA (mRNA) message is translated into a sequence of amino acids. Before any of this can take place, however, the RNA molecules made by transcription must be modified to generate the active forms of transfer RNA (tRNA), ribosomal RNA (rRNA), and mRNA.

POSTTRANSCRIPTIONAL MODIFICATIONS OF RNA

Ribonucleic acid is created by the transcription of genes as discussed in chapter 8. In any cell, most of the RNA molecules are derived from single-copy genes that provide information for the sequence of amino acids in a polypeptide. However, there are multiple copies of genes that provide the information for the formation of RNA molecules destined to become rRNA and tRNA.

Formation of mRNA Molecules

Following transcription of genes coding for polypeptides, the product is known as the primary transcript or *pre-mRNA*. Because in any nucleus, many hundreds of genes are being transcribed simultaneously, there will be a huge variety of pre-mRNA and other RNA molecules. Each pre-mRNA or other RNA molecule is associated with protein right from the time it is freed from its RNA polymerase until when it is transported out of the nucleus. The collection of all of the transcripts in the nucleus is described as *heterogeneous nuclear RNA* or hnRNA. The combination of the hnRNAs and the nuclear proteins with which they are associated is known as *heterogeneous ribonucleoprotein particles* or hnRNPs.

The pre-mRNA molecules undergo three kinds of changes before they become mature mRNA molecules ready to be exported from the nucleus. These alterations are referred to as RNA processing. First, each pre-mRNA has a cap added to its 5' end. Next, it has a tail added that consists of multiple copies of adenine nucleotides. Finally, it has its introns removed. This last step is very important because it means that the genes in eukaryotic organisms are interrupted—that these genes are in pieces. Parts of the gene, known as **exons,** contain information that will appear in the mature mRNA; these can be called coding regions. Other parts contain base sequences, known as *introns* or intervening sequences, that will not appear in the mature mRNA. As shown in figure 9.1, a gene can be illustrated as a segment of DNA with a transcribed section bounded at the 5' end by the start site and the poly A tail. Just upstream of the start site is the promoter. Within the transcribed portion are introns and exons. Figure 9.1*b* shows a typical way in which exons and introns can be illustrated. The primary transcript or pre-mRNA contains nucleotides corresponding to both introns and exons. The pre-mRNA in figure 9.1 also has a 5' cap and a poly A tail, which we will discuss next. The final processing

of the pre-mRNA molecule involves splicing out the introns so that between the cap and poly A tail is a section composed only of exons.

As mentioned in the previous chapter, we refer to the sense strand of DNA and refer to the size of the gene in terms of nucleotides or bases in the sense strand. The average gene in the human genome has about 3,000 nucleotides (bases). The largest is the gene that codes for the protein dystrophin; it contains 2.4 million nucleotides (bases). With introns as long as 100,000 to 200,000 bases, the dystrophin gene may take 24 h to be transcribed.

Capping the 5' End of the RNA

As you will recall, transcription of genes proceeds as the template strand of DNA is read in the 3' to 5' direction. The RNA transcript is generated in the 5' to 3' direction. While the RNA transcript initially stays with the RNA polymerase and loosely binds to the template strand of DNA by hydridization between complementary bases, the 5' end becomes free before transcription proceeds very far. Capping of the free 5' nucleotide of RNA involves a reaction between the 5'-phosphate group on the first ribonucleotide and a 7-methyl GTP (guanosine triphosphate). This creates a cap consisting of 7-methylguanine bound to sugar ribose, which in turn is attached to the first nucleotide of the RNA via three phosphate groups. This process is generally described as capping and is shown in figure 9.2*a*.

Figure 9.1 *(a)* A gene shown as the familiar double helix (top) and as a bar (bottom). Most eukaryotic genes have alternating sections containing coding information (exons—solid) and noncoding sequences (introns—open) downstream of the transcription start site (bent arrow) and the promoter region (hatched region). *(b)* The pre-mRNA molecule has introns and exons as well as a 5' cap and a 3' poly A tail (shaded region). The final processing of the pre-mRNA involves removal of sections of RNA that represent the introns.

Figure 9.2 Processing of pre-mRNA. *(a)* A cap is added to the 5' end of the transcript. The cap consists of a guanine base (modified by the addition of a methyl group) that is attached by three phosphate groups to the 5' carbon of the ribose on the initial nucleotide of the transcript. *(b)* The transcript illustrates the introns (open) and exons (solid). *(c)* A site where a sequence of 200 to 250 adenine nucleotides will be added subsequent to removal of a section of the 3' end of the transcript, leading to the pre-mRNA with both a cap and a poly A tail (gray).

Forming the Poly A Tail

All mRNA molecules in eukaryotes, except those for the histone proteins, have a poly A tail. This consists of 200 to 250 adenine nucleotides added to the 3' end of the pre-mRNA. A schematic of the process is shown in figure 9.2. Transcription proceeds past a consensus sequence (AAUAAA) and the poly A site to a region that generates a G- and U-rich sequence in the pre-mRNA. Two major processes now take place. First, the pre-mRNA is cleaved at the poly A site, releasing a small RNA particle with the G- and U-rich sequences. Then an enzyme known as *poly A polymerase* adds about 200 to 250 adenine nucleotides to the poly A site, which is some 20 to 35 nucleotides downstream (in the 3' direction) from a key signal sequence (AAUAAA). This signal sequence is absolutely essentially for polyadenylation to take place.

Splicing to Remove Introns

The next step is to remove the introns from the capped, polyadenylated, pre-mRNA via a splicing process in which the junction between introns and exons is cleaved at both the 5' and 3' ends. The introns are then removed and contiguous exons joined up. Generally, the splicing out of introns occurs following the addition of the cap and poly A tail to the pre-mRNA. For very long transcripts, splicing can begin before the transcription of the long gene has been completed.

The locations of the intron-exon splice sites have been studied in detail, and we know that there are consensus nucleotides that clearly delineate splice sites. The details of this process are complicated, involving small nuclear RNA molecules known as snRNAs in combination with the pre-mRNA molecule, forming a *spliceosome*. The actual process of splicing out the intron is shown in figure 9.3. The hypothetical pre-mRNA molecule starts with a 5' cap and 3' poly A tail and contains a single intron. There is a phosphoester bond exchange between the consensus nucleotides at either end of the intron, forming a lariat structure. This is removed from the second exon, and then the two exons join together by their consensus terminal guanine nucleotides.

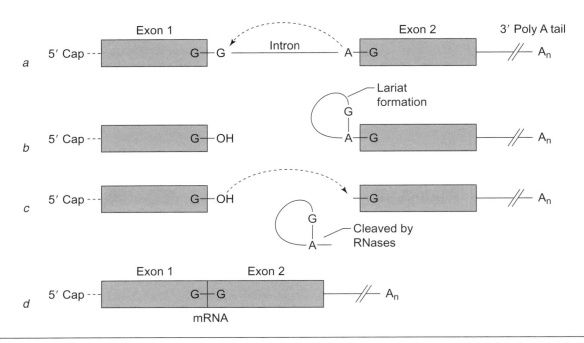

Figure 9.3   Splicing out of an intron in a simple pre-mRNA molecule containing a 5' cap and 3' poly A tail. *(a)* Exchange of the phosphoester bond of the **A** nucleotide residue at the 3' end of the intron with the **G** nucleotide residue at the 5' end of the same intron. *(b)* This forms a lariat structure in which the **A** nucleotide residue at the 3' end of the intron is still attached to the 5' **G** nucleotide residue of exon 2. *(c)* The intron, now a freed lariat, is attacked by RNA-cleaving RNases. *(d)* The 3' terminal **G** nucleotide residue of exon 1 forms a phosphoester bond with the 5' terminal **G** nucleotide residue of exon 2, linking the two exons together and forming the mRNA molecule.

The lariat is then attacked by RNA-digesting enzymes (RNases), releasing the nucleotides as nucleotide monophosphates.

As a result of the 5' capping, the 3' polyadenylation, and intron removal by cutting and splicing, we now have what has been called a mature mRNA that is ready to leave the nucleus for the cytosol, where it acts as a template on which a polypeptide (protein) will be made, employing ribosomes. During all the RNA processing, the RNA molecules are associated with proteins in a complex identified as heterogeneous ribonucleoprotein particles. Transport out of the nucleus takes place through *nuclear pores*, or openings in the nuclear membrane. Within the nuclear pores is a complex of filaments (the *nuclear pore complex*) that are involved in allowing the transport of ions and proteins (e.g., transcription factors) into the nucleus and the export of molecules such as mRNA, rRNA, and tRNA from the nucleus. Newly formed mRNA molecules, associated with their nuclear proteins, are exported through the nuclear pore complex using a specific mRNA exporter protein.

For many genes, every intron is removed and every exon is incorporated into the mature mRNA. However, it is not uncommon for one gene to give rise to two or more mRNA molecules and thus two or more final proteins. An example of the alternate splicing of a primary gene transcript is shown in figure 9.4. There is a gene, the calcitonin-CGRP (calcitonin gene-related peptide), that is expressed in thyroid gland and in brain. The primary transcript calcitonin-CGRP pre-mRNA contains six exons, but is treated in different ways in the two tissues. In thyroid, exons 5 and 6 are spliced out, creating an mRNA with just four exons. The protein product, *calcitonin*, is released as a hormone from the thyroid gland in response to elevated blood calcium ion levels. In brain, exon 4 is spliced out of the calcitonin-CGRP primary transcript, producing a different mRNA that contains exons 1, 2, 3, 5, and 6. The CGRP peptide has several functions, one of which is to act as a vasodilator. Alternate splicing of genes is common for some of the contractile proteins in skeletal, cardiac, and smooth muscle and can occur in these instances:

1. When transcription is initiated at different promoters, resulting in different 5' exons in the mRNA.

2. When transcription terminates differently because of more than one site of polyadenylation, resulting in different 3' exons.

Figure 9.4   An example of alternate pre-mRNA splicing to generate two different mRNA products. The calcitonin-CGRP gene is expressed in thyroid gland and brain as a primary transcript with six exons. In thyroid, exons 5 and 6 are spliced, and the resulting mRNA provides the amino acid sequence information to make the hormone calcitonin. In brain, the calcitonin-CGRP primary transcript is spliced a different way, removing exon 4. The resulting mRNA codes for the CGRP polypeptide.

3. When different internal exons in the gene are included or not included in the splicing process, giving rise to different mRNA molecules (thus two or more different polypeptides) in a process known as alternate mRNA splicing. This alternative splicing can occur in the same cell at the same time, in the same cell at different times during cell differentiation, or in different cells. Control of alternative splicing occurs through the action of specialized proteins.

Formation of rRNA and tRNA Molecules

Modification of primary transcripts from genes for rRNA and tRNA also occurs. As already mentioned, two different genes are needed to produce the rRNA molecules that make up much of the ribosomes. There are multiple copies of each rRNA gene; these are located in a part of the nucleus known as the *nucleolus*. The two different rRNA transcripts are so large that their mass is represented in Svedberg units (S), which describe their velocity of sedimentation in an ultracentrifuge.

For very large molecules or molecular complexes, size is measured by how far they travel in an ultracentrifuge, expressed in Svedberg units, rather than by molecular weight. The larger the molecule, the faster it sediments in a centrifuge and the larger the value of S. Unlike units such as kilodaltons (kD) or base pairs

(bp), typically used to describe the size of protein or DNA molecules, respectively, Svedberg units are not linearly related to size. Thus a whole ribosome, with a mass of 80S, is made up of two subunits of size 60S and 40S because the sedimentation behavior of particles in the centrifuge is not linearly related to size.

The two primary rRNA transcripts are described as 45S and 5S. The smaller transcript, 5S (120 bases), remains unchanged, whereas the larger transcript, 45S, undergoes some splitting and modification, resulting in three new rRNA molecules: 18S, 28S, and 5.8S. The 18S rRNA molecule combines with 33 protein molecules to make the small ribosomal subunit called the 40S ribosomal subunit. The 5.8S, the 28S, and the 5S rRNA molecules combine 50 protein molecules to make the large ribosomal subunit

> **KEY POINT**
>
> Although there are only two different genes for rRNA, multiple copies of these exist because so many rRNA molecules are needed to make up the hundreds of thousands of ribosomes in a cell. Whereas each mRNA message gets amplified because it results in thousands of polypeptide chains, there is no such amplification of the four rRNA molecules, as each is the final product. There is more rRNA by weight in the cell than all the mRNA and tRNA molecules combined.

known as the 60S ribosomal subunit. During protein synthesis, the 40S and 60S ribosomal subunits combine to make the complete ribosome, termed the 80S ribosome.

Primary transcripts for genes for the various tRNAs undergo some modifications. For example, nucleotides are removed from the 5' and 3' ends, and, if present, an intron is removed. Then the 3' end has a nucleotide sequence CCA added. Finally, some of the bases are modified. Figure 9.5 shows the main features of a fully functional tRNA molecule. At one end is a 5'

phosphate group. The 3' end has the sequence CCA with a free 3' hydroxyl group to which the amino acid becomes attached. Notice the hydrogen bonding that helps maintain the rough cloverleaf structure, characteristic of tRNA molecules. Three bases at the bottom of tRNA represent the **anticodon** bases. These will correspond to the codon for the particular amino acid, except that they will be antiparallel and complementary. For example, if the codon is UCG (always specified in the 5' to 3' direction) as in figure 9.5, the anticodon will be AGC, in the 3' to 5' direction.

Figure 9.5   An example of a tRNA molecule that has been modified from its primary transcript. The finished product has a cloverleaf-type structure that is maintained by hydrogen bonding between complementary bases (dotted lines). Solid lines represent the covalent attachment of adjacent nucleotides shown by the letter representing their base. The **X** refers to one of several different types of bases that are generated after transcription. The 3' end contains a terminal adenosine (A) whose 3' ribose OH group binds a specific amino acid. Each tRNA has a three-base sequence at the bottom, known as the anticodon, that will be complementary to the codon on mRNA representing the particular amino acid. The anticodon and complementary codon are for the amino acid serine.

Because there are 61 codons in mRNA specifying the 20 amino acids, we would expect there to be 61 tRNA molecules, each with one of 61 different anticodons. However, there is a bit of flexibility in the anticodon–codon binding because the first base in the anticodon (the one at the 5' end that pairs with the third base at the 3' end in the codon) can often recognize two bases. For example, the codons CUA and CUG could be recognized by the anticodon GAU. This flexibility in codon–anticodon recognition is known as **wobble.** As a result of wobble, fewer than 61 different tRNA molecules are needed to ensure that all 20 amino acids are able to participate in protein synthesis. Indeed, on a theoretical basis, considering all the possibilities created by wobble, only 31 different tRNA molecules would be needed. The actual number of different tRNA molecules found in eukaryotic cells is 32.

TRANSLATION

In the process of translation, which takes place on ribosomes in the cytosol of the cell, the mRNA message is converted into a sequence of amino acids in a polypeptide chain. This process can be described as a number of steps, beginning with the attachment of each specific amino acid to its respective tRNA molecule.

Formation of Aminoacyl-tRNA

Each amino acid has at least one tRNA to which it will be attached. Each tRNA will have an anticodon that will match one of the codons (likely more, due to wobble) for that amino acid, if it has more than one codon. Each amino acid will also have a specific enzyme to catalyze the joining of its alpha carboxyl group to the 3' OH group of the terminal adenosine of the single tRNA (or duplicate tRNAs if there are more than one for that amino acid) using an ester bond. The enzyme that joins the amino acid to its

tRNA is known as *aminoacyl-tRNA synthetase*. It is the responsibility of each synthetase to match the correct amino acid and its tRNA, as shown in the following equation:

$$\text{amino acid} + \text{tRNA} + \text{ATP} \rightarrow \text{aminoacyl-tRNA} + \text{AMP} + \text{PPi}$$

The energy needed for the formation of the ester bond between the amino acid and its tRNA is provided by the hydrolysis of ATP. The products of this hydrolysis reaction are AMP and inorganic pyrophosphate (PPi). Hydrolysis of PPi by inorganic pyrophosphatase ensures that the reaction is driven to the right. Amino acids attached to their respective tRNA molecules are often characterized as charged amino acids. Some aminoacyl-tRNA synthetases attach their amino acid directly to the 3' OH group of the terminal ribose on their tRNA, whereas others attach their amino acid to the 2' OH group of ribose. This latter attachment is only temporary, as subsequently the amino acid is moved to the 3' position. The role of the individual aminoacyl-tRNA synthetase enzymes, to recognize the correct amino acid and the correct tRNA, places these enzymes in a critical position to ensure that correct proteins are produced by the cell. Errors are extremely rare.

The Role of mRNA

Figure 9.6 illustrates the essential components of a functional mRNA molecule. The coding region, headed by the AUG start codon, contains the information to indicate the amino acid sequence for the polypeptide chain, which is followed by a stop codon (UAG, UGA, or UAA). The term **open reading frame** (ORF) is often used to describe a region of triplet coding, bordered on one end by the start codon and at the other end by a stop codon. We could use ORF to refer to a gene in DNA or, more likely, to refer to an mRNA molecule. Flanking the coding

Figure 9.6 A model of a mature mRNA molecule, ready for translation to form a polypeptide chain. At the left end is the 5' cap and at the 3' end, the poly A tail. The region that contains the information for the sequence of amino acids in a polypeptide chain (coding region) begins with the start codon AUG and ends with a stop codon (UAG is shown, but UAA or UGA also signal stop). Flanking the coding region are the 5' and 3' untranslated regions. N refers to nucleotides whose designation is not shown.

region are a 5' noncoding region and a 3' region that is also noncoding. Because the base sequence in these regions is not translated into an amino acid sequence, they are also known as the 5' UTR (untranslated region) and 3' UTR, respectively. Base sequences in the 5' UTR or 3' UTR can play a role in the regulation of translation. Except for histone proteins, virtually all mRNA molecules contain the poly A tail.

Initiation of Translation

Like transcription, the translation process has three stages: initiation, elongation, and termination. This section does not provide great detail but instead focuses on the overall process. The major players in the initiation of translation are the two ribosomal subunits 40S and 60S, the mRNA molecule, a number of protein factors to control the initiation process (eukaryotic initiation factors, or eIFs, such as eIF1A and eIF2), and a source of energy from the hydrolysis of ATP and GTP. Initiation cannot take place without methionine, attached to its tRNA. However, while methionine has a single codon (AUG) and a single methionine tRNA synthetase, it has two different tRNAs. One of these, tRNA$_i$, is used only for the initiation of translation at the first start codon of

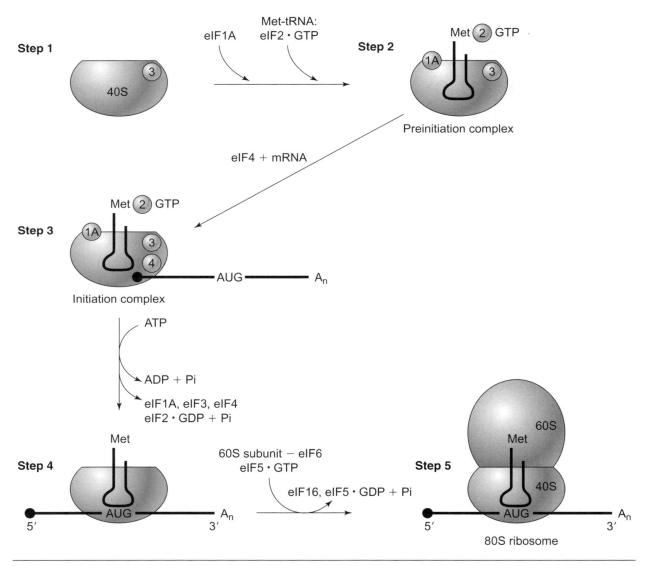

Figure 9.7 Stages in the formation of an 80S ribosome particle. A 40S ribosome subunit combines with eukaryotic initiation factor 3 (shown circled). Subsequent additions include the initiation tRNA (Met-tRNA$_i$) attached to GTP by EIF2 and eIF1A, forming a preinitiation complex (step 2). Messenger RNA is positioned at its 5' cap on the 40S subunit aided by eIF4. This generates the initiation complex (step 3). The mRNA is scanned for an initiation codon (AUG), and the ribosome moves along the mRNA in a process driven by the hydrolysis of ATP and GTP (step 4). Addition of the large ribosome subunit (60S) to the 40S subunit, forming an 80S ribosome, is facilitated by addition of initiation factors (eIF5 and eIF6) and the energy from GTP hydrolysis (step 5).

the open reading frame. The other tRNA, which binds methionine, does so when it is incorporated into the peptide chain beyond the initial methionine.

Figure 9.7 illustrates the initiation part of translation, which can be considered to take place as five discrete steps. In the first step, ribosome subunit 40S binds to eIF3. In step 2, a preinitiation complex is formed with the methionyl-tRNA (Met-tRNA$_i$), eIF1A, and a complex of eIF2 bound to GTP. In step 3, eIF4 and the mRNA bind, with the cap on mRNA recognized by and located at eIF4. This creates the initiation complex. With the help of the initiation factors and energy released from ATP and GTP hydrolysis, the mRNA is scanned in the 3' direction to find the initiation codon (step 4). During step 4, the initiation factors and the hydrolysis products of ATP and GTP are released. Finally, the 80S ribosome complex is formed in step 5, when the 60S ribosome subunit bound to eIF6 joins with the 40S subunit. Energy is provided by hydrolysis of GTP. Subsequently eIF6, eIF5, GDP, and Pi are released. The fact that GTP hydrolysis is needed to form the 80S ribosome complex helps to maintain the attachment of the 40S and 60S subunits throughout translation of the mRNA molecule.

Elongation of Translation

The elongation phase involves the addition of amino acids, one at a time, to the carboxy terminal end of the existing polypeptide chain (see figure 9.8). Let us start with the 80S ribosome as shown in step 5 of figure 9.7. For clarification, anticodons are shown on the tRNAs, as well as relevant codons on mRNA. Recall that each amino acid is attached to the 3' end of its respective tRNA. In step 1 of figure 9.7, the next aminoacyl-tRNA (Phe-tRNA) comes in, its anticodon (AAA) recognizing the mRNA codon (UUU) on the 3'

side of the initiator (AUG) codon. In step 2, a peptide bond is formed between the methionine carboxyl and the free amino group of the next amino acid (Phe), still attached to its tRNA. The formation of this peptide bond between Met and Phe means the methionine is released from its tRNA. The methionine tRNA leaves, and we now have a dipeptide attached to the second tRNA. The 80S ribosome complex then slides three bases along the mRNA molecule in the 3' direction. Then the next aminoacyl-tRNA (Val-tRNA) comes in, its anticodon (GUG) recognizing the codon CAC. Another peptide bond is formed, the Phe-tRNA is released, and we now have a tripeptide. Again, the 80S ribosomal complex slides along the mRNA. This process continues with the growing polypeptide chain still attached to the last incoming tRNA. The energy needed to make these peptide bonds comes from the hydrolysis of GTP to GDP and Pi. Eukaryotic elongation factors (eEFs) participate in the elongation process creating the polypeptide chain.

Termination of Translation

The process of elongation continues, with the ribosome and growing polypeptide chain moving along the mRNA three bases at a time until a stop codon is reached (UAA, UGA, or UAG). At this point, a release factor causes the release of the completed polypeptide chain from the last tRNA. The whole complex dissociates, and the 80S ribosome dissociates into its 40S and 60S subunits. As with peptide chain elongation, energy for the termination of translation is provided by GTP hydrolysis.

Because most mRNA molecules are large, there is room for more than one ribosome on a single mRNA molecule—each containing a growing polypeptide chain. We call the mRNA and the two or more ribosomes, each with its growing polypeptide chain, a

Figure 9.8 The elongation phase of translation. (*a*) The second aminoacyl-tRNA comes to the 80S ribosome, its anticodon AAA recognizing the second codon UUU, coding for the amino acid phenylalanine (Phe). (*b*) A peptide bond is formed between the carboxyl group of Met and the Phe, still attached to its tRNA. The 80S ribosome moves three bases along the mRNA toward the 3' end. Not shown are elongation factors and the hydrolysis of GTP, which provides energy for peptide bond formation and movement of the 80S ribosome along the mRNA.

Figure 9.9   A polyribosome, composed of five ribosomes. Most mRNA molecules are long compared to the dimensions of an 80S ribosome complex, allowing for more than one 80S ribosome complex with its growing polypeptide chain. Here there are five ribosomes, each with its polypeptide chain (shown as small connected circles). The complex of a number of ribosomes on one mRNA molecule is known as a polyribosome or polysome. Note that the farther along the mRNA toward the 3' end, the longer the polypeptide chain.

polyribosome or polysome (see figure 9.9). The overall process of translation occurs with remarkably few errors, roughly one error per 10,000 codons translated under normal conditions.

REGULATION OF TRANSLATION

When gene expression regulation is discussed, terms such as *transcription control* and *posttranscription control* are often used. In the previous chapter, we discussed transcription control primarily from the perspective of regulation of the initiation of gene transcription by RNA polymerase II. Here, we look at posttranscriptional control because this refers principally to the regulation of translation of mRNA. Translation of muscle mRNA, or the formation of polypeptides in muscle, is exquisitely sensitive to a variety of nutritional and exercise states. Moreover, while protein synthesis is a continuous activity in cells, it does require energy expenditure since the equivalent of four ATP is needed for the synthesis of each peptide bond.

Protein Synthesis Measurement

Many of the early studies measuring protein synthesis involved lab animals. Protein synthesis rate was measured as the incorporation of a radioactively labeled amino acid into muscle protein over a specific time period, expressed, for example, as nanomoles of amino acid incorporated per gram of muscle tissue per hour. Typically, phenylalanine is used for incorporation studies, as it is not metabolized by muscle; thus its only fate is to be used to make proteins. One can obtain the rate of protein synthesis in specific fractions of muscle proteins by isolating mitochondria, contractile proteins, or soluble cytosolic proteins and determining the incorporation rate of labeled phenylalanine in each fraction per hour. One can assess the efficiency of translation by mea-

suring the rate of protein synthesis and dividing this by the total content of RNA. In this way it is possible to obtain an indication of the relative use of RNA, which points to how effectively the mRNA, tRNA, and rRNA are being used in protein synthesis.

In studies with humans, ethical considerations preclude the use of radioactive amino acids for nonmedical studies. Therefore, scientists have employed isotopically labeled amino acids that are not radioactive but that can be measured because of differences in their masses. These are known as **stable isotopes.** The amino acid enriched with a particular stable isotope can be infused at a constant rate into a vein. We know that it will be used to make muscle proteins. Therefore, muscle samples are obtained by biopsy at specific intervals, and the amount of the amino acid with the stable isotope can be determined over time. Leucine, labeled at carbon 1 with an isotope of carbon ($1\text{-}^{13}\text{C}$-leucine) as opposed to the normal isotope of carbon (^{12}C), is often used. Recent experiments have also used phenylalanine labeled with deuterium.

Effects of Exercise and Diet

A number of studies using humans and animals have shown that the rate of protein synthesis is generally depressed during exercise. After exercise, protein synthesis increases for periods up to 48 h before declining to baseline values. This has been shown following both endurance and resistance exercise, although the increase in protein synthesis with resistance exercise tends to be greater than that observed following endurance exercise (Hornberger and Esser 2004). People who are new to resistance training show a greater response in muscle protein synthesis following a training bout compared to those accustomed to exercise (Karlsson et al. 2004). Therefore, regular training appears to create a situation in which the exercise stimulus has less influence on protein synthesis (Kimball and Jefferson 2004).

In the fasted state, muscle protein synthesis is increased following resistance exercise. This suggests that exercise by itself stimulates synthesis. If the resistance or endurance training is followed by ingestion of carbohydrates or carbohydrates and protein, a dramatic improvement in muscle protein synthesis occurs (Hornberger and Esser 2004). This points to the important synergistic effect of food intake and exercise, particularly if the food intake closely follows the end of exercise. One of the explanations for the effect of food is that the blood insulin concentration is elevated. We know that insulin has an independent effect on stimulating muscle protein synthesis. Recent studies have also demonstrated that ingestion of a mixture of amino acids after resistance exercise stimulates muscle protein synthesis (Karlsson et al. 2004); this was accompanied by an increase in blood insulin levels. The branched chain amino acids, leucine, isoleucine, and valine, or just leucine alone, can increase translation initiation (Baum et al. 2005).

How do exercise and diet increase muscle protein so quickly following the end of exercise? We might suggest that they increase the rate of transcription of a number of genes, leading to a general increase in the content of a variety of mRNA molecules. Alternatively, the exercise and diet could increase the transcription of genes for rRNA or tRNA, leading to increased capacity for protein synthesis. This explanation is highly unlikely given the short period of time in which the exercise (or exercise and diet) effect is noted—too short to transcribe genes, modify RNA, and get it into functional form in the cytosol where protein synthesis takes place. Moreover, these changes in protein synthesis take place without a measurable increase in total RNA or specific mRNA molecules. Therefore, the rate of protein synthesis increases without a change in total RNA, indicating an increased efficiency of translation. It has also been demonstrated that immobilizing a muscle leads to a dramatic decrease in muscle protein synthesis, without a change in total RNA content. Apparently, the overall translation process is very sensitive to muscle activity.

In chapter 5, we discussed the mechanism by which insulin increases glucose uptake into muscle. In this mechanism, insulin binding to its receptor leads to a number of changes in the activities of kinases within the muscle cell, one outcome of which is an increased translocation of GLUT-4 transporters from intracellular storage sites into the muscle cell membrane. In the postexercise period, muscle is particularly sensitive to carbohydrate uptake, in part to restore muscle glycogen used during the previous exercise. Insulin potentiates this effect by increasing the activity of a particular kinase, phosphatidylinositol 3-kinase (PI3K). One of the downstream targets of PI3K is another protein kinase known as the mammalian target of rapamycin (mTOR). The mTOR protein kinase activity is enhanced by insulin, insulin-like growth factor-1 (IGF-1), ingestion of amino acids, and, most likely, exercise—especially resistance exercise. Insulin-like growth factor-1 can be increased by exercise stress and likely acts via a membrane receptor that is coupled to PI3K. An increase in mTOR activity leads to an increase in the overall protein synthesis rates in muscle. Figure 9.10 summarizes a

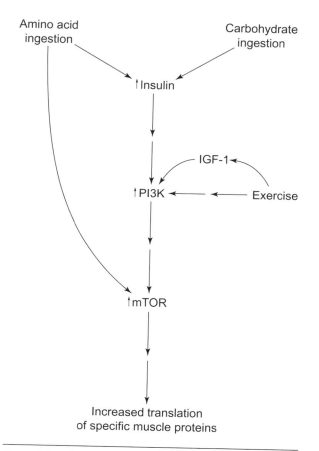

Figure 9.10 The ability of exercise, amino acid, and carbohydrate ingestion to increase the translation of specific muscle proteins. Two key kinases, phosphatidylinositol 3-kinase (PI3K) and the mammalian target of rapamycin (mTOR), are known to play key roles. The figure suggests how these interact. Carbohydrate and amino acid ingestion are known to increase insulin concentration, which leads to an increase in PI3K activity. Amino acid ingestion, particularly of leucine, may have an independent effect on mTOR activity. Exercise can function through several routes, one of which is to induce an increase in insulin-like growth factor-1 (IGF-1) production that can lead to PI3K activation.

proposed mechanism by which exercise, amino acids, and carbohydrates can lead to increased synthesis of specific muscle proteins by activating two particular kinases, PI3K and mTOR. This figure omits many of the complicated intermediates that can be found in other sources (Bolster et al. 2004; Kimball and Jefferson 2004; Tidball 2005).

Response Elements in mRNA and Translation

Control regions in some mRNA molecules are found in both the 5' and 3' untranslated regions (UTR); these can regulate the ability of the mRNA message to be initiated. A classic example of this type of control of mRNA translation involves the regulation of two key proteins in iron metabolism in the body. The average adult male has a total of only 4 to 5 g of iron in his body. For adult women this value is 2 to 3 g. As we have seen in previous chapters, iron plays essential roles in transporting and storing oxygen as a component of hemoglobin and myoglobin, respectively. Iron is also a critical component in oxidative phosphorylation in mitochondria, as it is a constituent of the cytochromes as well as iron-sulfur clusters in other mitochondrial complexes.

Because we have so little iron in our bodies, its uptake into tissues and the storage of any excess are rigidly controlled. The *transferrin receptor* is a membrane protein that allows iron to be taken up by cells. *Ferritin* is an iron storage protein. In times of an excess of cellular iron, it would be helpful to depress iron uptake and to increase the storage of any excess. Thus we would expect the transferrin receptor protein to be downregulated during iron excess, whereas we would expect ferritin, the storage protein, to increase. For the opposite situation, in which there is a scarcity of iron in the blood and cells, increasing the content of the transferrin receptor protein and decreasing the storage protein ferritin would be useful adaptations. Indeed this is exactly what the body does, and the regulation is based on the translation of each of the respective mRNAs.

Figure 9.11 summarizes the translation regulation of both transferrin receptor and ferritin. Both ferritin mRNA and transferrin receptor mRNA have *iron response elements* (IRE) that influence the translation and stability of each mRNA. For ferritin mRNA, its IRE is in the 5' UTR; for transferrin receptor mRNA, it has 5 IREs in the 3' UTR. The IRE is an AU-rich sequence in the mRNA that creates binding sites for a protein, *iron response element binding protein* (IRE-BP). Under conditions of low cellular iron, IRE-BP has little bound iron and can therefore bind to the IREs on each of the ferritin and transferrin receptor mRNAs. When cellular iron content is elevated, IRE-BP binds iron, changing its conformation so that it can no longer bind to IREs. When IRE-BP binds to the 5' UTR of ferritin mRNA, it blocks the ability of the 40S ribosome subunit to bind to the 60S ribosome subunit so an actively translating 80S ribosome cannot be created; thus ferritin mRNA cannot be translated, and its protein product, ferritin, is not made. On the other hand, the transferrin receptor mRNA is unstable and is easily degraded by ribonucleases unless it has IRE-BP bound, and this occurs only when cellular iron is low. As summarized in figure 9.11, the net effect is that ferritin mRNA is translated at an increased rate when body iron stores are high and it is appropriate for the excess iron to be stored by its product, ferritin. Conversely, transferrin mRNA is translated at an increased rate when cell iron need is increased (Papanikolaou and Pantopoulos 2005).

Messenger RNA Turnover

The amount of a specific polypeptide product generated in a cell by translation depends on the number of molecules of its mRNA and the rate at which translation is initiated and terminated for each mRNA. The number of mRNA molecules in a cell depends on how fast the gene template for the mRNA is transcribed and the mRNA modified and transported to the cytosol where translation takes place. We must also consider the average length of time during which each mRNA is capable of being transcribed. The lifetime of mRNA molecules, measured as a half-life of mRNA, can vary from a few minutes (for immune system signaling molecules such as cytokines) to many hours. If the rate of translation of mRNA molecules is fixed, then the amount of polypeptide product generated from each mRNA depends on how long it remains capable of being translated. Obviously, then, the lifetime of mRNA must be considered as a level of control. As we have seen, mRNA is generated by transcription. How long the mRNA lasts in the cytosol is controlled by degradative enzymes known as **ribonucleases.** These enzymes cleave the sugar-phosphate backbone of RNA, releasing individual nucleoside

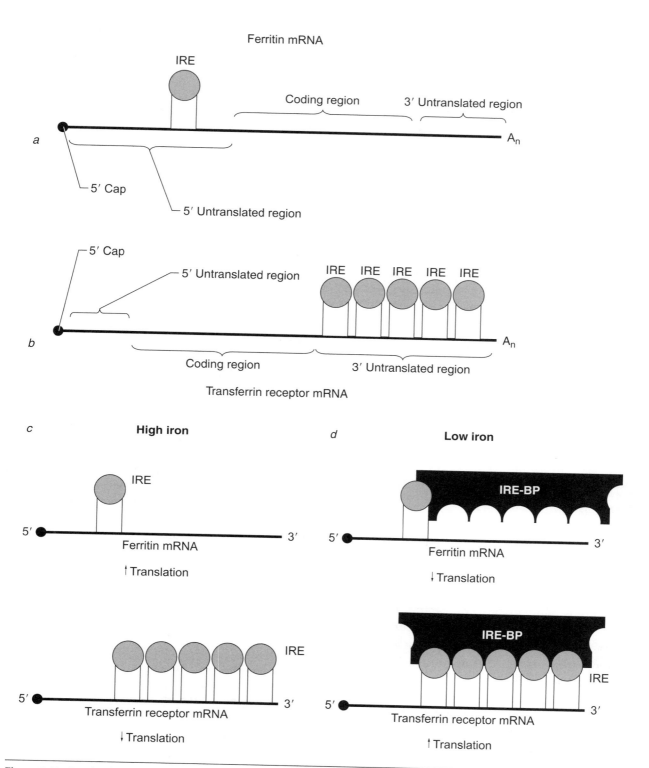

Figure 9.11 Regulation of the content of ferritin and the transferrin receptor, at the level of the translation of their respective mRNAs. *(a, b)* Each mRNA contains an iron response element (IRE), in the 5' untranslated region for ferritin and multiple copies in the 3' untranslated region for the transferrin receptor. *(c, d)* With low cellular iron, an iron response element binding protein (IRE-BP) can bind to IRE. When iron levels are elevated and bound to IRE-BP, the latter is incapable of binding to IREs. When IRE-BP is bound to the ferritin IRE, it blocks translation of the ferritin mRNA. The transferrin receptor mRNA is unstable unless IRE-BP binds to its IREs. The net effect is that ferritin protein is low under conditions of low cellular iron, but elevated when iron concentration is elevated. With transferrin receptor, its concentration is elevated under conditions of low cellular iron and depressed when cell iron content is high.

monophosphates. The length of time an individual mRNA molecule lasts in the cytosol depends on its ability to escape being degraded by the cytosolic ribonucleases. This is a very complicated area, since there are proteins that stabilize and proteins that destabilize mRNA. Regulation of these proteins can therefore indirectly influence the lifetime of the mRNA as well as regulate the ability of the mRNA to be translated.

During its lifetime, the poly A tail of most mRNA molecules is gradually shortened by a *deadenylase*. Since most mRNA molecules are stabilized in the cytosol by a poly A-binding protein, reduction of the length of the poly A tail reduces the ability of the poly A-binding protein to stabilize the mRNA, including the 5' cap, and it becomes susceptible to decapping and attack from the 5' end. Some mRNA molecules lose their cap without polyadenylation. Apparently nucleotide sequences in the 3' untranslated region render the cap more or less susceptible to removal and subsequent attack from the 5' end. Signals in the 5' and 3' UTRs of mRNA molecules can regulate not only the initiation of translation of the message, but also the length of time the mRNA lasts.

▶ K E Y   P O I N T ◀

Regulation of translation is important to many athletes who are attempting to increase muscle size. This means increasing the frequency of translation of each mRNA and increasing the lifetime of each mRNA. Simple strategies such as adopting an appropriate exercise training stimulus (which is not the same for everyone) and adjusting the timing and content of meals can play dominant roles. Such simple approaches may not be satisfying to the impatient athlete, and a host of alternative strategies are possible such as supplementing with essential amino acids, the essential branched chain amino acids (leucine, isoleucine, and valine), or simply leucine (Karlsson et al. 2004; Paddon-Jones et al. 2004). Many athletes have resorted to drugs and hormones. Some of these practices are successful, but many have little or no influence or may be illegal, unethical, or harmful.

POSTTRANSLATIONAL PROCESSING OF POLYPEPTIDES

Translation of the mRNA message generates a one-dimensional amino acid sequence in a polypeptide. As we saw in chapter 1, there are higher levels of structure of polypeptide chains that must be created before the polypeptide can function either as a single unit, or combined with one or more cofactors, or combined with other polypeptides in a protein with quaternary structure. All the information needed to make a complete three-dimensional structure for an amino acid is found in its amino acid sequence. However, the cytosol, where proteins are synthesized, is a very crowded space; and improper associations could certainly be made. Even if a polypeptide could spontaneously assume its correct native shape after it is released from the translation apparatus, it would likely need to bind to other molecules (cofactors or other polypeptide chains) or be in a different location in the cell to function properly. A group of proteins known as *chaperones* bind and stabilize completed polypeptide chains or partially folded polypeptides, or participate in the process of complete folding of proteins. They are often characterized as *molecular chaperones*, if they bind to and stabilize newly synthesized and partially folded polypeptides, or **chaperonins** if they function to directly fold proteins. Many of these proteins are **heat shock proteins** (HSP), a class of protein molecules initially found to be synthesized in response to a heat stress. HSP70 is particularly important as a molecular chaperone to stabilize partially folded proteins. The number following the designation HSP identifies the molecular weight in kilodaltons. We now know that the HSP proteins are synthesized in response to an exercise stress as well (Milne and Noble 2002).

Chaperonins are a subgroup of chaperone proteins that function primarily in protein folding and assembly of polypeptide subunits into molecular aggregates. Remember that some polypeptides must also bind with other polypeptides to generate the active oligomeric protein. For example, myosin is a hexamer, so six polypeptides must combine in the proper way to generate a functional myosin molecule. Moreover, once formed, myosin hexamers must be inserted into thick filaments in order to be able to hydrolyze ATP and generate a force (see chapter 3). Some polypeptides need to have a prosthetic group attached to them in order to be effective. For example, myoglobin, the protein in muscle that aids in storing and moving oxygen in the cytosol of cardiac and skeletal muscle, needs to have a heme group attached to its polypeptide.

Other polypeptides must be packaged for export because they function in the blood; liver is actively involved in synthesizing and secreting a host of plasma proteins. Others must be incorporated into membranes because much of the mass of a membrane is made up of a variety of proteins. Mitochondria represent an important organelle that needs more than 1,000 different protein molecules to function. Since the mitochondrial genome codes for only 13 polypeptide subunits, at least 99.9% of all proteins in mitochondria are coded by nuclear genes and imported. Moreover, a functional mitochondrion has four separate areas where specific proteins must be inserted: the outer and inner membranes, the matrix, and the intermembrane space. To facilitate the proper positioning of proteins in the various mitochondrial locations, cytosolic-generated polypeptides contain N-terminal signal sequences that, much like shipping labels, target a given polypeptide to its proper location (Stojanovski et al. 2003). These signals are recognized by membrane protein translocases TOM and TIM (translocase of the outer membrane and inner membrane, respectively). The signals are typically 15 to 40 amino acid N-terminal sequences that are removed after the mitochondrial protein is properly located. Mitochondria represent one important example of the complexity of a cell, in that protein synthesis on ribosomes is only one key step in a process to create a functional mitochondrial protein product.

PROTEIN DEGRADATION

In chapters 1, 5, and 7, the concept of protein turnover was discussed. Turnover relates to the balance between the opposing processes of protein synthesis and protein degradation. The content of any protein in a cell is based on the relative rates of protein synthesis and protein degradation. This can be described by the equation

cellular protein content =
synthesis rate – degradation rate

This tells us that the moment-to-moment change in the content of any protein is a reflection of the differences in rates of its synthesis and degradation.

We have already seen how proteins are synthesized. The opposing process is carried out by three major systems of protein-degrading enzymes, known as *proteinases*.

For years, the content of a particular protein was thought to be regulated primarily by changes in the transcription of its gene and subsequent translation of this message. The degradation side of the preceding equation was considered to be relatively constant. We now know that protein degradation is extremely important, involved in regulating cellular processes such as growth and atrophy, transcription, rates of metabolic pathways, development, and some disease states such as cancer. Three main protein-degrading systems operate in healthy skeletal muscle: the ubiquitin-proteasome pathway, the lysosomal system, and the calpain system. As these are complex and incompletely understood mechanisms, we will briefly summarize how they function.

Ubiquitin-Proteasome Pathway

The ubiquitin-proteasome pathway plays a major role in targeting and degrading cellular proteins. This pathway must recognize and label a specific protein for degradation, then carry out the process. The recognition part is based on marking a protein by covalent attachment of a small ubiquitous protein known as **ubiquitin.** This is a protein containing 76 amino acids that is found in all eukaryotes, virtually unchanged. Ubiquitinated proteins are subsequently degraded to small polypeptides in a 26S proteasome complex (figure 9.12). The 26S proteasome complex is formed by the union of two 19S regulatory particles and the 20S core particle. The barrel-shaped core particle is arranged as four stacked rings with proteolytic enzymes inside that can break down polypeptides to small peptide subunits.

Proteins are marked for degradation by the addition of ubiquitin units, termed polyubiquitination. Three conjugating enzymes are involved in this process, designated as E1, E2, and E3. Using the free energy released from ATP hydrolysis and the three conjugating enzymes, at least four ubiquitins are added to the target protein, generating a polyubiquitinated protein. Thus modified by ubiquitin addition, the protein binds to a 19S regulatory particle of the 26S proteasome complex. The target protein is unfolded, and its polypeptide chain is fed into the 20S core particle where the polypeptide will be degraded by proteases generating and releasing small

Figure 9.12 Summary of the steps involved in the degradation of an intracellular protein by the ubiquitin-proteasome system. Proteins are tagged for degradation by the attachment of multiple ubiquitin subunits to the protein using three enzymes, identified as E1, E2, and E3, and the energy from ATP hydrolysis. The polyubiquitin chain binds to the 19S regulatory particle and feeds the polypeptide chain into the 20S core particle, where the protein is hydrolyzed and small peptides are released. The polyubiquitin chain is not degraded by the proteasome but released, and the individual ubiquitin subunits are freed. Peptidases cleave the small peptides released from the proteasome to free amino acids.

peptides. These are subsequently degraded to free amino acids by cytosolic peptidases. The polyubiquitin marker is freed by the regulatory particle, and free ubiquitin units are subsequently released by hydrolysis of the peptide bonds holding them together. Many forms of exercise result in decreased activity of the ubiquitin-proteasome system, whereas disuse as a result of illness or immobilization is associated with increased activity (Taillandier et al. 2004).

Lysosomal System

Lysosomes are organelles, found in most cells, that are responsible for degrading a variety of cell constituents. In particular, lysosomes contain a battery of protein-degrading enzymes known as *cathepsins* that break peptide bonds in the interior of the protein molecule. In addition, they contain proteinases that cleave amino acids, one at a time, from the amino and carboxy ends of the polypeptide. To be degraded by lysosomes, proteins must enter via a process known as endocytosis.

Calpain System

The calpains are a family of calcium-activated proteinases in cell cytosol. They are found in two major forms in skeletal muscle, each activated by specific levels of intracellular calcium. Calpain activation has been implicated in a number of conditions, including muscular dystrophy, diabetes, multiple sclerosis, and some cancers (Zatz and Starling 2005). It is believed that calpains are activated by the rise in calcium with exercise and that they are responsible in part for initiating the increase in protein degradation associated with exercise. Moreover, severe exercise stress, known to produce muscle soreness and evidence of muscle damage at the microscopic level, is attributed to disturbances in internal calcium homeostasis, leading to activation of calpains. Because calpain acts on structural proteins in muscle, it is believed that its activation may be a first step in the response to exercise-induced muscle injury (Belcastro et al. 1998).

SUMMARY

The information for the sequence of amino acids in a polypeptide chain is based on the sequence of nucleotides (bases) in the messenger RNA molecule. Created during the transcription of a protein-coding gene, the pre-mRNA molecule undergoes extensive modifications, including the addition of a cap to the 5' end, the addition of multiple adenine nucleotides at the 3' end (poly A tail), and the removal of sequences not involved in the gene message (introns) from those that do provide amino acid sequence information (exons). The sequence of nucleotides (bases) in the mRNA, read in groups of three, specifies amino acids. The coding region or open reading frame describes the area between the start codon (AUG), where protein synthesis will begin, and a stop codon indicating that the message is terminated.

Protein synthesis occurs in the cell cytosol using mRNA, two ribosome subunits (described on the basis of their mobility in a centrifugal field as 40S and 60S), amino acids, energy sources, and a large number of protein factors. The modified rRNA molecules, created from two different kinds of RNA gene transcripts, are major components in the structure of ribosome subunits. A significant first step is the attachment of amino acids to specific tRNA molecules by a specific aminoacyl-tRNA synthetase. Each tRNA contains a three-base sequence called an anticodon; the complementary base pairing between the anticodon on tRNA (containing its attached amino acid) with the codon on mRNA provides fidelity in amino acid sequence in a protein.

Translation initiation is the most complex step in protein synthesis. It involves the formation of a complete ribosome subunit (80S) at the start (AUG) codon on mRNA with a methionyl-tRNA. Energy provided from GTP and the assistance of a number of eukaryotic initiation factors (eIFs) are also necessary.

Elongation is the step at which individual aminoacyl-tRNA molecules come to the ribosome complex and form a peptide bond between amino acids attached to their respective tRNAs. Individual ribosomes then move along the mRNA molecules three bases at a time. Each time the ribosome complex moves along the mRNA, a new amino acid is added to the growing polypeptide chain. Elongation continues the building of a polypeptide chain until a stop codon on mRNA is encountered. The ribosome complex then dissociates from the mRNA, and the completed polypeptide is released. Each mRNA molecule may have a number of ribosomes moving along independently, each with a growing polypeptide chain. Since the quantity of individual proteins in a cell can dictate rates of cell metabolism, control of the translation process must be carefully regulated. Translation, mainly initiation, is controlled at a number of different sites and may be based on nucleotide sequences in the 5' or 3' untranslated regions of the mRNA. The lifetime of its mRNA molecules also determines the amount of a cell's protein.

The amount of a specific protein in a cell is related to its rate of breakdown as well as its synthesis. Protein degradation is an important process that plays a significant role in the function of a cell. Three major systems are involved in normal protein degradation in muscle. The ubiquitin-proteasome pathway involves targeting proteins for degradation by attaching multiple units of a small protein known as ubiquitin. Subsequent degradation to small peptides is carried out by a proteasome complex known as the 26S proteasome. Lysosomes, which are internal organelles, degrade endocytosed proteins using a battery of proteinases. Calpains are calcium-activated proteinases that may play a significant role in the response of muscle to severe exercise stress.

REVIEW QUESTIONS

1. If a functional protein contains 200 amino acids, what is the minimum number of nucleotides in the coding region of the mature mRNA molecule for this protein? How many nucleotides would there be in the open reading frame for this protein?

2. If we assume that there are 30,000 genes coding for polypeptide subunits in the human genome and that the average number of amino acids in each polypeptide is 400, estimate the percentage of the human genome (3.1647×10^9 bases) that is actually involved in coding for amino acid sequence in a polypeptide. Assume that only one strand of DNA codes for all of the polypeptides.

Answers
to Review Questions

CHAPTER 1

1.

   a. 11

   b. N-terminus is glycine and C-terminus is leucine.

   c. Zero: arg and lys have +1 each, and asp and glu have net −1 each.

   d. cysteine only

   e. Conpolar amino acids are gly, phe, val, leu, trp.

   f. phe, val, leu, trp

   g. Yes, there are two cys residues.

   h. GRCEDKFVWCL

2. For cysteine to carry a net negative charge, the side chain sulfhydryl group must have a net negative charge while the ammonium group carries its charge. This happens only when the pH is greater than the pK_a for the sulfhydryl group but less than the pK_a for the amino group, which would be about pH 10. Thus the pH must between 8.2 and 10.0.

3. At pH of 6.0, the $H_2PO_4^-$ (acid form) predominates, whereas at pH 8.0, it is HPO_4^{2-}.

CHAPTER 2

1. Plot the data using a Lineweaver-Burk plot. These are actual data, so your line will need to be adjusted to be a fair reflection of the data. Make sure you extend the line into the other quadrant to get an intercept on the X-axis. K_m should be about 15 to 16 mmol/L and Vmax should be about 33 mmol · ml⁻¹ · min⁻¹.

2. Assuming that the enzyme behaves traditionally, then a 10 °C rise in temperature should double the Vmax.

3. This was a common problem in early studies because labs were not careful about temperature control. Too often, samples sat at room temperature and then were measured in the spectrophotometer or fluorometer. Different labs could have different temperatures. A second possibility lies in calibration using solutions of NADH, especially with the use of a fluorometric assay. Finally, samples may have less or more water depending on how they are treated. Since water is about 77% of a wet muscle sample, this could introduce a large error.

4.

   a. 10 mmol · kg⁻¹ · min⁻¹

   b. 10 IU/g

5. 40 IU/g dry weight

6. The K_m was the same, but the Vmax was lower (27 mmol · ml⁻¹ · min⁻¹). This means that the inhibitor was noncompetitive.

7. Your research should tell you that LDH is a tetramer. To express the activity per active site, you would divide the Vmax you determined by 4 (four active sites per enzyme). Hopefully you would have reservations because you would assume that the enzyme is active only as a tetramer, and thus the exercise is wasted. Some might also suggest that there are two different kinds of subunits in LDH and that these could have different activities.

CHAPTER 3

1. Γ rest is 5.85×10^{-3} mM; Γ mild fatigue is 3.53×10^{-2} mM; Γ severe fatigue is 0.629 mM; and Γ following marathon is 3.75×10^{-2} mM.

2.

   a. 4.0 μmol/g wet weight

   b. 17.4 mmol/kg dry weight

   c. 5.7 mM

3. K_{eq} is 25.6.

4. ΔG for ATP hydrolysis = 10.5 kcal/mol.

5. Approximately 10.2 mM for total magnesium concentration

6. ΔG for ATP hydrolysis = 7.0 kcal/mol.

7.

$$ATP + H_2O \rightarrow ADP + Pi$$
$$PCr + ADP \rightarrow ATP + Cr$$
$$\overline{PCr + H_2O \rightarrow Cr + Pi}$$

CHAPTER 4

1. The number of ATP formed maximally from the complete oxidation of a pyruvate would be 12.5. Four NADH are generated (4×2.5), one $FADH_2$ (1.5), and one GTP (equivalent to an ATP). Using the earlier numbers, the ATP yield would be 15.

2. To reduce one oxygen molecule requires four electrons. To reduce four would require 16 electrons. Forty protons would be pumped.

3. The same, 2.0 L/min

4. Rotenone inhibits complex I. Therefore any substrate that generates NADH will not work. A good substrate would be succinate because you would bypass complex I.

5. The $\dot{V}O_2$ of 0.4 L/min would represent the person's resting metabolism. Since the subject is sitting and not doing much in terms of muscular work, the metabolism would represent mostly metabolism of the heart, respiratory muscles, kidney, brain, and liver.

6. Use the relationship that a $\dot{V}O_2$ of 1.0 L/min is equivalent to an energy expenditure of 5 kcal/min. Then multiply by 120 min and use 3.0 L/min. You should get 1,800 kcal or 7,525 kJ.

7. The level-off phenomenon occurs with people who are trained or well trained. It means that they continue to exercise past the point of any increase in $\dot{V}O_2$, a fact attributable to their $\dot{V}O_2$ being limited by the ability to pump oxygen-rich blood from the heart. Untrained people often stop without a plateau in $\dot{V}O_2$. This is taken to mean that their weak link lies in the ability to use the oxygen in blood pumped from the heart. The limitation would be in the citric acid cycle, in the electron transport chain, or both.

8. Using table 4.4, determine that the $\Delta E^{\circ\prime}$ for the algebraic sum of $FADH_2$ oxidation and reduc-

tion of oxygen is 1.04 V. The standard free energy change from this would be 200 kJ/mol.

9. Dehydroascorbic acid can donate electrons only to substances below it in table 4.4. Therefore, the answer is both oxygen and the oxidized form of cytochrome c.

10. The $\Delta E^{\circ\prime}$ is –0.14 V.

11. The ΔE is –0.13 V.

CHAPTER 5

1. Glycogen by itself has a fuel value of 4 kcal/g (17 kJ/g). However, since each gram of glycogen is associated with 3 g of water, the fuel value is diluted to 1 kcal/g or 4.2 kJ/g.

2. Activities that involve vertically lifting your body weight may require anaerobic glycolysis, but if the activity is relatively brief, there is no advantage to having a huge glycogen store. Rock climbing or rope climbing would use glycogen at a high rate, but for brief periods of time, so large stores would be just added weight. Storing a lot of glycogen could also add extra weight, so wrestlers should be careful.

3. A lactate concentration of 990 mg/dL is equivalent to 11 mM. A glucose concentration of 108 mg/dL is equivalent to 6 mM.

4. Using the calorie equivalents of 4 kcal/g for carbohydrate and 9 kcal/g for fat, the runner's rate of energy expenditure is 20.5 kcal/min (86 kJ/min). The proportion of energy from CHO is 16/20.5 or 0.78. The total time he runs is total energy expended divided by rate of energy expended. The answer is 39.9 min. At 4 min/km, he covers just under 10 km.

5. A glycogen concentration of 500 mmol/kg dry weight is equivalent to 115 mmol/kg wet weight if the water content is 77%. Converting this to millimolar units, you would divide 115 mmol/kg wet weight by 0.7, because each kilogram of muscle would contain 0.7 kg of water or 0.7 L of intracellular water.

6. The FT fiber is metabolically endowed to produce lactate, whereas the ST fiber is better endowed to use lactate as a fuel. An elevated concentration of extracellular lactate leads to fatigue, whereas intracellularly, lactate may help maintain force. Therefore, the FT fiber should have the high K_m transporter, while the ST fiber should have the low K_m transporter.

CHAPTER 6

1.

Arachidonic acid, showing all carbon atoms and cis orientation for each double bond.

Elaidic acid, showing all carbon atoms and the orientation about the double bond.

2. A 3 to 1 ratio tells us that 100% of fatty acids released by lipolysis are released. Therefore a 1.5:1 ratio means that 50% of the fatty acids are being re-esterified.

3. Energy expended during 60 min is 60 min × 2.4 L O_2/min × 5.043 kcal/L O_2 = 726 kcal (3,037 kJ). The number of grams of carbohydrate oxidized is 0.585 × 726 kcal divided by 4 kcal/g = 106.2 g carbohydrate and 33.4 g fat. Because this is a nonprotein RER, we are excluding the contribution made by protein oxidation. This is usually less than 5%.

4. In the section on regulation of lipolysis, it was mentioned that adenosine is an antilipolytic substance. Caffeine is believed to bind to adenosine receptors to block the antilipolytic action. The net effect of this is to increase blood FFA concentration. It was once suspected that caffeine acted to inhibit the phosphodiesterase that breaks down cyclic AMP in fat cells, but this does not seem to be the case with the concentrations of caffeine produced from regular coffee drinking.

5. Each mole of triacylglycerol has 1 mol of glycerol. The molecular formula for glycerol is $C_3H_8O_3$, and its molecular weight is 92. During the formation of a triacylglycerol, three water molecules are released for each fatty acid attached. If we ignore that glycerol contributes to this water loss, we can determine that glycerol contributes 92/860 × 100 or 10.7%.

6. The $\dot{V}O_2$ due entirely to the active kicking muscles was 0.75 L/min (three times that of rest, which must then be 0.25 L/min). The kilocalories of energy devoted entirely to kicking are 60 min × 0.75 L O_2/min × 5.043 kcal/L O_2 = 227 kcal. Weight of intramuscular TAG used is first calculated using the relationship that if there are 860 g/mol of TAG, then there are 860 mg/mmol. Now, intramuscular triacylglycerol used = (12 mmol/kg – 10 mmol/kg) × 3 kg of muscle used × 860 mg/mmol TAG = 5,160 mg or 5.16 g. The energy equivalent would be 5.16 g × 9 kcal/g or 46.4 kcal. The mass of fat oxidized is given from the RQ relationship as described in question 3. With an RER of 0.87, 41.5% of the energy comes from fat. This amounts to 227 kcal × 0.415 = 94 kcal. We can account for 46 kcal from IMTG. Thus the remainder is from FFA and perhaps VLDL. Obviously the balance of the remaining energy would be derived from carbohydrate, notably muscle glycogen.

CHAPTER 7

1. Energy balance means that food energy intake is equal to total daily energy expenditure. Protein balance means that there is no net protein

stored in the body, so that protein intake is matched by the complete degradation of protein.

2. Nitrogen balance means that nitrogen intake and nitrogen excretion are equivalent. If excretion of nitrogen is 16 g/day, then intake is 100/16 × 16 g, or 100 g/day.

3. Reviewing the chemical structure of urea, you can determine that its molecular formula is CH_4N_2O. Its molecular weight is 60, and the contribution to this molecular weight based on nitrogen is 28/60. Therefore, the nitrogen content in 40 g of excreted urea is 40 × 28/60 or 18.7 g. To convert this to a protein value, simply multiply by 100/16, which gives us 117 g of protein intake per day.

4. You would need to measure the total volume of urine produced over 24 h.

5. As you did in question 2, you would determine nitrogen excretion as 28/60 × 30 g urea per day or 14 g N per day. Multiply this by 100/16 to get total protein lost = 87.5 g protein. Because she is in nitrogen balance (protein balance), she must take in 87.5 g of protein. At 18 kJ/g, the energy equivalent of protein is 18 kJ/g × 87.5 g = 1,575 kJ. This represents 1,575/8,000 × 100% or 19.7% protein.

6.
   a. A negative arterial concentration difference means that the venous value is greater than the arterial value.

   b. 8 L/min × 60 min = 480 L of blood.

   c. Fold increase = 8 L/min divided by 0.3 L/min or 26.7-fold greater.

   d. Glutamate exchange would be uptake of 40 mmol/min. Multiply this by 60 min to get total exchange during exercise.

   e. Muscle glycogen and glucose taken up from the blood

   f. Release of alanine is 100 mmol/min × 60 min × 89 mg/mmol. Divide by 1,000 to get grams = 534 g.

   g. The MW of alanine is 89, so 534 g represents 6 mol of alanine. Now, you need to know that each mole of alanine is converted to 1 mol of pyruvate. Further, you need to know that it takes 2 mol of pyruvate to make 1 mol of glucose (from chapter 5). Thus the 6 mol of alanine would be converted into 6 mol of pyruvate, which would be converted into 3 mol of glucose or 3,000 mmol of glucose.

   h. 100 mmol of alanine released per minute would be 6,000 mmol released for 1 h. As shown in answer g, it takes 2 mmol of alanine to make 1 mmol of glucose. Therefore there would be the potential to make 3,000 mmol of glucose from the alanine.

CHAPTER 8

1. Complementary strand: 3'-TCAGGTCG-CAATCTGGCTTCA-5'

2. Base sequence of mRNA: 5'-AGUCCAGC-GUUAGACCGAAGU-3'

3. Amino acid sequence: N-Ser-Pro-Ala-Leu-Asp-Arg-Ser-C

4. The sequence of the sense strand would be the same as the mRNA sequence except that you would have a T instead of a U. Therefore, go to the genetic code and work from amino acid to base sequence. You would get some redundancy, but this is also correct: 5'-GAAAAAATG-GCTGGTCATACT-3'.

CHAPTER 9

1. This is a bit of a tricky question because you are asked two different things. The coding region would begin with AUG, the start codon, and proceed to a stop codon. Thus the minimum number of nucleotides would be 3 × 200 if the protein began with methionine or 3 × 201 if the protein did not begin with methionine because the initial methionine coded by AUG would be removed. The open reading frame is the area between the start and stop codon and so would contain a minimum of 600 nucleotides.

2. If only one DNA strand codes for polypeptides, then a total of 3.1647 × 10^9 bases can be involved. Omitting the fact that most protein-coding genes have introns, and neglecting stop and start signals, we would multiply 30,000 genes × 400 amino acids per gene × 3 bases to specify one amino acid. When you do the arithmetic, the answer is 1.1375% of the genome.

Glossary

acceptor control—Regulation of the rate of oxidative phosphorylation by the availability of ADP.

acid—A compound that can donate a proton.

acid dissociation constant—The constant K_a, which describes the ability of an acid to donate a proton. The larger the value, the stronger the acid.

acidosis—A metabolic condition in which the production of protons is elevated and the capacity to buffer these is diminished. The blood pH may be reduced.

actin—The principal protein of the thin filament in muscle. It can bind with myosin and enhance myosin's ability to hydrolyze ATP.

activator—A DNA-binding protein that enhances the expression of a gene, or an allosteric effector that increases the activity of an enzyme.

active site—The region of an enzyme protein that binds substrate and converts it to product.

active transport—Transport across a membrane against a concentration gradient so that energy is needed for the transport.

adenosine—The molecule formed when the base adenine binds to the sugar ribose.

adenosine diphosphate—See ADP.

adenosine triphosphate—See ATP.

adenylyl (adenylate) cyclase—A membrane-bound enzyme that converts ATP into cyclic AMP.

adipocyte—A single fat cell.

adipokine—A cytokine secreted from adipose tissue.

adipose tissue—A collection of adipocytes embedded in a connective tissue network.

ADP—Adenosine 5'-diphosphate, an adenine nucleotide produced when ATP is hydrolyzed, and used as a substrate in reactions producing ATP.

adrenergic receptors—Membrane-bound receptors, categorized as alpha or beta, that can bind epinephrine and norepinephrine.

aerobic—Requiring or taking place in the presence of oxygen.

alkalosis—A condition in which the blood or intracellular pH becomes too high due to a decrease in the proton concentration.

allosteric enzyme—An enzyme that plays a regulatory role in metabolism because, in addition to binding its substrate or substrates at its active site, it binds small molecules at allosteric sites, which modulates the enzyme activity.

allosteric site—The site on an allosteric enzyme where effector molecules bind.

alpha-helix—A type of secondary structure of a protein in which the polypeptide backbone forms coils with maximum hydrogen bonding between amino acids.

amino acid pool—The collection of free amino acids in a compartment such as the bloodstream.

aminotransferases—A class of enzymes that remove an amino group from one amino acid and transfer it to another, usually α-ketoglutarate.

AMP—Adenosine 5'-monophosphate, an adenine nucleotide.

amphipathic—Referring to a molecule that has both hydrophilic and hydrophobic properties.

amphoteric—Capable of accepting or donating protons and thus able to act as a base or acid, respectively.

anabolism—The part of metabolism in which larger molecules are made from smaller molecules.

anaerobic—Occurring without air or oxygen.

anaplerosis—Reactions that can generate TCA cycle intermediates.

anhydride bond—A type of energy-rich bond formed when two acids are joined together with the elimination of water.

anticodon—A specific sequence of three bases in tRNA that is complementary to an mRNA codon.

antiparallel—Referring to two polynucleotide chains that are opposite in polarity.

antiport—A membrane transport protein that simultaneously moves two molecules in opposite directions across the membrane.

apoptosis—Programmed cell death in which a regulated process leads through a series of steps to the ultimate death of a cell.

ATP—Adenosine 5'-triphosphate, an adenine nucleotide used as the energy currency in metabolism. The free energy released when ATP is hydrolyzed is used to drive reactions in cells.

ATP-ADP antiport (adenine nucleotide translocase)—An inner membrane protein that transports one ATP out of the matrix while simultaneously transferring an ADP into the matrix.

ATPase—An enzyme that can hydrolyze ATP, producing ADP and inorganic phosphate.

ATP synthase—An enzyme complex in the inner mitochondrial membrane that uses the free energy release in oxidative phosphorylation to combine Pi with ADP to make ATP.

autocrine—Pertaining to the effect exerted on a cell by a substance produced by that cell.

base pair—Two nucleotides in polynucleotide chains that can be paired through hydrogen bonding between their bases, for example base A with base T.

beta-oxidation—The process through which fatty acids attached to coenzyme A are broken down into a sequence of four steps to produce acetyl CoA units.

beta-sheet—A type of secondary structure of a polypeptide chain in which hydrogen bonding between components of the chain creates a zigzag structure.

binding site—The part of the active site of an enzyme where substrate is bound, prior to its conversion to product.

biotin—A B vitamin involved as a coenzyme in some carboxylation reactions.

branched chain amino acids—The name given to three essential, hydrophobic amino acids, leucine, isoleucine, and valine.

Brønsted-Lowry—The surnames of scientists who described acids as proton donors and bases as proton acceptors.

brown adipose tissue (BAT)—A form of adipose tissue in which much of the reduction of oxygen to make water is not coupled to ADP phosphorylation to make ATP. Therefore most of the energy release appears as heat.

buffer—A system that can resist changes in the pH of a solution.

calmodulin—A ubiquitous calcium-binding protein in cells, capable of binding four calcium ions.

carboxy terminus—The end of a polypeptide chain with a free alpha carboxate group.

carnitine—A small molecule that aids in transport of long chain fatty acids across the inner mitochondrial membrane.

catabolism—That part of metabolism in which larger molecules are broken down to smaller components and energy is released.

catalase—Enzyme found in peroxisomes that degrades hydrogen peroxide.

catalytic site—That part of the active site of an enzyme molecule that actually participates in the bond making or breaking (or both) that converts a substrate into a product.

chaperonins—A class of proteins involved in post-translational processing of polypeptide chains.

chemiosmotic hypothesis—The proposal currently accepted to describe how the free energy released during electron transport from reduced coenzymes to oxygen is coupled to ATP formation in oxidative phosphorylation.

chromatin—A complex of DNA, histone, and nonhistone proteins from which chromosomes emerge during cell division.

chromosome—A discrete structure, observed as such during cell division, consisting of a long DNA molecule with associated proteins.

chylomicrons—Lipoproteins formed in intestinal cells and released into the bloodstream from the lymphatic system. These contain lipids obtained from the diet and some proteins to maintain their particulate form.

citric acid cycle—A sequence of reactions taking place in mitochondria where acetyl units attached to CoA are degraded to carbon dioxide and the electrons produced are transferred to the coenzymes NAD^+ and FAD. Also known as TCA or Krebs cycle.

codon—Three adjacent nucleotides that specify a particular amino acid. We generally consider these to be in mRNA.

coenzyme—A molecule generally derived from a B vitamin that is essential to a particular enzyme in its catalytic role.

coenzyme A (CoA)—A particular coenzyme, formed in part from the B vitamin pantothenic acid, used to carry acyl groups such as acetyl groups or fatty acyl groups from fatty acids.

coenzyme Q (also known as ubiquinone)—A small molecule involved in the electron transport chain, leading to ATP formation.

competitive inhibitor—An enzyme inhibitor that resembles the normal substrate and blocks the enzyme by reversibly binding and releasing from the active site. Its effects can be overcome by an excess of substrate.

complementary—Referring to two bases or two nucleotide chains containing purine and pyrimidine bases that recognize each other by their unique structures and that can form stable hydrogen bonds with each other.

complementary DNA (cDNA)—A DNA molecule formed when an mRNA is copied by reverse transcriptase to make a complementary DNA strand. Creation of a new complementary DNA strand produces double-stranded cDNA.

cooperativity—Property of some proteins with multiple subunits in which binding of a substrate or ligand to one subunit alters the affinity of the other subunits for the substrate or ligand.

Cori cycle—A cycle in which glucose is broken down to lactate in a muscle fiber. The lactate is released to the blood, taken up by the liver, and converted back to glucose, then released again to be used by muscle to form lactate.

cristae—Infoldings of the inner mitochondrial membrane.

cross bridge—A part of the myosin molecule with ATPase- and actin-binding activity. During ATP hydrolysis, a myosin cross bridge can go through a cycle of attachment to actin, force generation, and then detachment.

crossover concept—The point during graded exercise when the contribution of fat to ATP production is matched by that of carbohydrate. At intensities beyond this, carbohydrate becomes increasingly important.

cyclic AMP (cAMP)—Adenosine 3', 5'-monophosphate in which the phosphate group forms an ester bond with OH groups attached to both the 3' and 5' carbon atoms of the sugar ribose.

cytochromes—Heme-containing proteins in the electron transport chain that can be alternately in an oxidized or a reduced state.

cytokines—Polypeptide hormones secreted by a variety of tissues that bind to cell membrane receptors and signal such cell activities as proliferation or differentiation.

deamination—An enzymatic process in which an amino acid loses an amino group.

degenerate—A term to describe the fact that there is more than one codon specifying each amino acid.

degradation—Breakdown of something, as when a protein is changed to its component amino acids.

denaturation—Disruption of secondary and tertiary structure of a biological polymer such as a protein. With this comes a loss of normal activity.

dephosphorylation—Removal of a phosphate group from a molecule to which it is covalently attached.

diabetes mellitus—Type 1 or insulin-dependent diabetes, caused by destruction of the insulin-producing and -secreting beta cells of the pancreas.

diacylglycerol (also known as a diglyceride)—A glycerol molecule with two fatty acids attached via ester bonds.

disaccharide—A molecule composed of two monosaccharides joined together by a covalent bond.

dismutation reaction—Reaction that occurs when two identical chemical species react together and one species donates an electron to the other.

dissociation—The process in which an acid dissolved in water gives up a proton.

disulfide bond—A covalent bond joining two sulfhydryl (SH) groups on cysteine side chains in the same or different polypeptide chains.

DNA polymerase—The enzyme involved in replicating nuclear DNA prior to cell division (mitosis).

domain—A distinct section of a protein molecule with a distinct three-dimensional structure. It may or may not have a special function.

downstream—For a gene, representing the direction in which RNA polymerase moves during transcription, considered to be in the 3' direction.

electrochemical gradient—The energy required to separate a charge and concentration difference across a membrane.

electron transport chain—A group of four protein-lipid complexes in the inner mitochondrial membrane that transfer electrons from reduced coenzymes to oxygen.

endergonic reaction—A reaction or process that requires free energy in order to occur.

enhancers—Regulatory base sequences in DNA to which specific protein transcription factors may bind to influence the expression of a gene. Generally these lie a considerable distance from the gene they regulate.

enthalpy change—The total energy liberated in a reaction or process.

entropy change—A change in the degree of randomness or disorder in a system.

epinephrine—Hormone released from the adrenal medulla in response to physical stress.

equilibrium—The state of a system when there is no net change in energy.

equilibrium constant (K_{eq})—The ratio of the product of the concentrations of products of a reaction to the product of the concentrations of reactants when the system is in equilibrium.

essential amino acids—Amino acids that must be obtained in the diet because the body cannot synthesize them or cannot synthesize them in adequate amounts. Examples are leucine, isoleucine, valine, threonine, phenylalanine, histidine, tryptophan, lysine, and methionine.

essential fatty acids—Fatty acids that are needed by the body but cannot be synthesized by the body in amounts sufficient to meet our physiological needs. Examples are linoleic acid and linolenic acid.

estrogens—Female sex hormones that stimulate and maintain female secondary sex characteristics.

euglycemia—A normal range of blood glucose concentration.

exergonic reaction—A reaction that can proceed spontaneously with the release of free energy.

exon—Part of the primary transcript that will be retained in the functional mRNA molecule.

exothermic reaction—A reaction that liberates energy when it proceeds from reactants to products.

facilitated diffusion—Diffusion in which a substance crosses a membrane, down its concentration gradient, with the aid of a specific carrier.

FAD (flavin adenine dinucleotide)—An oxidation–reduction coenzyme that oxidizes a class of substrate molecules, transferring their electrons to coenzyme Q.

Faraday constant (F)—The number (96.5 kJ per mole per volt) that allows one to convert a change in redox potential, expressed in volts, into units of energy.

feedback inhibition—Inhibition of an allosteric enzyme by a product of its reaction or a product of a pathway in which the enzyme participates.

Fenton reaction—The nonenzymatic reaction of ferrous iron with hydrogen peroxide to produce hydroxyl radical.

FFA (free fatty acid)—A fatty acid, bound noncovalently to albumin, circulating in the blood.

free energy change—The fraction of the total energy released in a reaction or process that can be used to do useful work.

free radical—A chemical species containing an unpaired electron.

gene—The segment of DNA that provides the information for the sequence of amino acids in a polypeptide chain.

genome—The total genetic information carried within a cell nucleus.

ghrelin—A polypeptide secreted by gastric cells. It can signal the need to form new fat cells.

glucagon—Polypeptide hormone secreted from the alpha cells of the pancreas in response to a decrease in blood glucose concentration.

glucogenic amino acids—Those amino acids whose carbon skeletons can be used to make glucose in the liver.

glucose transporters—A family of membrane proteins that transport glucose across cell membranes, down their concentration gradient.

glutathione—A tripeptide, containing a sulfhydryl group (–SH), that participates in a variety of intracellular reduction reactions.

glutathione peroxidase—An enzyme that breaks down hydrogen peroxide or other peroxides and uses the reducing power of glutathione.

glyceroneogenesis—The formation of glycerol from pyruvate or carbon skeletons of some amino acids.

glycogen—A large branched polymer, formed entirely of glucose, that is the principal means of storing carbohydrate in liver and muscle.

glycogenesis—The formation of glycogen from glucose.

glycogenin—A self-glycosylating polypeptide that is the precursor for glycogen synthesis.

glycogenolysis—The process in which glycogen is broken down in liver and muscle to yield glucose 6-phosphate.

glycolysis—A catabolic pathway that breaks down glucose 6-phosphate, derived from glucose or glycogen, into lactate and in the process generates ATP.

glycosome—A molecule of glycogen along with the enzymes involved both in synthesizing and in degrading glycogen.

G protein—A GTP-binding membrane protein that participates in the process of signal transduction, linking the binding of an external ligand, its membrane receptor, into an action within the cell.

growth factors—A family of extracellular polypeptides that can bind to specific receptors on the cell membrane, influencing the growth and differentiation of that cell.

growth hormone—A polypeptide hormone synthesized in and secreted from the anterior portion of the pituitary gland.

half-life—The time taken for the disappearance or decay of one-half of the population of chemical substances.

heat shock proteins—A family of proteins originally shown to be synthesized in response to cellular or organism stress. Among the members of this expanded family are the chaperonins, which participate in posttranslational processing of new polypeptides.

heme—A prosthetic group containing a central iron ion.

Henderson-Hasselbalch equation—The equation relating the solution pH, the pK_a, and the ratio of undissociated to dissociated acid concentrations in a solution.

histones—A family of nuclear proteins carrying net positive charges because of a preponderance of basic amino acids. These are associated with the polynucleotide backbone of DNA in structures known as nucleosomes.

holoenzyme—A catalytically active enzyme containing the polypeptide subunit and any necessary cofactors.

hormone receptor—A protein located in a cell membrane or within a cell that recognizes and binds a specific hormone.

hydridization—Association of two complementary polynucleotide chains because of base pairing through hydrogen bonding between complementary bases.

hydride ion—A hydrogen atom with an additional electron so that it carries a negative charge.

hydrogen bond—A weak, noncovalent bond between a hydrogen atom, covalently attached to an electronegative atom, and another electronegative atom in the same or a different molecule.

hydrophilic—Referring to polar molecules or charged ions that can interact with the polar water molecule.

hydrophobic—Referring to nonpolar groups or molecules that cannot interact with water.

hydroxyl radical—The most reactive free radical normally produced in our bodies.

hyperglycemia—Elevated blood glucose concentration.

hypoglycemia—Very low level of blood glucose concentration.

inorganic phosphate—Ionized phosphate group that contains a net negative charge at neutral pH.

inorganic pyrophosphate—An anhydride containing two phosphate groups joined together; it is hydrolyzed by the enzyme inorganic pyrophosphatase.

insulin—A polypeptide hormone produced in and secreted from the beta cells of the pancreas in response to an elevation in blood glucose concentration.

intermembrane space—The space between the outer and inner mitochondrial membranes.

intermyofibrillar mitochondria—Rows of mitochondria located between adjacent myofibrils in a muscle fiber.

international unit—The amount of enzyme that converts 1 μm of substrate to product in 1 min.

isoelectric point (pI)—The pH at which a molecule or ion has no net charge but contains at least one positive and one negative charge.

isozymes (isoenzymes)—Different molecular forms of the same enzyme, catalyzing the same reaction with the same mechanism but having different kinetic parameters.

ketogenic amino acids—The carbon skeletons of amino acids that cannot be converted into glucose in the liver. Two examples are lysine and leucine.

ketone bodies—A group of small molecules, including acetoacetate, 3-hydroxybutyrate, and acetone that are formed in the body during carbohydrate deprivation or uncontrolled diabetes mellitus.

ketonemia—A condition of elevated blood ketone body concentration.

ketonuria—Loss of ketone bodies in the urine.

kinetic parameters—The Vmax and K_m are considered to define parameters of an enzyme.

Krebs cycle—See citric acid cycle.

leptin—A polypeptide synthesized in and secreted by adipose tissue especially when the fat cells are enlarged from storing more triacylglycerol.

ligand—A small molecule that binds to a specific larger one, such as a receptor.

Lineweaver-Burk plot—A transformation of the Michaelis-Menten equation in which reciprocals of substrate concentration are plotted, yielding a straight line whose intercepts can produce values for the kinetic parameter.

lipoic acid—An organic coenzyme, made in the body, that works with two mitochondrial dehydrogenase enzymes.

lipolysis—The enzymatic breakdown of triacylglycerol molecules, catalyzed by a class of enzymes known as lipases.

lipoproteins—A family of protein-lipid particles found in the blood.

long chain carboxylic acids—Fatty acids with 16 to 20 carbon atoms; may be saturated or unsaturated.

macrophage—Tissue-based phagocytic cell, derived from blood monocyte cells, that plays a role in immune responses within tissues.

MAP kinases—A family of protein kinases that respond to cell stimulation by mitogens.

mass action ratio—The ratio of the product of the concentrations of the products of a reaction to the product of the concentrations of reactants at any time during the course of a chemical reaction.

matrix—That part of the mitochondrion that is bounded by the inner membrane.

maximum velocity (V_{max})—The maximum velocity of an enzyme-catalyzed reaction.

medium chain triacylglycerol—Triacylglycerol molecules containing fatty acids with a carbon chain length of 6 to 10.

messenger RNA (mRNA)—A class of RNA molecules, complementary to a portion of a strand of DNA, that provides the information for the synthesis of a polypeptide chain.

metabolon—A cluster of enzymes in which the product of one enzyme is immediately passed to the next enzyme because of the physical association.

Michaelis constant (K_m)—The concentration of substrate needed to produce one-half the maximum velocity for an enzyme-catalyzed reaction.

monocarboxylate transporter—A membrane protein that transports small anions against their concentration gradient. Usually refers to transport of lactate and pyruvate.

monosaccharides—Simple sugars that cannot be digested or broken down further without having their structure destroyed.

mutation—A change in the nucleotide sequence of DNA that can be passed on.

myeloperoxidase—An enzyme found in a variety of phagocytic cells (neutrophils, macrophages) that reacts hydrogen peroxide with a chloride ion to generate hypochlorite.

myogenic regulatory factor—A class of transcription factors that, when bound to their cognate cis elements, regulate the expression of genes specific for development and maintenance of muscle properties.

myosin—The principal protein of the thick filament of muscle fibers. It has ATPase- and actin-binding activity.

myosin heavy chain—The largest subunit in the oligomeric protein, myosin.

NAD$^+$ (nicotinamide adenine dinucleotide)—The oxidized form of the more common coenzyme involved in redox reactions in the cell.

NADH—Reduced form of NAD$^+$. Transfers its electrons to oxygen in the electron transport chain.

NADPH (nicotinamide adenine dinucleotide phosphate)—The reduced form for the coenzyme involved in redox reactions in which larger molecules are built out of smaller components. The oxidized form is NADP$^+$.

NADPH oxidase—Enzyme found in immune cells and endothelial cells that can use electrons from NADPH to reduce oxygen to superoxide.

Nernst equation—An equation to determine the redox potential of a reaction using known values for standard redox potential, temperature, and the concentrations of the oxidized and reduced forms of the substrates and products.

neutrophils—A class of white blood cells that attack foreign cells such as bacteria.

NFAT (NF-AT, nuclear factor of activated T cells)—A transcription factor first identified in T lymphocytes, but now known to operate in other cells.

nitric oxide (NO)—Nitrogen monoxide, a free radical signaling molecule, produced by nitric oxide synthase and derived from the amino acid arginine.

nitric oxide synthase—A class of enzyme that converts arginine to nitric oxide and citrulline.

nitrogen balance—The relationship between nitrogen intake into the body in the form of food protein and nitrogen loss from the body in urine, feces, and sweat.

noncompetitive inhibitor—An agent that causes irreversible inhibition of an enzyme and loss of enzyme function. This kind of inhibition decreases Vmax, but has no effect on K_m.

norepinephrine (also known as noradrenaline)—The chief neurotransmitter in the sympathetic nervous system.

nucleoside—Product that results when a purine or pyrimidine base is attached to a ribose sugar.

nucleoside triphosphate—A sugar plus base (nucleoside) with one or more attached phosphate groups.

nucleosome—A basic structural component of chromatin, consisting of a DNA strand wound around a core of histone proteins.

nucleotide—A nucleoside with one or more attached phosphate groups.

oligomer—A small polymer, usually of nucleotides (oligonucleotide), amino acids (oligopeptide), or monosaccharides (oligosaccharide).

open reading frame—A sequence of nucleotide codons that code for amino acids in the absence of stop codons.

osmotic pressure—The pressure that must be applied to a solution to prevent solvent (water) from flowing into it.

oxidation—A chemical reaction in which a substrate loses electrons.

oxidative phosphorylation—The formation of ATP from ADP and Pi in association with the transfer of electrons from substrate to coenzymes to oxygen.

oxidative stress—The condition in a cell or organism in which the formation of reactive oxygen and nitrogen species exceeds the capacity to quench them.

oxygen consumption ($\dot{V}O_2$)—The utilization of oxygen (measured in L/min) to accept electrons and produce ATP in oxidative phosphorylation.

pancreatic lipase—A lipase produced in and secreted by the pancreas. Its function is to break down dietary triacylglycerol.

pantothenic acid—A B vitamin that forms a major part of coenzyme A.

paracrine—A term to describe the mechanism of signaling between neighboring cells in which a growth factor diffuses from one cell to its close neighbors.

pentose phosphate pathway—A pathway to interconvert hexoses and pentoses; a source of NADPH for synthetic reactions. It is also called the phosphogluconate pathway.

peptide bond—A specialized form of amide bond that joins two amino acids together.

perilipins—A family of proteins that coat adipose tissue droplets in fat cells so that the triacylglycerol molecules are inaccessible to enzymes unless the cell is stimulated.

phosphagen—A molecule such as phosphocreatine that can transfer its phosphate group to ADP to make ATP.

phosphate transporter—Protein that transfers a phosphate group across the inner membrane while simultaneously transferring a hydroxide group the opposite way, to the intermembrane space.

phosphatidic acid—A diglyceride with a phosphate group attached in ester linkage to the third OH of glycerol.

phosphatidylinositols—A family of phospholipids in which inositol or phosphorylated derivatives are attached to phosphatidic acid.

phosphocreatine—A phosphagen capable of rapidly transferring its phosphate group to ADP to quickly regenerate ATP.

phospholipid—A lipid containing one or more phosphate groups.

phosphoprotein phosphatase—An enzyme that removes a phosphate group attached in ester linkage to a protein.

phosphorylation—Covalent attachment of a phosphate group to a molecule.

phosphorylation potential—The ratio of the concentration of ATP, divided by the concentrations of ADP and Pi (i.e. [ATP]/[ADP] [Pi]).

phytochemicals—Name given to a wide variety of compounds from plant sources that can have an antioxidant effect in our bodies.

polypeptide—A molecule composed of 20 or more amino acids joined together by peptide bonds.

polyribosome—An mRNA molecule with two or more ribosomes that carry out protein synthesis on it.

P/O ratio—The ratio of the number of ATP molecules produced per pair of electrons transferred from a substrate to oxygen to reduce the oxygen to water.

porphyrin—A heterocyclic molecule, made in the body, that forms the basic structure for heme groups in hemoglobin, myoglobin, and the cytochromes.

primary gene transcript—The initial product of transcription.

primary structure—The sequence of amino acids in a peptide or protein, starting from the N-terminus.

promoter—The region upstream of the start site of a gene to which transcription factors can bind to influence the initiation of transcription.

prosthetic group—A cofactor (metal ion or organic compound) that is covalently attached to a protein and is necessary for that protein to function.

proteasome—A cylindrical-shaped protein complex responsible for much of the degradation of cellular proteins.

protein balance—The relationship between protein synthesis and protein degradation, which affects the net content of protein in an organism. This is often inferred from nitrogen balance.

protein kinase—An enzyme that attaches a phosphate group from ATP to the side chain of a serine, threonine, or tyrosine residue in the protein.

protein tyrosine kinase—An enzyme that attaches the terminal phosphate of ATP to the hydroxyl group in the side chain of tyrosine residues in a protein.

proteomics—The study of large compilations of proteins, such as those expressed in a single cell.

proton motive force (pmf)—The combination of the membrane electric potential difference and the pH gradient that provides the driving force for protons to move from the intermembrane space to the matrix of mitochondria.

purine (pyrimidine)—A nitrogen-containing heterocyclic base found in nucleic acids. Principal examples are adenine and guanine, cytosine, thymine, and uracil.

quaternary structure—The three-dimensional organization of a protein that is composed of subunits, dealing with the way the subunits interact with each other.

rate limiting—Referring to that step in a reaction or pathway that limits the overall rate of the reaction or pathway.

reactive oxygen species—A molecule or ion of oxygen that is highly reactive in accepting or donating one or more electrons, or an oxygen species with one or more unpaired electrons.

redox reaction—A reaction in which electrons are transferred from one substrate (oxidation) to another substrate (reduction).

reducing equivalents—A pair of electrons that can be used to reduce something. Generally transferred as a hydride ion or a pair of hydrogen atoms.

reduction—What occurs when a substrate gains one or more electrons.

reduction potential—The tendency of a substance to be reduced, measured in volts.

repressor—A protein that binds to a regulatory element associated with a gene and blocks its transcription.

respiratory burst—A process by which a reactive oxygen species such as superoxide is produced in macrophages or other phagocytic cells when stimulated.

respiratory exchange ratio (RER)—The ratio of carbon dioxide produced to oxygen consumed as measured by collection of expired air at the mouth of an individual.

respiratory quotient (RQ)—The ratio of carbon dioxide produced to oxygen consumed as measured in an isolated system or across a tissue or organ.

ribonuclease—A class of enzymes that can hydrolyze the sugar-phosphate backbone of RNA.

RNA polymerase—The enzyme that transcribes a section of DNA, producing an RNA molecule complementary to the copied strand of DNA.

sarcolemma—The cell membrane of a muscle cell or fiber.

sarcoplasmic reticulum—A specialized form of endoplasmic reticulum, found in striated muscle, that holds, releases, and takes up calcium.

secondary structure—A regular, repeating structure assumed by the polypeptide backbone of a polypeptide. Examples are the alpha-helix and the beta-sheet.

second messenger—A molecule or ion produced in or released from the interior of a cell in response to the binding of a ligand (first messenger) to a cell membrane receptor.

semiconservative—Referring to the replication of DNA in which each strand acts as a template upon which a complementary strand is produced.

signal transduction—The process through which an extracellular signal is amplified and converted to a response within the cell.

small nuclear RNA (snRNA)—Small RNA molecules in the nucleus that play a role in RNA processing, such as splicing out introns.

sodium-calcium antiport—A membrane protein that transports calcium against its gradient using the energy produced when sodium flows down its gradient.

stable isotopes—Isotopes of common elements that are not radioactive.

standard free energy change ($\Delta G°'$)—The free energy change for a reaction under standard conditions at pH 7.0 and with all reactants and products at 1 M concentration.

standard redox potential ($\Delta E°'$)—The tendency for a substance to be reduced, measured in volts and based on standard conditions.

start codon—The codon AUG, which signals the start of translation; also the codon for methionine.

start site—The nucleotide on the sense strand of DNA that indicates the beginning of transcription.

state 3 respiration—Maximally stimulated oxidative phosphorylation when all substrates and components are present in adequate amounts.

state 4 respiration—The level of electron transport in mitochondria when the substrate ADP is inadequate.

steady state—A nonequilibrium condition of a system in which there is a constant flux, yet the intermediates within the system are at constant concentration.

stereoisomers—Isomers that have the same composition of components yet have different molecular arrangements.

steroid hormones—A class of hormones based on the structure of their precursor, cholesterol.

stop (termination) codons—Three codons, UAG, UAA, and UGA, that signal the end to translation.

stroke volume—The volume of blood ejected from the left ventricle at each beat of the heart.

subsarcolemmal mitochondria—Mitochondria located immediately beneath the cell membrane (sarcolemma) of a muscle fiber.

substrate—The specific compound that is chemically modified by an enzyme.

substrate-level phosphorylation—Process whereby ADP is converted into ATP by any mechanism other than oxidative phosphorylation.

superoxide—The product of the addition of a single electron to an oxygen molecule.

superoxide dismutase—An enzyme that catalyzes the reaction between two superoxide anions, in which an electron is donated by one to the other, creating hydrogen peroxide.

TATA box—A sequence of bases located about 25 to 35 base pairs upstream of the start of many eukaryotic genes that helps to locate RNA polymerase.

TCA (tricarboxylic acid) cycle—See citric acid cycle.

template DNA strand—The strand of DNA that actually is the strand copied in a complementary way by RNA polymerase.

thiamine pyrophosphate (TPP)—A coenzyme involved in several mitochondrial oxidative decarboxylation reactions.

thick filament—A filament in the middle of the sarcomere of a striated muscle fiber, composed almost entirely of myosin molecules.

thin filament—A filament in striated muscle consisting of the proteins actin, tropomyosin, and troponin.

thymine—A pyrimidine base found in DNA.

total adenine nucleotide (TAN)—The sum of the concentrations of ATP, ADP, and AMP.

trans (transcription) factor—A protein transcription factor that binds to a cis element near a gene and increases the frequency of initiation of transcription.

transamination—Transferal of an amino group from an amino acid to a keto acid to make a new amino acid and a new keto acid.

transcription—The enzymatic reaction whereby a segment of a DNA strand is copied via the making of RNA.

translation—The process whereby the sequence of codons on an mRNA molecule is used to direct the sequence of amino acids in a polypeptide chain.

triacylglycerol (triglyceride)—A triester consisting of the alcohol glycerol and three fatty acid molecules.

troponin—A calcium-binding protein in the thin filament of striated muscle.

T-tubules—Invaginations in the sarcolemma of striated muscle.

turnover number—The number of molecules of substrate that an enzyme active site can convert to product in 1 s.

ubiquinol (reduced coenzyme Q or QH$_2$)—The reduced form of an electron carrier in the mitochondrial inner membrane.

ubiquinone (oxidized coenzyme Q or Q)—The oxidized form of an electron carrier in the mitochondrial inner membrane.

ubiquitin—Small protein found in all cells that becomes covalently attached to proteins destined for degradation by the proteasome complex.

ubisemiquinone—An intermediate in the reduction of ubiquinone or the oxidation of ubiquinol in which there is an unpaired electron that can be passed to oxygen.

uncoupling protein—Inner membrane protein that allows protons to flow down their electrochemical gradient without ADP phosphorylation.

uniport—A membrane protein that allows a polar molecule or ion to flow across the membrane down a concentration gradient.

upstream—Representing a direction with respect to a gene, that is, in the 5' direction, based on the sense strand of DNA.

uracil—A pyrimidine base found in RNA.

urea—Small molecule produced in the liver in the urea cycle; the major form in which the body expels nitrogen groups from amino acids.

urea cycle—A sequence of reactions in the liver in which urea is synthesized.

very low density lipoprotein (VLDL)—A triacylglycerol-rich lipoprotein fraction, synthesized in liver.

wobble—The relatively loose base pairing between the third (3') base in a codon and the first base (5') of an anticodon.

xanthine oxidase—An enzyme that can convert hypoxanthine to xanthine and xanthine to uric acid; in the process, generates superoxide.

xenobiotics—Man-made foreign molecules that may or may not be harmful to us if they enter our bodies.

zwitterion—A molecule with one positive and one negative charge.

References

Alberts, B., A. Johnson, J. Lewis, M. Raff, K. Roberts, and P. Walter. 2002. *Molecular Biology of the Cell.* New York: Garland Science.

Allen, D.A., and H. Westerblad. 2004. Lactic acid—the latest performance-enhancing drug. *Science* 305: 1112-1113.

Angus, D.J., M. Hargreaves, J. Dancey, and M.A. Febbraio. 2000. Effect of carbohydrate or carbohydrate plus medium chain triglyceride ingestion on cycling time trial performance. *Journal of Applied Physiology* 88: 113-119.

Arkinstall, M.J., C.R. Bruce, S.A. Clark, C.A. Rickards, L.M. Burke, and J.A. Hawley. 2004. Regulation of fuel metabolism by preexercise muscle glycogen content and exercise intensity. *Journal of Applied Physiology* 97: 2275-2283.

Armitage, J.A., I.Y. Khan, P.D. Taylor, P.W. Nathanielsz, and L. Poston. 2004. Developmental programming of the metabolic syndrome by maternal nutritional imbalance: How strong is the evidence from experimental models in mammals? *Journal of Physiology* 561: 355-377.

Bailey, D.M., B. Davies, I.S. Young, M.J. Jackson, G.W. Davison, R. Isaacson, and R.S. Richardson. 2003. EPR spectroscopic detection of free radical outflow from an isolated muscle bed in exercising humans. *Journal of Applied Physiology* 94: 1714-1718.

Baldwin, J., R.J. Snow, M.J. Gibala, A. Garnham, K. Howarth, and M.A. Febbraio. 2003. Glycogen availability does not affect the TCA cycle or TAN pools during prolonged, fatiguing exercise. *Journal of Applied Physiology* 94: 2181-2187.

Barazzoni, R., A. Bosutti, M. Stebel, M.R. Cattlin, E. Roder, L. Visinin, L. Cattin, G. Biolf, M. Zanetti, and G. Guarnieri. 2005. Ghrelin regulates mitochondrial-lipid metabolism gene expression and tissue fat distribution in liver and skeletal muscle. *American Journal of Physiology* 288: E228-E235.

Battram, D.S., J. Shearer, D. Robinson, and T.E. Graham. 2004. Caffeine ingestion does not impede the resynthesis of proglycogen and macroglycogen after prolonged exercise and carbohydrate supplementation. *Journal of Applied Physiology* 96: 943-950.

Baum, J.I., J.C. O'Connor, J.E. Seyler, T.G. Anthony, G.G. Freund, and D.K. Layman. 2005. Leucine reduces the duration of insulin-induced PI 3-kinase activity in rat skeletal muscle. *American Journal of Physiology* 288: E86-E91.

Belcastro, A.N., L.D. Shewchuk, and D.A. Raj. 1998. Exercise-induced muscle injury: A calpain hypothesis. *Molecular and Cellular Biochemistry* 179: 135-145.

Berg, J.M., J.L. Tymoczko, and L. Stryer. 2002. *Biochemistry, Fifth Edition.* New York: Freeman.

Boeger, H., D.A. Bushnell, R. Davis, J. Griesenbeck, Y. Lorch, J.S. Strattan, K.D. Westover, and R.D. Kornberg. 2005. Structural basis of eukaryotic gene transcription. *FEBS Letters* 579: 899-903.

Bolster, D.R., L.S. Jefferson, and S.R. Kimball. 2004. Regulation of protein synthesis associated with skeletal muscle hypertrophy by insulin-, amino acid-, and exercise-induced signaling. *Proceedings of the Nutrition Society* 63: 351-356.

Booth, F.W., and P.D. Neufer. 2005. Exercise controls gene expression. *American Scientist* 93: 28-35.

Boveris, A., L.E. Cosa, E. Cadenas, and J.J. Poderoso. 1999. Regulation of mitochondrial respiration by adenosine diphosphate, oxygen, and nitric oxide. *Methods in Enzymology* 301: 188-198.

Brooks, G.A. 1997. Importance of the "crossover" concept in exercise metabolism. *Clinical and Experimental Pharmacology and Physiology* 24: 889-895.

Brooks, G.A., M.A. Brown, C.E. Butz, J.P. Sicurello, and H. Dubouchaud. 1999. Cardiac and skeletal muscle mitochondria have a monocarboxylate transporter MCT1. *Journal of Applied Physiology* 87: 1713-1718.

Brooks, G.A., H. Dubouchaud, M. Brown, J.P. Sicurello, and C.E. Butz. 1999. Role of mitochondrial lactate dehydrogenase and lactate oxidation in the intercellular lactate shuttle. *Proceedings of the National Academy of Science USA* 96: 1129-1134.

Brown, J.M., and M.K. McIntosh. 2003. Conjugated linoleic acid in humans: Regulation of adiposity and insulin sensitivity. *Journal of Nutrition* 133: 3041-3046.

Bülow, J. 2003. Lipid mobilization and utilization. In: *Principles of Exercise Biochemistry, Third Edition,* ed. J.R. Poortmans. Basel: Karger.

Burtis, C.A., and E.R. Ashwood, Eds. 2001. *Tietz Fundamentals of Clinical Chemistry, Fifth Edition.* New York: Saunders.

Cesari, M., M. Pahor, B. Bartali, A. Cherubini, B.W.J.H. Penning, G. Rhys Williams, H. Atkinson, A. Martin, J.M. Guralnik, and L. Ferrucci. 2004. Antioxidants and physical performance in elderly persons: The Invecchiare in Chianti (InCHIANTI) study. *American Journal of Clinical Nutrition* 79: 289-294.

Chabowski, A., S.L.M. Coort, J. Calles-Escandon, N.N. Tandon, J.F.C. Glatz, J.J.F.P. Luiken, A. Bonen. 2004. Insulin stimulates fatty acid transport by regulating the expression of FAT/CD36 but not FABPpm. *American Journal of Physiology* 287: E781-E789.

Chakravarthy, S.,Y.-J. Park, J. Chodaparambil, R.S. Edayathumangalam, and K. Luger. 2005. Structure and dynamic properties of nucleosome core particles. *FEBS Letters* 579: 895-898.

Chanutin, A. 1926. The fate of creatine when administered to man. *Journal of Biological Chemistry* 67: 29-43.

Chinetti-Gbaguidi, G., J.-C. Fruchart, and B. Staels. 2005. Role of the PPAR family of nuclear receptors in the regulation of metabolic and cardiovascular homeostasis: New approaches to therapy. *Current Opinion in Pharmacology* 5: 177-183.

Coles, L., J. Litt, H. Hatta, and A. Bonen. 2004. Exercise rapidly increases expression of the monocarboxylate transporters MCT1 and MCT4. *Journal of Physiology* 561: 253-261.

Conaway, J.W., L. Florens, S. Sata, C. Tomomori-Sato, T.J. Parmely, T. Yao, S.K. Swanson, C.A.S. Banks, M.P. Washburn, and R.C. Conaway. 2005. The mammalian mediator complex. *FEBS Letters* 579: 904-908.

Costa, L.E., G. Mendez, and A. Boveris. 1997. Oxygen dependence of mitochondrial function measured by high-resolution respirometry in long-term hypoxic rats. *American Journal of Physiology* 273: C852-C858.

Crowther, G.J., M.F. Carey, W.F. Kemper, and K.E. Conley. 2002a. Control of glycolysis in contracting skeletal muscle. II. Turning it off. *American Journal of Physiology* 282: E74-E79.

Crowther, G.J., W.F. Kemper, M.F. Carey, and K.E. Conley. 2002b. Control of glycolysis in contracting skeletal muscle. I. Turning it on. *American Journal of Physiology* 282: E67-E73.

de Glisezinski, I., C. Moro, F. Pillard, F. Marion-Latard, I. Harant, M. Meste, M. Berlan, F. Crampes, and D. Rivière. 2003. Aerobic training improves exercise-induced lipolysis in SCAT and lipid utilization in overweight men. *American Journal of Physiology* 285: E984-E990.

Didier, L., R.S. Hundal, A. Dresner, T.B. Price, S.M. Vogel, F. Falk Petersen, and G.I. Shulman. 2000. Mechanism of muscle glycogen autoregulation in humans. *American Journal of Physiology* 278: E663-E668.

Duchen, M.R. 2004. Mitochondria in health and disease: Perspectives on a new mitochondrial biology. *Molecular Aspects of Medicine* 25: 365-451.

Duteil, S., C. Bourrilhon, J.S. Raynaud, C. Wary, R.S. Richardson, A. Leroy-Willig, J.C. Jouanin, C.Y. Guezennec, and P.G. Carlier. 2004. Metabolic and vascular support for the role of myoglobin in humans: A multiparametric NMR study. *American Journal of Physiology* 287: R1441-R1449.

Elosua, R., L. Molina, M. Fito, A. Arquer, J.L. Sanchez-Quesada, M.I. Covas, J. Ordoñez-Llanos, and J. Arruga. 2003. Response of oxidative stress biomarkers to a 16-week aerobic physical activity program and to acute physical activity, in healthy young men and women. *Atherosclerosis* 167: 327-334.

Flaus, A., and T. Owen-Hughes. 2004. Mechanisms for ATP-dependent chromatin remodeling: Farewell to the tuna-can octamer. *Current Opinion in Genetics and Development* 14: 165-173.

Frøsig, C., S.B. Jørgensen, D.G. Hardie, E.A. Richter, and J.F.P. Wojtaszewski. 2004. 5'-AMP-activated protein kinase activity and protein expression are regulated by endurance training in human skeletal muscle. *American Journal of Physiology* 286: E411-E417.

Fueger, P.T., S. Heikkinen, D.P. Bracy, C.M. Malabanan, R.R. Pencek, M. Laakso, and D.H. Wasserman. 2003. Hexokinase II partial knockout impairs exercise-stimulated glucose uptake in oxidative muscles of mice. *American Journal of Physiology* 285: E958-E963.

Furukawa, S., T. Fujita, M. Shimabukuro, M. Iwaki, Y. Yamada, Y. Nakajima, O. Nakayama, M. Makishima, M. Matsuda, and I. Shimomura. 2004. Increased oxidative stress in obesity and its impact on metabolic syndrome. *Journal of Clinical Investigation* 114: 1752-1761.

Grassi, B. 2001. Regulation of oxygen consumption at exercise onset: Is it really controversial? *Exercise and Sport Sciences Reviews* 29: 134-138.

Gropper, S.S., J.L. Smith, and J.L. Groff. 2005. *Advanced Nutrition and Human Metabolism.* Belmont, CA: Thomson Wadsworth.

Guerre-Millo, M. 2003. Extending the glucose/fatty acid cycle: A glucose/adipose tissue cycle. *Biochemical Society Transactions* 31: 1161-1164.

Halliwell, B., and J.M.C. Gutteridge. 1999. *Free Radicals in Biology and Medicine, Third Edition.* Oxford: Clarendon Press.

Hamilton, M.T., D.G. Hamilton, and T.W. Zderic. 2004. Exercise physiology versus inactivity physiology: An essential concept for understanding lipoprotein lipase regulation. *Exercise and Sport Sciences Reviews* 32: 161-166.

Hardie, D.G. 2004. AMP-activated protein kinase: A key system mediating metabolic responses to exercise. *Medicine and Science in Sports and Exercise* 36: 28-34.

Hewitt, S.C., B.J. Deroo, and K.S. Lorach. 2005. A new mediator for an old hormone? *Science* 307: 1572-1573.

Hickner, R.C., J.S. Fisher, P.A. Hansen, S.B. Racette, C.M. Mier, M.J. Turner, and J.O. Holloszy. 1997. Muscle glycogen accumulation after endurance exercise in trained and untrained individuals. *Journal of Applied Physiology* 83: 897-903.

Holm, C. 2003. Molecular mechanisms regulating hormone-sensitive lipase and lipolysis. *Biochemical Society Transactions* 31: 1120-1124.

Hood, D.A. 2001. Contractile activity-induced mitochondrial biogenesis in skeletal muscle. *Journal of Applied Physiology* 90: 1137-1157.

Hornberger, T.A., and K.A. Esser. 2004. Mechanotransduction and the regulation of protein synthesis in skeletal muscle. *Proceedings of the Nutrition Society* 63: 331-335.

Horowitz, J.F., A.E. Kaufman, A.K. Fox, and M.P. Harber. 2005. Energy deficit without reducing dietary carbohydrate alters resting carbohydrate oxidation and fatty acid availability. *Journal of Applied Physiology* 98: 1612-1628.

Horowitz, J.F., R. Mora-Rodriguez, L.O. Byerley, and E.F. Coyle. 1999. Substrate metabolism when subjects are fed carbohydrate during exercise. *American Journal of Physiology* 276: E828-E835.

Houston, M.E. 1999. Gaining weight: The scientific basis of increasing skeletal muscle mass. *Canadian Journal of Applied Physiology* 24: 305-316.

Hughson, R.L. In press. Regulation of VO$_2$ on kinetics by O$_2$ delivery. In: *Oxygen Uptake Kinetics in Health and Disease*, ed. A.M. Jones and D.C. Poole. London: Routledge.

Hughson, R.L., M.E. Tschakovsky, and M.E. Houston. 2001. Regulation of oxygen consumption at the onset of exercise. *Exercise and Sport Sciences Reviews* 29: 129-133.

Ivy, J.L., H.W. Goforth Jr., B.M. Damon, T.R. McCauley, E.C. Parsons, and T.B. Price. 2002. Early postexercise muscle glycogen recovery is enhanced with a carbohydrate-protein supplement. *Journal of Applied Physiology* 93: 1337-1344.

Jeukendrup, A.E. 2003. Modulation of carbohydrate and fat utilization by diet, exercise and environment. *Biochemical Society Transactions* 31: 1270-1273.

Jeukendrup, A.E., W.H.M. Saris, and A.J.M. Wagenmakers. 1998a. Fat metabolism during exercise: A review. Part I: Fatty acid mobilization and muscle metabolism. *International Journal of Sports Medicine* 19: 231-244.

Jeukendrup, A.E., W.H.M. Saris, and A.J.M. Wagenmakers. 1998b. Fat metabolism during exercise: A review. Part II: Regulation of metabolism and the effects of training. *International Journal of Sports Medicine* 19: 293-302.

Karlsson, H.K.R., P.-A. Nilsson, J. Nilsson, A.V. Chibalin, J.R. Zierath, and E. Blomstrand. 2004. Branched-chain amino acids increase p70[S6k] phosphorylation in human skeletal muscle after resistance exercise. *American Journal of Physiology* 287: E1-E7.

Katz, A., and K. Sahlin. 1990. Role of oxygen in regulation of glycolysis and lactate production in human skeletal muscle. *Exercise and Sport Sciences Reviews* 18: 1-28.

Kimball, S.R., and L.S. Jefferson. 2004. Amino acids as regulators of gene expression. *Nutrition and Metabolism* 1: 1-10.

Kjaer, M., K. Howlett, J. Langfort, T. Zimmerman-Belsing, J. Lorentsen, J. Bülow, J. Ihlemann, U. Feldt-Rasmussen, and H. Galbo. 2000. Adrenaline and glycogenolysis in skeletal muscle during exercise: A study in adrenalectomised humans. *Journal of Physiology* 528: 371-378.

Leary, S.C., C.N. Lyons, A.G. Rosenberger, J.S. Ballantyne, J. Stillman, and C.D. Moyes. 2003. Fiber-type differences in muscle mitochondrial profiles. *American Journal of Physiology* 285: R817-R826.

LeBlanc, P.J., K.R. Howarth, M.J. Gibala, and G.J.F. Heigenhauser. 2004. Effects of 7 wk of endurance training on human skeletal muscle metabolism during submaximal exercise. *Journal of Applied Physiology* 97: 2148-2153.

Louis, M., J.R. Poortmans, M. Francaux, J. Berré, N. Boisseau, E. Brassine, D.J.R. Cuthbertson, K. Smith, J.A. Babraj, T. Waddell, and M.J. Rennie. 2003. No effect of creatine supplementation on human myofibrillar and sarcoplasmic protein synthesis after resistance exercise. *American Journal of Physiology* 285: E1089-E1094.

Louis, M., J.R. Poortmans, M. Francaux, E. Hultman, J. Berré, N. Boisseau, V.R. Young, K. Smith, W. Meier-Augenstein, J.A. Babraj, T. Waddell, and M.J. Rennie. 2003. Creatine supplementation has no effect on human muscle protein turnover at rest in the post-absorptive or fed states. *American Journal of Physiology* 284: E764-E770.

Lyon, C.J., R.E. Law, and W.A. Hsieh. 2003. Adiposity, inflammation, and atherogenesis. *Endocrinology* 144: 2195-2200.

Machida, S., and F.W. Booth. 2005. Changes in signaling molecule levels in 10-day hindlimb immobilized rat muscles. *Acta Physiologica Scandinavica* 183:171-179.

McComas, A.J. 1996. *Skeletal Muscle Form and Function.* Human Kinetics, Champaign, IL.

Michel, R.N., S.E. Dunn, and E.R. Chin. 2004. Calcineurin and skeletal muscle growth. *Proceedings of the Nutrition Society* 63: 341-349.

Milne, K., and E.G. Noble. 2002. The exercise-induced elevation of Hsp70 is intensity dependent. *Journal of Applied Physiology* 93: 561-568.

Mittendorfer, B., D.A. Fields, and S. Klein. 2004. Excess body fat in men decreases plasma fatty acid availability and oxidation during endurance exercise. *American Journal of Physiology* 286: 354-362.

Mooren, F.C., A. Lechtermann, and K. Völker. 2004. Exercise-induced apoptosis of lymphocytes depends on training status. *Medicine and Science in Sports and Exercise* 36: 1476-1483.

Morio, B., U. Holmbäck, D. Gore, and R.R. Wolfe. 2004. Increased VLDL-TAG turnover during and after acute moderate intensity exercise. *Medicine and Science in Sports and Exercise* 36: 801-806.

Mourtzakis, M., and T.E. Graham. 2003. Glutamate ingestion and its effects at rest and during exercise in humans. *Journal of Applied Physiology* 93: 1251-1259.

Moyes, C.D., and D.A. Hood. 2003. Origins and consequences of mitochondrial variation in vertebrate muscle. *Annual Reviews of Physiology* 65: 177-201.

Nguyen, H.X., and J.G. Tidball. 2003. Expression of a muscle-specific, nitric oxide synthase transgene prevents muscle membrane injury and reduces muscle inflammation during modified muscle use in mice. *Journal of Physiology* 550: 347-356.

Nicholls, D.G., and S.J. Ferguson. 2002. *Bioenergetics 3*. London: Academic Press.

Nielsen, O.B., F. dePaoli, and K. Overgaard. 2001. Protective effects of lactic acid on force production in rat skeletal muscle. *Journal of Physiology* 536: 161-166.

Nielsen, S.Z., K. Guo, J.B. Albu, S. Klein, P.C. O'Brien, and M.D. Jensen. 2003. Energy expenditure, sex, and endogenous fuel availability in humans. *Journal of Clinical Investigation* 111: 981-988.

Paddon-Jones, D., M. Sheffield-Moore, X.-J. Zhang, E. Volpi, S.E. Wolf, A. Aarsland, A.F. Ferrando, and R.R. Wolfe. 2004. Amino acid ingestion improves muscle protein synthesis in the young and elderly. *American Journal of Physiology* 286: E321-E328.

Papanikolaou, G., and K. Pantopoulos. 2005. Iron metabolism and toxicity. *Toxicology and Applied Pharmacology* 202: 199-211.

Phillips, S.M., H.J. Green, M.A. Tarnopolsky, G.J.F. Heigenhauser, and S.M. Grant. 1996. Progressive effect of endurance training on metabolic adaptations in working skeletal muscle. *American Journal of Physiology* 270: E265-E272.

Price, T.B., D. Laurent, K.F. Petersen, D.L. Rothman, and G.I. Shulman. 2000. Glycogen loading alters muscle glycogen resynthesis after exercise. *Journal of Applied Physiology* 88: 698-704.

Pruchnic, R., A. Katsiaras, J. He, D.E. Kelley, C. Winters, and B.H. Goodpaster. 2004. Exercise training increases intramyocellular lipid and oxidative capacity in older subjects. *American Journal of Physiology* 287: E857-E862.

Randle, P.J., P.B. Garland, C.N. Hales, and E.A. Newsholme. 1963. The glucose fatty-acid cycle: Its role in insulin sensitivity and the metabolic disturbances of diabetes mellitus. *Lancet* 1: 785-789.

Randle, P.J., E.A. Newsholme, and P.B. Garland. 1964. Regulation of glucose uptake by muscle. *Biochemistry Journal* 93: 652-665.

Rasmussen, H.N., G. van Hall, and U.F. Rasmussen. 2002. Lactate dehydrogenase is not a mitochondrial enzyme in human and mouse vastus lateralis muscle. *Journal of Physiology* 541: 575-580.

Reid, M.B. 2001. Redox modulation of skeletal muscle contraction: What we know and what we don't. *Journal of Applied Physiology* 90: 724-731.

Rennie, M.J., and K.D. Tipton. 2001. Protein and amino acid metabolism during and after exercise and the effects of training. *Annual Reviews of Nutrition* 20: 457-483.

Reshef, L., Y. Olswang, H. Cassuto, B. Blum, C.M. Croniger, S.C. Halhan, S.M. Tilghman, and R.W. Hanson. 2003. Glyceroneogenesis and the triglyceride/fatty acid cycle. *Journal of Biological Chemistry* 278: 30413-30416.

Richardson, R.S., J.S. Leigh, P.D. Wagner, and E.A. Noyszewski. 1999. Cellular PO_2 as a determinant of maximal mitochondrial O_2 consumption in trained human skeletal muscle. *Journal of Applied Physiology* 87: 325-331.

Richardson, R.S., E.A. Noyszewski, L.J. Haseler, S. Blumi, and L.R. Frank. 2002. Evolving techniques for the investigation of muscle bioenergetics and oxygenation. *Biochemical Society Transactions* 30: 232-237.

Richter, E.A., B. Vistisen, S.J. Maarbjerg, M. Sajan, R.V. Farese, and B. Kiens. 2004. Differential effect of bicycling exercise intensity on activity and phosphorylation of atypical protein kinase C and extracellular signal-regulated protein kinase in skeletal muscle. *Journal of Physiology* 560: 909-918.

Rider, M.H., L. Bertrand, D. Vertommen, P.A. Michels, G.G. Rousseau, and L. Hue. 2004. 6-Phosphofructo-2-kinase/fructose-2,6-bisphosphatase: Head-to-head with a bifunctional enzyme that controls glycolysis. *Biochemical Journal* 381: 561-579.

Roberge, R.A., F. Ghiasvand, and D. Parker. 2004. Biochemistry of exercise-induced metabolic acidosis. *American Journal of Physiology* 287: R502-R516.

Roberge, R.A., and S.O. Roberts. 1997. *Exercise Physiology: Exercise Performance and Clinical Applications*. St. Louis: Mosby.

Rocchini, A.P. 2002. Childhood obesity and the diabetes epidemic. *New England Journal of Medicine* 346:854-855.

Roden, M. 2004. How free fatty acids inhibit glucose utilization in human skeletal muscle. *News in Physiological Sciences* 19: 92-96.

Roeder, R.G. 2005. Transcriptional regulation and the role of diverse coactivators in animal cells. *FEBS Letters* 579: 909-915.

Roepstorff, C., N. Halberg, T. Hillis, A.K. Saha, N.B. Ruderman, J.F.P. Wojtaszewski, E.A. Richter, and B. Kiens. 2005. Malonyl-CoA and carnitine in regulation of fat oxidation in human skeletal muscle during exercise. *American Journal of Physiology* 288: E133-E142.

Roepstorff, C., B. Vistisen, M. Donsmark, J.N. Nielsen, H. Galbo, K.A. Green, D.G. Hardie, J.F.P. Wojtaszewski, E.A. Richter, and B. Kiens. 2004. Regulation of hormone-sensitive lipase activity and Ser[563] and Ser[565] phosphorylation in human skeletal muscle during exercise. *Journal of Physiology* 560: 551-562.

Rousset, S., M.-C. Alves-Guerra, J. Mozo, B. Miroux, A.-M. Cassard-Doulcier, F. Bouillaud, and D. Ricquier. 2004. The biology of mitochondrial uncoupling proteins. *Diabetes* 53 (suppl 1): S130-S135.

Rush, J.W.E., and L.L. Spriet. 2001. Skeletal muscle glycogen phosphorylase kinetics: Effects of adenine nucleotides and caffeine. *Journal of Applied Physiology* 91: 2071-2078.

Safdar, A., N. Yardley, R.J. Snow, S. Melov, and M.A. Tarnopolsky. 2004. Genomic and protein expression in skeletal muscle of men following acute creatine monohydrate supplementation. *Canadian Journal of Applied Physiology* 29: S77.

Sahlin, K., M. Fernström, M. Svensson, and M. Tonkonogi. 2002. No evidence of an intracellular lactate shuttle in rat skeletal muscle. *Journal of Physiology* 541: 569-574.

Sambandamm, N., M. Steinmetz, A. Chu, J.Y. Altarejos, J.R.B. Dyck, and G.D. Lopaschuk. 2004. Malonyl CoA decarboxylase (MCD) is differentially regulated in subcellular compartments by 5' AMP-activated protein kinase (AMPK). *European Journal of Biochemistry* 271: 2831-2840.

Shearer, J., I. Marchand, P. Sathasivam, M.A. Tarnopolsky, and T.E. Graham. 2000. Glycogenin activity in human skeletal muscle is proportional to muscle glycogen concentration. *American Journal of Physiology* 278: E177-E180.

Snow, R.J., and R.M. Murphy. 2003. Factors influencing creatine loading into human skeletal muscle. *Exercise and Sport Sciences Reviews* 31: 154-158.

Spencer, M.K., Z. Yan, and A. Katz. 1992. Effect of low glycogen on carbohydrate and energy metabolism in human muscle during exercise. *American Journal of Physiology* 262: C975-C979.

Stich, V., I. de Glisezinski, F. Crampes, J. Hejnova, J.M. Cottet-Emard, J. Galitzky, M. Lafontan, D. Rivière, and M. Berlan. 2000. Activation of α_2 adrenergic receptors impairs exercise-induced lipolysis in SCAT of obese subjects. *American Journal of Physiology* 279: R499-R504.

Stojanovski, D., A.J. Johnston, I. Streimann, N.J. Hoogenraad, and M.T. Ryan. 2003. Import of nuclear-encoded proteins into mitochondria. *Experimental Physiology* 88: 57-64.

Strawford, A., F. Antero, M. Christiansen, and M.K. Hellerstein. 2004. Adipose tissue triglyceride turnover, de novo lipogenesis, and cell proliferation in humans measured with ^2H_2O. *American Journal of Physiology* 286: E577-E588.

Taillandier, D., L. Combaret, M.-N. Pouch, S.E. Samuels, D. Béchet, and D. Attaix. 2004. The role of ubiquitin-proteasome-dependent proteolysis in the remodeling of skeletal muscle. *Proceedings of the Nutrition Society* 63: 357-361.

Tordjman, J., G. Chauvet, J. Quette, E.G. Beale, C. Gorest, and B. Antoine. 2003. Thiazolidinediones block fatty acid release by inducing glyceroneogenesis in fat cells. *Journal of Biological Chemistry* 278: 18785-18790.

Tidball, J.G. 2005. Mechanical signal transduction in skeletal muscle growth and adaptation. *Journal of Applied Physiology* 98: 1900-1908.

Trimmer, J.K., J.-M. Schwarz, G.A. Casazza, M.A. Horning, N. Rodriguez, and G.A. Brooks. 2002. Measurement of gluconeogenesis in exercising men by mass isotopomer distribution analysis. *Journal of Applied Physiology* 93: 233-241.

Trumpower, B.L. 1990. The protonmotive Q cycle. Energy transduction by coupling of proton translocation to electron transfer by the cytochrome bc1 complex. *Journal of Biological Chemistry* 265: 11409-11412.

Tucker, R.M., C. May, R. Bennett, J. Hymer, and B. McHaney. 2004. A gym-based wellness challenge for people with type 2 diabetes:Effect on weight loss, body composition and glycemic control. *Diabetes Spectrum* 17: 176-180.

Tunstall, R.J., K.A. Mehan, G.D. Wadley, G.R. Collier, A. Bonen, M. Hargreaves, and D. Cameron-Smith. 2002. Exercise training increases lipid metabolism gene expression in human skeletal muscle. *American Journal of Physiology* 283: E66-E72.

Vandenboom, R. 2004. The myofibrillar complex and fatigue: A review. *Canadian Journal of Applied Physiology* 29: 330-356.

Van de Poll, M.C.G., P.B. Soeters, N.E.P. Deutz, K.C.H. Fearon, and C.H.D. Dejong. 2004. Renal metabolism of amino acids: Its role in interorgan amino acid exchange. *American Journal of Clinical Nutrition* 79: 185-197.

van Loon, L.J.C., V.B. Schrauwen-Hinderling, R. Koopman, A.J.M. Wagenmakers, M.K.C. Hesselink, G. Schaart, M.E. Kooi, and W.H.M. Saris. 2003. Influence of prolonged endurance cycling and recovery diet on intramuscular triglyceride content in trained males. *American Journal of Physiology* 285: E804-E811.

Venables, M.C., J. Achten, and A.E. Jeukendrup. 2005. Determinants of fat oxidation during exercise in healthy men and women: A cross-sectional study. *Journal of Applied Physiology* 98: 160-167.

Venugopal, J., K. Hanashiron, Z-Z. Yang, and Y. Nagamine. 2004. Identification and modulation of a caveola-dependent signal pathway that regulates plasminogen activator-receptor-1 in insulin-resistant adipocytes. *Proceedings of the National Academy of Sciences USA* 101: 17120-17125.

Vincent, H.K., J.W. Morgan, and K.R. Vincent. 2004. Obesity exacerbates oxidative stress levels after acute exercise. *Medicine and Science in Sports and Exercise* 36: 772-779.

Vistisen, B., L. Nybo, X. Xu, C.-E. Høy, and B. Kiens. 2003. Minor amounts of plasma medium-chain fatty acids and no improved time trial performance after consuming lipids. *Journal of Applied Physiology* 95: 2434-2443.

Vogt, M., A. Puntschart, H. Howard, B. Mueller, C. Mannhart, L. Feller-Tuescher, P. Mullis, and H. Hoppeler. 2003. Effects of dietary fat on muscle substrates, metabolism, and performance in athletes. *Medicine and Science in Sports and Exercise* 35: 952-960.

Wasserman, K., B.J. Whipp, S.N. Koyal, and W.L. Beaver. 1973. Anaerobic threshold and respiratory gas exchange during exercise. *Journal of Applied Physiology* 35: 236-245.

Watson, T.A., R. Callister, R.D. Taylor, D.W. Sibbritt, L.K. Macdonald-Wicks, and M.L. Garg. 2005. Antioxidant restriction and oxidative stress in short-duration exhaustive exercise. *Medicine and Science in Sports and Exercise* 37: 63-71.

Watt, M.J., G.J.F. Heigenhauser, and L.L. Spriet. 2002. Intramuscular triacylglycerol utilization in human skeletal muscle during exercise: Is there a controversy? *Journal of Applied Physiology* 93: 1185-1195.

Watt, M.J., A.G. Holmes, G.R. Steinberg, J.L. Mesa, B.E. Kemp, and M.A. Febbraio. 2004. Reduced plasma FFA availability increases net triacylglycerol degradation, but not GPAT or HSL activity, in human skeletal muscle. *American Journal of Physiology* 287: E120-E127.

Watt, M.J., P. Krustrup, N.H. Secher, B. Saltin, B.K. Pedersen, and M.A. Febbraio. 2004. Glucose ingestion blunts hormone-sensitive lipase activity in contracting human skeletal muscle. *American Journal of Physiology* 286: E144-E150.

Watt, M.J., T. Stellingwerff, G.J.F. Heigenhauser, and L.L. Sprite. 2003. Effects of plasma adrenaline on hormone-sensitive lipase at rest and during moderate exercise in human skeletal muscle. *Journal of Physiology* 550: 325-332.

Wellen, K.E., and G.S. Hotamisligil. 2003. Obesity-induced inflammatory changes in adipose tissue. *Journal of Clinical Investigation* 112: 1785-1788.

Weltan, S.M., A.N. Bosch, S.C. Dennis, and T.D. Noakes. 1998. Preexercise muscle glycogen content affects metabolism during exercise despite maintenance of hyperglycemia. *American Journal of Physiology* 274: E83-E88.

Westerblad, H., D.C. Allen, and J. Lännergren. 2002. Muscle fatigue: Lactic acid or inorganic phosphate the major cause? *News in Physiological Sciences* 17: 17-21.

Widegren, U., J.W. Ryder, and J.R. Zierath. 2001. Mitogen-activated protein kinase signal transduction in skeletal muscle: Effects of exercise and muscle contraction. *Acta Physiologica Scandinavica* 172: 227-238.

Wolf, G. 2004. Insulin resistance and obesity: Resistin, a hormone secreted by adipose tissue. *Nutrition Reviews* 62: 389-395.

Yeaman, S.J. 2004. Hormone-sensitive lipase—new roles for an old enzyme. *Biochemical Journal* 379: 11-22.

Zatz, M., and A. Starling. 2005. Calpains and disease. *New England Journal of Medicine* 352: 2413-2423.

Zderic, T.W., C.J. Davidson, S. Schenk, L.O. Byerley, and E.F. Coyle. 2004. High fat diet elevates resting intramuscular triglyceride concentration and whole body lipolysis during exercise. *American Journal of Physiology* 286: E217-E225.

Index

Note: The letters *f* and *t* after page numbers indicate figures and tables, respectively.

About the Author

Michael E. Houston, PhD, is a professor in the Department of Human Nutrition, Foods, and Exercise at Virginia Tech in Blacksburg, Virginia. Besides writing the first two editions of this book, Houston has written more than 100 refereed journal publications dealing with biochemical and nutritional implications to exercise. He has taught biochemistry of exercise for more than 35 years and has conducted extensive research in exercise. In 2003 he received the Honour Award from the Canadian Society for Exercise Physiology. In his leisure time he enjoys weight training and road and mountain biking.

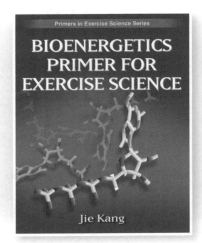